THE MISSOURI RIVER JOURNALS

OF JOHN JAMES AUDUBON

THE MISSOURI RIVER JOURNALS OF

JOHN JAMES AUDUBON

JOHN JAMES AUDUBON

Edited and with original commentary by Daniel Patterson

UNIVERSITY OF NEBRASKA PRESS

Lincoln and London

Publication of this volume was assisted by the Virginia Faulkner Fund,
established in memory of Virginia Faulkner, editor in chief of the
University of Nebraska Press.

Library of Congress Cataloging-in-Publication Data
Audubon, John James, 1785–1851.
The Missouri River journals of
John James Audubon / John James Audubon;
edited and with original commentary
by Daniel Patterson.
pages   cm
Includes bibliographical references and index.
ISBN 978-0-8032-4498-6 (cloth: alkaline paper)
ISBN 978-0-8032-9483-7 (pdf)
1. Audubon, John James, 1785–1851—Diaries. 2. Audubon,
John James, 1785–1851—Travel—Missouri River Valley.
3. Scientific expeditions—Missouri River Valley—History—
19th century. 4. Naturalists—United States—Diaries. 5. Wildlife artists—
United States—Diaries. 6. Natural history—Missouri River Valley.
7. Missouri River Valley—Description and travel. 8. Missouri
River—Description and travel. 9. River steamers—Missouri River—
History—19th century. 10. Audubon, John James, 1785–1851—
Ethics. I. Patterson, Daniel, 1953– II. Title.
QL31.A9A3 2016
508.092—dc23
2015019950

Set in Arno Pro by L. Auten.
Designed by Rachel Gould.

To Victor Morris Tyler
Descendant and Friend

Our Boilers have to be cleaned every Evening &c &c and
when we come to the cutting of our own Wood, there will be our
Traps &c and our Guns a going.—May it please our God that I
may procure a good round number of animals.
—JOHN JAMES AUDUBON, APRIL 25, 1843

[We saw] a very large band of buffalo, and as we came opposite to
them they got the wind of us, and started to pass in front of us. We then
rode on until we parted them and passed through. After about 300 had
passed us, the others then passed behind us, in a line, and they kept comeing
and comeing from the River and running to the hills. They formed a complete
line, one, two, and sometimes three deep, for more than two miles long
with scarcely an opening between them. They are like sheep, when
one starts the others will all follow. To me it was a most beautiful
sight, to see them passing within 150 yards of me running as
fast as they could, with their tongues hanging out.
—JOHN G. BELL, AUGUST 5, 1843

We now regretted having destroyed these noble beasts for no earthly
reason but to gratify a sanguinary disposition which appears to be inherent
in our natures. We had no means of carrying home the meat and after cutting
out the tongues we wended our way back to camp, completely disgusted with
ourselves and with the conduct of all white men who come to this country.
—EDWARD HARRIS, JULY 20, 1843

This was good shooting, to kill two buffaloes weighing 2000 lbs each—
with a single ball each—but the sight of them as they lay bleeding
on the earth seemed to me too much like butchering—I immediately
made a sketch of the nearest one as he lay, afterwards cut
off the tails as trophys and left them for the wolves!
·   —ISAAC SPRAGUE, AUGUST 20, 1843

CONTENTS

List of Illustrations   xi

Preface   xv

Acknowledgments   xix

List of Abbreviations   xxv

Editorial Principles   xxvii

*Part I. Maria Rebecca Audubon, Her Grandfather's
1843 Missouri River Journals, and the "Great Auk Speech"*   1

Part II. Audubon's Missouri River Expedition of 1843

His Eminence   27

Preparations   31

Minnie's Land to St. Louis, March 11–28   47

St. Louis, March 28–April 24   55

St. Louis to the Yellowstone River and Fort Union,
April 25–June 12   63

Fort Union and the Prairies, June 13–August 15   83

Fort Union to St. Louis, August 16–October 19   115

St. Louis to Minnie's Land, October 22–November 7   127

Part III. The Three Forgotten Manuscript Journals

The Beinecke Partial Copy   133

The Original Field Notebook and the Newberry Partial Copy   159

Part IV. Audubon's Conservation Ethic Reconsidered

Audubon's Hunting and Conservation Ethic as
Represented in the Biographies   211

The Lived Ethic   235

The Written Ethic   267

Epilogue   299

Part V. Other Materials from the 1843 Expedition

The 1843 Diary of John Graham Bell   307

The 1843 Diary of Isaac Sprague   375

Audubon's "George Catlin" Powder Horn from
the Missouri River Expedition   423

Appendix: "The Pet Bear," an Unpublished Episode   427

Works Cited   433

Index   439

ILLUSTRATIONS

Following page 162

1. Map of the Missouri River expedition

2. *John James Audubon,* by John Woodhouse Audubon
and Victor Gifford Audubon, 1841

3. Front view of Maria Audubon's Salem, New York, home

4. Rear view of the Salem home

5. The sisters Maria Rebecca and Florence Audubon

6. Edward Harris

7. Isaac Sprague, self-portrait

8. Isaac Sprague, *John James Audubon,* 1843

9. J. C. Wild, *View of Front Street, St. Louis,* 1840

10. "Pseudostoma Bursarius, Canada Pouched Rat," plate 44, 1845

11. Karl Bodmer, *Snags on the Missouri River,* 1833

12. "Cervus Macrotis, Mule Deer," plate 78, 1846

13. "Spermophilus Ludovicianus, Prairie Dog," plate 99, 1846

14. "Bos Americanus, American Bison or Buffalo," plate 57, 1845

15. Isaac Sprague, sketch of Missouri River, June 1, 1843

16. Isaac Sprague, *Fort Union, U. Mo.,* 1843

17. "Castor Fiber Americanus, American Beaver," plate 46, 1845

18. Edward Harris's bison tail trophies, 1843

19–20. Audubon's field sketch of the camp at Three Mammelles, 1843

21. Audubon's finished sketch of the camp at
Three Mammelles, July 27, 1843

22. Isaac Sprague, sketch of encampment on
Missouri River, September 2, 1843

23. "Vulpes Velox, Swift Fox," plate 52, 1845

24. Isaac Sprague, sketch of encampment on Missouri
River, May 26, 27, and September 16, 1843

25. John Woodhouse Audubon, *John James Audubon*, 1843

26. First page of the Beinecke partial copy, bearing
journal entry for August 5, 1843

27. Pages from the Beinecke partial copy bearing
journal entry for August 8, 1843

28. Audubon's original field notebook, 1843, front cover

29. Audubon's original field notebook, 1843, back cover

30. First page of the Newberry partial copy

31. Audubon's field sketch of "The Coupe," July 2, 1843

32. First of Audubon's two field sketches of dead bison, July 31, 1843

33. Second of Audubon's two field sketches of dead bison, July 31, 1843

34. Audubon's powder horn from Missouri River expedition

35–38. Detail of Audubon powder horn

39. Edward Harris's powder horn

40–42. Detail of Harris powder horn

43. Photo of the Catlin powder horn

44. Sketch of engravings on the Catlin powder horn

45. George Catlin, *Buffalo Chase*

46. The fourth St. Louis–style powder horn

PREFACE

Out of the trunk grow the branches.
—HERMAN MELVILLE

This project began with a trip to the Audubon Center at Mill Grove (Audubon, Pennsylvania) in June 2008. My wife and I made the drive so that I could study the Alice Ford Papers housed there in hopes of finding answers to questions that had arisen while I was working on my edition of Audubon's journal of 1826. Since Ford's work on Audubon spanned half a century, the six archival boxes and one large rubber storage tub contained many hundreds of documents: correspondence, transcriptions, notebooks, photographs, exhibit catalogs, brochures, and books. One item especially intrigued me, though, when I lifted it from its box (box 3, item 14, of the inventory I prepared). In an envelope from the Newberry Library in Chicago was a photocopy of a manuscript, across the top of the front page of which Audubon had written, "Copy of my Journal from Fort Union homeward Commenced Augt 16$^{\underline{th}}$ at 12 o'clock, the moment of our departure." So far as I knew at that time, no portion of his 1843 Missouri River journals was known to have survived. His granddaughter Maria Rebecca Audubon was said to have destroyed the original journals after publishing her rather freely edited *Audubon and His Journals* in 1897 (more on this in part 1). Yet here was a copy of a portion of the original journals that Audubon ordered his scribe to make. Differences between the Newberry partial copy and the granddaughter's version of the same dates motivated me to keep searching.

I next found that the Beinecke Rare Book and Manuscript Library, Yale University, held in plain sight—that is, in the Morris Tyler Family Collection of John James Audubon—another partial copy of the Missouri River journals. A bit later I learned that one of Audubon's original field notebooks from the 1843 expedition had also survived and was in a private collection. All together, then, the three "forgotten" journals contained Audubon's entries for 65 of the expedition's 178 days (counting from the day they left St. Louis until they returned). And the differences between these journal entries and those published by the granddaughter told me that here was an opportunity to tell the story of this expedition more fully and accurately than it had been possible to do before.

The momentum continued to build as I began to gather the primary documents relevant to the expedition and saw that there was sufficient material to support a full treatment of the story. Of Audubon's four traveling companions, three kept journals. A reliable edition of Edward Harris's had been published by John Francis McDermott in 1951 and so was readily available (EH). McDermott had also published an edition of letters Audubon wrote while on the expedition (AW). The journals of John G. Bell and Isaac Sprague were in the Beinecke and the Boston Athenaeum, respectively. Otherwise, an abundance of letters to and from Audubon, family members, colleagues, and friends—published and unpublished—lay scattered about in libraries, museums, and books.

As I studied the amassed primary documents, however, I saw that this supposedly scientific and artistic expedition that Audubon had mounted to advance his work in progress *The Viviparous Quadrupeds of North America* was rather a ceaselessly bloody hunting expedition—and that John James Audubon had been granted a status as conservationist that these documents did not support. It was his granddaughter's spurious edition of his journals that had led many to regard Audubon as a remarkably early conservationist, despite his killing of untold thousands of birds and other animals, which he himself reported in his massive prose work *Ornithological Biography* (1831–39). When the recovered partial copies of the original expedition journals showed me that his granddaughter had essentially manufactured in 1897 the man she wished her grandfather had been in 1843, I understood that all discussions of his thought about conservation and hunting were based upon tainted evidence, but no one had known that. With the recovered journals in hand, then, I undertook to reread all the available evidence in hopes of arriving at a more accurate understanding of Audubon and his thought than had been possible before. What I learned finally—the evidence and argument for which I present fully in the pages that follow—is that Audubon was a great sportsman as well as the earliest advocate for the ethical considerability of birds in this country. In the history of American conservationist thought, Audubon is one of the earliest voices for human restraint on the land, but the evidence for that claim is not in how he lived his life, for he rarely exercised restraint, but in the words he published, particularly the *Ornithological Biography*. Thus I learned that Maria Rebecca Audubon set up a misleading beacon that lured scholars away from the true and reliable evidence of what her grandfather actually believed about how humans should live in relation to the natural world.

I begin, then, with the story of the granddaughter's destruction of the 1843 journals and an analysis of the most significant changes she made in them (part 1). Then I tell, for the first time, the full story of the travels and adventures of Audubon and his four companions up the Missouri River to the confluence with the Yellowstone over some seven months (part 2). My narrative is based upon extensive primary evidence, the most important of which are the "forgotten" journals. In part 3 I present the full text of those expedition journals for others to enjoy and mine. In part 4 I address the questions that Maria Audubon's edition and destruction of the original 1843 journals raised and present my case for a revised and more accurate understanding of Audubon's conservationist thought. In part 5 I present the full diaries of Audubon's traveling companions John Bell and Isaac Sprague, which I drew upon heavily for my narrative of the expedition, and a consideration of Audubon's "Catlin" powder horn from the expedition. An appendix contains a previously unpublished "episode" that Audubon wrote about a pet bear, which he may have "collected" originally with publication in the *Quadrupeds* in mind.

My hope is that a new understanding of Audubon emerges from all this, and that it is based upon reliable evidence and sound reasoning. Another hope is that readers will enjoy the numerous good stories included herein.

ACKNOWLEDGMENTS

My debts are great; I now gladly acknowledge them.

Amy T. Montague, the director of and creative force behind the Museum of American Bird Art at Mass Audubon in Canton, Massachusetts, provided lodging for a visiting scholar and introduced me to an Audubon family descendant who subsequently aided this project significantly. I thank Amy and Mass Audubon for permission to publish Isaac Sprague's portrait of Audubon.

Matt Bokovoy, senior acquisitions editor at the University of Nebraska Press, continues to be a wise, savvy, and cordial editor and advisor. This is the second project on which we have worked together, and my debts to Matt are many and complex, but my gratitude is simple: thank you.

Also at Nebraska, Heather Stauffer and Sabrina Stellrecht kept complex communications about complex processes coherent, efficacious, and pleasant. Joy Margheim served as copyeditor and saved me from much embarrassment.

Both John R. Knott, emeritus professor of English of the University of Michigan, and William Benemann, archivist of the School of Law, University of California, Berkeley, worked their way through the rather mazy manuscript of this book and guided me toward greater coherence and away from errors. The degree of improvement was not small.

I thank Martha Briggs and John Powell of the Newberry Library, Chicago, for a copy of the partial copy of Audubon's 1843 journal and for permission to publish my transcription of it and an image of its first page. Similarly, Anne

Marie Menta and Adrienne Sharpe at Yale University's Beinecke Rare Book and Manuscript Library supplied copies of the partial copy of Audubon's original 1843 journal and John G. Bell's diary from the same expedition in the Beinecke's holdings. I am grateful to both of these institutions for the privilege of working in their reading rooms.

Another institution with great reading rooms is the Boston Athenaeum, where Stanley Cushing, Anne C. and David J. Bromer Curator of Manuscripts and Rare Books, welcomed me and permitted me to transcribe and publish the diary held in the Athenaeum's Special Collections that Isaac Sprague kept on the 1843 expedition.

In Montgomery, Alabama, at the State of Alabama Department of Archives and History, I enjoyed a cordial welcome and an opportunity to study the journals and correspondence of Edward Harris, Audubon's best friend and fellow traveler on the Missouri River expedition. I thank Norwood Kerr and his colleagues for all their courtesies and assistance. For permission to publish the daguerreotype of Harris, I thank Meredith McDonough. For arranging for photographs to be made of Harris's powder horn and trophy bison tails from the expedition, I thank Bob Bradley, curator of archives, and Bob Cason, senior curator, retired. Bob Bradley was especially generous with his time and allowed me to study the artifacts Harris brought home from the expedition, including several bird drawings supposedly made by Harris.

At Auburn University's Jule Collins Smith Museum of Fine Art, Director Marilyn Laufer and Scott Bishop, director of educational programs, extended a generous welcome and the privilege of examining the plates of their superbly preserved Imperial Folio *Viviparous Quadrupeds of North America* in the Louise Hauss and David Brent Miller Audubon Collection. With their kind permission, I include six of those plates in this volume.

In the Special Collections of the Research Library at the American Museum of Natural History in Manhattan, I was welcomed by Gregory Raml and Mai Qaraman, both of whom I thank for permission to publish from the correspondence between Maria Rebecca Audubon and Frank

Chapman in their holdings as well as one sketch made by Audubon on the expedition and three sketches made by Isaac Sprague. I thank them also for permission to publish a portion of my transcription of Audubon's manuscript essay "American Woodcock." Mary LeCroy took time out of her day to talk with me about the history of Audubon studies and to show me the artifacts in their holdings that Audubon acquired on the 1843 expedition. It was an illuminating conversation with an important part of the American Museum of Natural History's institutional memory.

I thank Eleanor Gillers and her assistant Robert Delap in the Department of Rights and Reproductions of the New-York Historical Society for permission to publish Isaac Sprague's sketch of Fort Union.

For gracious hospitality and generous assistance, I thank Sarah Boehme, curator, and Jenniffer Connors, librarian, at the Stark Museum of Art, Orange, Texas. I thank Sarah also for permission to publish passages from Maria Audubon's letters to the book dealer Patrick F. Madigan in the Stark's holdings.

For permission to publish extracts of my transcription of Audubon's journal for 1820–21, I thank Constance Rinaldo at the Ernst Mayr Library at Harvard University's Museum of Comparative Zoology.

The venerable American Philosophical Society has supplied copies of and granted permission to publish manuscripts in their holdings, for which I am grateful to Roy E. Goodman, curator of printed materials; Charles B. Greifenstein, associate librarian and curator of manuscripts; and Iren L. Snavely, NEH archivist.

The curator of the Missouri History Museum, Jeff Meyer, responded to wide-ranging queries. I am grateful to him and to Amanda Claunch, associate archivist, photographs and prints, for permission to publish J. C. Wild's view of Front Street, St. Louis, and for permission to publish passages from manuscripts in the archive of the museum.

I thank Anne Crouchley, assistant registrar at the Joslyn Art Museum, Omaha, for permitting publication of Karl Bodmer's *Snags on the Missouri River* (1833). My thanks to Princeton University Library for permission to publish from the John James Audubon Collection 1788–1969, Don C. Skemer, curator of manuscripts.

Lee Burke, an independent historian of Dallas, Texas, concentrating on the American fur trade in general and Kit Carson in particular, led me to the "Catlin" powder horn in Wilkes-Barre, Pennsylvania, and to the 1930 publication describing it. He was always generous with his time and shared his expertise freely. Similarly, Rick Sheets, journeyman horner of the Honourable Company of Horners, generously informed me of yet another powder horn of the St. Louis style bearing the same engraved scenes as Audubon's horn. This horn is in a private collection, and I thank Rick for permission to publish his photograph of it.

Erin Greb, the superb cartographer who created the map for this volume, was a pleasure to work with and, fortunately for me, patient with my ability to discover yet another location I wanted to be indicated on the map.

At the Wellesley Historical Society, Kathleen Fahey, curator, kindly arranged to have the self-portrait of Isaac Sprague photographed and granted permission to publish it here for, I believe, the first time ever.

My friend Eugene Beckham, naturalist-illustrator, always generous with his time and talents, made all the photographs of Audubon's original field notebook and powder horn from the 1843 expedition presented herein.

For reading and commenting on drafts of several chapters, I thank Henry Fulton, Matthew Spady, and Eric Russell. I especially thank Eric for many hours of good Audubon talk at the "Audubon Bar."

Matthew Spady, the historian of Audubon Park, has deepened my understanding of the Audubon family from the 1850s onward. With specific regard to this book, his knowledge of the Shufeldt affair significantly improved the chapter about Maria Audubon and the "Great Auk Speech." Our communications have been marked by the free and open exchange of information, ideas, and materials, from which always the best scholarship emerges.

Lance Spallholz of Round Lake, New York, kindly took the trouble to supply me with his grandfather's photographs of Maria Rebecca and Florence

Audubon at their home in Salem, New York. I thank Susie Crowfoot Davis for introducing me to Lance.

For gracious permission to publish passages from a letter by Hannah Mary Rathbone in her collection, I thank Rathbone family descendant Hannah Royle of Edinburgh.

At home, Central Michigan University, I thank Dean Pamela Gates of the College of Humanities and Social and Behavioral Sciences and James H. Hageman, interim vice provost for research, for time away from my "day job" to assist with this project.

I am blessed with family and friends who, even to this very day, are willing to listen to me go on and on about one Audubon thing or another. My humble thanks to you all.

Finally, the dedication page speaks quietly of great generosity of spirit and warm cordiality. Thank you, Vic.

ABBREVIATIONS

AB   Audubon, John James. Manuscript journal. "Journal No 5."
         August 5–13, 1843. Morris Tyler Family Collection of John James
         Audubon. Gen MS 85, box 8, folder 446. Beinecke Rare Book
         and Manuscript Library, Yale University.

AJ   Audubon, Maria Rebecca, ed. *Audubon and His Journals*. 2 vols.
         New York: Charles Scribner's Sons, 1897.

AN   Audubon, John James. Manuscript journal. August 16–October
         19, 1843. "Copy of my Journal from Fort Union homeward."
         Everett D. Graff Collection of Western Americana, Vault Graff
         109, Newberry Library, Chicago. [Reprinted here in part 3.]

AP   Audubon, John James. Manuscript journal: The original field
         notebook. July 26–28 and August 16–November 6, 1843. Private
         Collection. [Reprinted here in part 3.]

APS   Audubon, John James. Papers, 1821–45. American Philosophical
         Society, Philadelphia.

AW   Audubon, John James. *Audubon in the West*. Ed. John Francis
         McDermott. Norman: University of Oklahoma Press, 1965.

Beinecke Morris Tyler Family Collection of John James Audubon. Gen MS
         85. Beinecke Rare Book and Manuscript Library, Yale University.
         [Collections are listed under the individual authors' names in
         the works cited.]

EH   Harris, Edward. *Up the Missouri with Audubon: The Journal of Edward Harris*. Ed. John Francis McDermott. Norman: University of Oklahoma Press, 1951.

MH   Audubon, John James. Collection. Missouri History Museum Archives, St. Louis.

OB   Audubon, John James. *Ornithological Biography*. 5 vols. Edinburgh: Adam and Charles Black, 1831–39.

Princeton Audubon, John James. Collection. Department of Rare Books and Special Collections, Princeton University Library.

QU   Audubon, John James, and John Bachman. *The Quadrupeds of North America*. 3 vols. New York: Victor Gifford Audubon, 1854. [Royal Octavo edition.]

RO   Audubon, John James. *The Birds of America from Drawings Made in the United States and Its Territories*. 7 vols. New York: J. J. Audubon; Philadelphia: J. B. Chevalier, 1840–44. [Octavo edition.]

EDITORIAL PRINCIPLES

I stay as close as possible to the primary documents, doing my own transcriptions as much as possible, including Audubon's journal of 1820–21. The following practices are used for all five of the manuscript journals edited and published in this volume (three by Audubon and those by Bell and Sprague in part 5).

I retain the manuscript use of the ampersand (&).

Conjectural readings are indicated in curly brackets, thus: {indeed}.

I use four asterisks in square brackets [****] to indicate lost or illegible letters.

Words canceled in the manuscript are reported here as text with a line drawn through it (e.g., ~~de pouille~~).

Clear scribal errors are silently emended: e.g., "as it as it" becomes "as it."

I indicate in square brackets where I have omitted measurements of an animal killed.

To aid reading, I silently add punctuation and capitalization where it does not alter meaning, but I take a conservative approach. I also occasionally break a long paragraph to aid reading. Marginal subheadings in the manuscript are here included on a separate line.

I occasionally expand an abbreviation that might impede reading.

Occasional explanatory footnotes are indicated by a superscript number in the text.

For the sake of accuracy and readability, I employ the following editorial principles for the diaries of Bell and Sprague.

I regularize the placement of dates, indentations to begin paragraphs, and the spelling of the names of days and months.

Beginning on May 23, Bell briefly characterizes the weather of the day and notes temperatures within the bordered heading of each date. He appears to intend to record temperatures for morning, midday, and night separated by x's. A full temperature entry includes three temperatures separated by x's. Most entries provide only one or two temperatures.

I have arranged Bell's entries chronologically. In the manuscript journal, all Sunday entries are segregated from those for the other days of the week. The first volume contains mostly single-paragraph entries for each date. Later in the same volume and in the second volume, Bell expanded the single-paragraph entries with more information and observations. To eliminate pointless repetition, I report only the fuller entries. Where information in the shorter entry is not included in the expanded entry, however, I work that into the longer entry verbatim but without notation.

Since Sprague's paragraphing principle is inconsistent, I reproduce the manuscript paragraph breaks, but where a series of related lines seem each to be indented, I combine them into a single paragraph.

I retain Sprague's practice of using dashes for punctuation. I also retain Sprague's overuse of commas because of the possibility that he places them with an ear for cadence, rhetoric, or pace. See, for example, "I lost my ballance and fell, in, and in trying to get out dropped my gun" (April 5).

Where Sprague adds a note keyed to an asterisk, I insert it in parentheses following the location of the asterisk.

THE MISSOURI RIVER JOURNALS

OF JOHN JAMES AUDUBON

PART I

Maria Rebecca Audubon, Her
Grandfather's 1843 Missouri River
Journals, and the "Great Auk Speech"

The journals of John James Audubon would have offered one of the most fascinating and illuminating reading experiences in all of American history. This observant naturalist traveled extensively throughout the Ohio and Lower Mississippi River Valleys during the unprecedented acceleration of settlement activity through the first four decades of the nineteenth century. Later he sailed to England and traveled to Scotland and France. His further travels in the United States took him throughout the northeastern states, down through the Eastern Seaboard to Florida, and then to the Dry Tortugas and along the Gulf States to Texas. He also sailed from Maine to Nova Scotia and Labrador. And everywhere he wrote lengthy, expressive entries in his journals, recording and commenting upon what and whom he saw, occasionally revealing his inmost thoughts and feelings, often in his characteristic francophone syntax. No more extensive eyewitness testimony to the youthful United States and its rivers, forests, swamps, prairies, and estuaries was ever written. Indeed, while announcing to his friend Richard Harlan (November 16, 1834) his plan to write the story of his life from the amassed journals, Audubon could not estimate the length that work would reach: "As to Numbers of Volumes, I cannot say but I know this well that had you read the Journals now around me you would be not a little puzzled to condense them even by the assistance of steam" (Beinecke).

But history, or fortune, or fate did not intend for us to enjoy or benefit from this treasure trove of writing. On December 16, 1835, what began as a warehouse fire but quickly became a conflagration in Manhattan destroyed an unspecified number of Audubon's journals and field notebooks. Audubon noted the loss while struggling to describe a "fork-tailed owl" for his *Ornithological Biography*: "[B]ut I am unable to describe it more particularly, the Journal in which it was noticed having been, along with others, destroyed by the great fire which happened in New York some years ago" (5:334). Still, a large body of his journals survived the Great Fire of 1835 and have been attested to by several witnesses. Ultimately, however, most of the surviving journals were destroyed by fire around the end of the century, but this time it was no accident. After preparing her heavily edited, bowdlerized, and partially forged edition of her grandfather's journals, *Audubon and His Journals* (1897), Maria Rebecca Audubon burned the originals.[1]

Following Audubon's death on January 27, 1851, and before the final deliberate destruction, his surviving journals were in the possession of his widow, Lucy Bakewell Audubon. In the late 1850s, if the memory of George Bird Grinnell's friend John P. Holman was correct, Lucy allowed the young Grinnell, a boy of ten or twelve at the time, to peruse the journals in her home surrounded by the natural history collections and other artifacts brought home by Audubon from his final expedition. Grinnell was reportedly drawn especially to the journals from the Missouri River expedition, presumably because of their emphasis on the excitement of hunting on the already-legendary western prairies. Holman recalled that the young Grinnell was greatly inspired by the Missouri River journals, but Grinnell himself left behind no commentary about or impressions of those journals.[2]

1   Other than the Beinecke and Newberry partial copies and the 1843 field notebook, only three of Audubon's original journals are known to have survived: those for 1820–21 and 1840–42 (both at Harvard, Ernst Mayr Library) and for 1826 (at the Field Museum, Chicago). Subscription notebooks and business ledgers are held by the Stark Museum of Art (Orange, Texas), Harvard's Houghton Library, and the Beinecke, which also has a partial copy of the 1826 journal.

2   John Reiger reports young Grinnell's fascination with Audubon's 1843 journals (*Passing* 24). Grinnell describes his time as one of Lucy's pupils in his "Memoirs" and "Recollections." Reiger interviewed John P. Holman, a close friend of Grinnell in his later years,

In the 1860s, a devastating decade for Lucy, her sons, Victor Gifford and John Woodhouse, and their families, Lucy tried to benefit from her deceased husband's journals and to put the Audubon patriarch's fame in some sort of authoritative framework. Having sold her home, Minnie's Land, on the Hudson River because of financial losses and the deaths of both sons, she moved for a while into the home of the Reverend Charles Coffin Adams in Manhattanville, where Adams and she worked together in the summer of 1867 to prepare a large manuscript from the formidable assemblage of Audubon's journals and some of the published "episodes" for the purpose of producing a biography (A. Tyler 61). This hefty manuscript was shipped to the London publishing house Sampson Low, Son, and Marston, for whom Robert Buchanan cut, shaped, and edited a publishable narrative biography. Buchanan claimed that his volume equaled "in bulk to about one-fifth of the original manuscript" (v). After Buchanan's 366-page narrative was published in 1868, Lucy wanted an American edition (as well as a chance to remove a few "objectionable passages"), so she marked up a copy of Buchanan's work and handed it over to G. P. Putnam's Sons, the original manuscript not having been returned to her. Her treatment of the journals of 1843 is brief, covering some eighteen of the book's 443 pages (L. Audubon, *Life* 417–34). The Missouri River journals in her possession comprised "four folio volumes" and concluded on August 13, 1843, which means that the calf-bound field notebook (AP) whose entries run from August 16 to November 6 had already been misplaced and forgotten in the "back of an old secretary," where Maria Audubon reported having found it in August 1896 (AJ 1:449).

It appears that the journals Lucy used to prepare the manuscript she shipped to London were dispersed among various family members after 1867. The Audubon family was also dispersing as the surviving members struggled to make their way out of the financial chaos that resulted from

---

in 1969. Holman was recalling what Grinnell told him in his eighties (Grinnell died in 1938) about his time in Audubon Park in the late 1850s and early 1860s (Reiger, *Passing* 157n45; *Escaping* 157–58).

failed business decisions, the deaths of both sons, the Civil War, and sheer bad luck. Within the family of John Woodhouse, however, Maria Rebecca Audubon (1843–1925), John's eldest surviving child by his second wife, Caroline Hall, gradually emerged as the effective head of the family. Maria had become well situated with her employer, Mary Louise Comstock, in New York City. When Maria became a full-time resident in Comstock's home in Salem, New York, in the early 1880s, her dispersed family members were welcome to join her there if they wished. (See fig. 3.) Here too Maria began to gather the journals and other materials she would use to create her two-volume *Audubon and His Journals*. In her preface to the work, Maria thanks particularly her half sister Harriet, her sister Florence, and her cousin Mary Eliza: "The first and last have lent me of their choicest treasures; letters, journals, and other manuscripts they have placed unconditionally in my hands, besides supplying many details from other sources; and my sister Florence has been my almost hourly assistant in more ways than I can specify." Maria also notes that about 1885, "[t]here came into my hands [. . .] some of these journals,—those of the Missouri and Labrador journeys; and since then others have been added, all of which had been virtually lost for years" (1:viii).[3] Thus, by 1885 or so, Maria had a base of operations, thanks to the "loving generosity" of Mary Comstock (who bequeathed the Salem home to Maria), from which she could work on the vast project of editing the journals, as well as something like a family agreement that she would be in charge of the surviving journals and responsible for shaping the public image of the patriarch of "the Audubon family" (AJ 1:ix). Alice Jaynes Tyler, who married a son of Maria's first cousin Delia, characterized the family arrangement in this way: "And so both branches of the family gave full authority to Maria Audubon to use her sole discretion in selecting for publication those portions of John James Audubon's papers which she

3   After noting that an unspecified amount of the journals that Lucy Audubon had made use of for her volume were "destroyed by fire in Shelbyville," Kentucky (AJ 1:ix), Maria also makes the only reference I know of to her grandmother's "voluminous diaries." Did Lucy or Maria also destroy these, or do some survive, waiting to help us tell Lucy's story?

considered wise to preserve; and to eliminate those she deemed it expedient to conceal" (13).[4]

Maria recruited Dr. Elliott Coues, the nation's most accomplished and prolific ornithologist at the time, to assist with her complex undertaking by writing scientific notes and giving advice. In the summer of 1897 Coues and his wife were guests of Maria and her mother and sister while they worked on *Audubon and His Journals*. According to Coues, Audubon "left altogether about thirty private journals. Most of these have perished, by fire or otherwise, but I have examined nine of them, now carefully preserved by Miss Audubon, and to be extensively drawn upon for her forthcoming publication, some of them to be printed in full" (June 1897, 150). Six months or so earlier, Maria had accompanied Coues to the annual meeting of the American Ornithologists' Union in Cambridge, Massachusetts, in order to allow him to display the journals she had gathered, at least in part to promote her upcoming publication. According to Coues, "Among these are the personal diary of 1827, when Audubon was in England, on the threshold of his great work: the Labrador Journal of 1833: and especially the narrative of his voyage up the Missouri to the Yellowstone in 1843. All these records are of absorbing interest to those who desire closer acquaintance with the 'American Backwoodsman,' as he used to style himself" (July–August 1897, 135). I think of this image of the assembled surviving journals as a still moment before a storm. Never again would most of these be seen by anyone outside the family.

Surviving letters by Maria and several published testimonials reveal a consistent family desire over several decades to limit access to Audubon manuscripts and to control the public perception of the great American naturalist. The motivation that Maria most often expressed was the desire to keep personal family matters out of public view. She included only "a few extracts" from the 1822–24 journal in *Audubon and His Journals* because of the personal misery it reported: "As I turn over the pages of this volume […],

---

4   I thank Matthew Spady for sharing his knowledge of public land records, wills, and other official records.

well do I understand the mental suffering of which it tells so constantly."
She explains further: "It has been hard for me to keep from copying much
from this journal, but I have felt it too sacred. Some would see in it the very
heart of the man who wrote it, but to others—and the greater number—it
would be, as I have decided to leave it, a sealed book" (AJ 1:54–55), which
is her euphemism for a burned book. In a letter to Ruthven Deane twenty
years earlier (July 29, 1876), she shows that this family concern was a long-
standing one: "I enclose one of the few letters which yet remain to us, and
am sorry that I cannot send one more decidedly 'ornithological' but so
much of mere family matters are in all the longer letters that we cannot
let them yet pass into other hands" (MH). Maria was wary even of writing
about the journals in letters that she knew might be preserved and shown
to others. In a letter to the ornithologist Frank Chapman on March 27, 1901,
she writes, "There is much that I cannot put on paper about the journals,
of which I would willingly speak to *you*, and there are, of course, many
minor difficulties in doing *anything* with them, because of some feeling on
the part of the 'other Audubons,' but time often settles things for us."[5] If,
as that last ominous expression suggests, there was occasional dissension
about Maria's handling of the family manuscripts, she remained in charge. A
former neighbor of the Audubon sisters in their Salem home remembered
this about her: "The dominant character of the three was Maria, 'Deedie'
as she was known at home and in the village. It was Maria who edited the
Journals of John James Audubon, destroying as she worked such portions
of the manuscript as she thought best to cast to oblivion!" (Ashton 245).
Maria's older half sister Harriet probably captures the general family feeling about what should be published about their grandfather in a letter to
Grinnell, April 19, 1915: "I am very sorry that your friend Mr. Herrick is
writing a biography of my grandfather, for there are already half a dozen
or so, and Maria's is accurate and contains all that anyone would want to
know" (Beinecke).

5 [Maria R.] Audubon to Frank Chapman, March 27, 1901, Archives, Library, American
   Museum of Natural History.

Remarks Maria makes in a letter to Ruthven Deane in 1904 (specifically about the 1820–22 journal) may illuminate a powerful catalyst in 1895 to actively protect her family. Unambiguously she writes, "I was quite truthful as to the destruction of the journals by fire so far as this one goes for *I burned it myself* in 1895. I had copied from it all I ever meant to give to the public, and if you will go back to that bitter year, *you* will perfectly understand why my mother, the other members of the family, and Dr. Coues who read it *all*, thought that in view of the existing circumstances, fire was our only surety that many family details should be put beyond the reach of vandal hands" (qtd. in Arthur 243). (See fig. 4.) "[T]hat bitter year" was 1895, and "the existing circumstances" were her sister Florence's recent and sudden marriage to and divorce from the physician and ornithologist Robert W. Shufeldt, a founding member of the American Ornithologists' Union.[6] Shufeldt met the forty-two-year-old Florence in August 1895 after he had published a brief biographical sketch of her grandfather. A month later they married. Difficulties arose, and in November Florence returned to Maria's home in Salem announcing that she would sue for divorce on grounds of adultery. Shufeldt tried to prevent her from suing for divorce by threatening to publish a lurid (he claimed "scientific") case study of "female impotency" that, while not naming either the husband or the wife involved, clearly indicates Florence Audubon as the "impotent" one. Anyone who knew them could not miss the association despite the feigned ploy of anonymity. After describing the hypothetical wife's physiological incapacity for sexual arousal and the consequent failure of the husband to achieve erection, Shufeldt devotes a section of his argument (for it reads like a legal brief) to establishing a "neuropathic taint" in the wife's "*family history*" (*On a Case* 9). He then offers numerous examples of what he terms "both erotic and neurotic constitutions in the highest degree" among the wife's family members: an uncle who was a "drunkard" and a "man of

6   See Barrow 63–67 for a full, documented account of the "Shufeldt Affair." Here I thank Matthew Spady also for pointing out the relevance of this background story to an understanding of Maria's motives.

passion"; a female cousin who "never married, and was a great coquette in early life"; a half sister who "had a child five months after marriage"; a fifty-four-year-old sister who was "intensely erotic in passing the period of the menopause" and who, immediately after learning of the patient's (her sister's) engagement "would talk to her about little else save the 'horrors of copulation,' the 'pangs of labor,' etc., and later on desired her to read works upon 'The age of consent,' 'Prostitution within the bond of marriage,' etc." (10). When Elliott Coues learned about this, he led a movement to have Shufeldt removed from the American Ornithologists' Union's membership. (See fig. 5.)

It is certainly easy to understand that Shufeldt's threat to publish such a vile and inflammatory exposé would have been extremely alarming to Maria, Florence, and their mother, then living with them. And his strategy of arguing that Florence's problems should be seen in the context of larger Audubon family problems over two generations might very well have caused Maria and the other Audubons to see all the assembled journals of Audubon—possibly also Lucy's "voluminous diaries"—as offering far too much personal family material to be exploited, misconstrued, or opportunistically interpreted. Maria would have been extremely motivated to protect her damaged sister. That acute motivation likely made her hypersensitive to the mendacious purposes to which some of the private utterances and candid observations scattered about in her grandfather's journals could be turned. And Robert Shufeldt loomed nearby as a tangible and active menace.[7]

As Huck says, "It was a close place," but I'm not sure that, in Maria's place, I would not have burned them myself.

She might have destroyed some of the journals even without such extreme cause, and she might have used Shufeldt simply as an excuse for doing

7   Nearly four years after Florence's ordeal, Coues wrote about Shufeldt to Julia Stockton Robins, April 22, 1899: "Dr. Shufeldt is morally a cancer—the vilest & most depraved wretch I ever knew. His former wife committed suicide in an insane asylum to which his brutalities had consigned her. The horrors of poor Florence Audubon's situation I never saw surpassed" (qtd. in Cutright and Brodhead 404).

what some family members opposed, but Maria's urgent need to protect her sister and her family may well have been the spark that started the fire.

Following the publication of *Audubon and His Journals*, for at least sixteen years, 1901–17, Maria worked with the New York City bookshop owner and dealer in manuscripts Patrick Francis Madigan to retrieve personal family manuscripts that had made their way to public auction. When on May 31, 1901, she received from Madigan an auction listing that included a "Unique Audubon Relic," she responded urgently: "When your letter came this morning I had not very much time to reply & wrote—as you see—very hastily. I was utterly taken aback by the boldness of the sale of matter of so private a nature and which has evidently been stolen along with the books which once belonged to Audubon. As regards the item *40* which is the only one of value if the volumes are all *accounts of animals and birds*, or *accounts of money* I don't care for them at any price, if on the contrary they are personal journals I will give $50." Her ardor for this retrieval work, however, waned as she grew older and as ever fewer items were offered for sale. In a letter to Madigan of July 5, 1917, she thanks him for sending a "catalogue with the notice of one of Grandfather's letters in it" but does not refer again to the letter.[8]

How the Surviving Partial Copies Differ
from *Audubon and His Journals*

With regard specifically to the journals from the Missouri River expedition, Maria's creative editing and rewriting intended to represent Audubon as "a refined and cultured gentleman" and visionary conservationist become clear when we contrast the recovered partial copies to the granddaughter's published edition.[9] What we'll see is that the differences between the 1843

8   Maria Audubon to Patrick Francis Madigan, May 31, 1901, cat. no. 11.3.222, and Maria Audubon to Patrick Francis Madigan, July 5, 1917, cat. no. 11.3.268, Audubon Archives, Stark Museum of Art, Orange TX.

9   In Maria's preface to AJ, she indirectly exposes some of the motive behind her editorial practices: "In these journals [...] there is not one sentence, one expression, that is other than that of a refined and cultured gentleman. More than that, there is not one utterance of 'anger, hatred or malice'" (1:ix–x). She intended to create a man practically without flaws.

journals and the 1897 published edition parallel the general changes in the attitudes of Americans toward hunting, wild animal populations, and the need for conservation that occurred during those intervening decades. I'll begin this discussion by presenting the two differing versions of the two entries from the 1843 journals that have the greatest implications for a revised understanding of Audubon's thought, those for August 5 and August 8.

The entries for August 5 read:

Beinecke Partial Copy

One can hardly conceive how it happens, that notwithstanding their general deaths and the immence numbers that are murdered daily over the immence wastes, called prairies, that so many are yet every day to be found!—Sprague never spoke to me on my return, but that is his way. Bell related all his adventures to every one [****] and I am off to bed pretty well fatigued. Our boat is going [****] and I wish I had a couple more big horns Males.

AJ

One can hardly conceive how it happens, notwithstanding these many deaths and the immense numbers that are murdered almost daily on these boundless wastes called prairies, besides the hosts that are drowned in the freshets, and the hundreds of young calves who die in early spring, so many are yet to be found. Daily we see so many that we hardly notice them more than the cattle in our pastures about our homes. But this cannot last; even now there is a perceptible difference in the size of the herds, and before many years the Buffalo, like the Great Auk, will have disappeared; surely this should not be permitted. Bell has been relating his adventures, our boat is going on, and I wish I had a couple of Bighorns. God bless you all.

Having spent the previous day, August 4, seeking "the best promise of sport"—that is, chasing and killing bison cows—(EH 166), Audubon, in the company of his good friend Edward Harris, the Fort Union supervisor, and a few assistants made their way back to the fort by midday. There the famed trapper and trader Etienne Provost, who had just returned from an attempt to kill bighorns for Audubon, told him about the poor condition many bison reach in "hard winters when the snow covers the ground to the depth of 2 or 3 feet, that they lose their hair, become covered over with scales, on which the Magpies feed and the poor beasts die by hundreds" (AB). The Beinecke copy then shows that Audubon responded with an expression of disbelief and excitement, saying it was hard to believe that with so many causes of bison deaths, "so many are yet every day to be found!" The exclamation point is significant, and Maria removed it in order to change the tone of the passage from excitement to lamentation. In Maria's version, however, after noting how abundant bison are on the prairies, Audubon supposedly writes what I have come to refer to as the "Great Auk Speech": "But this cannot last; even now there is a perceptible difference in the size of the herds, and before many years the Buffalo, like the Great Auk, will have disappeared; surely this should not be permitted." The Beinecke copy includes no trace of this concern about the future disappearance of the bison. Audubon simply complains about Sprague's odd behavior and retires for the evening. Maria's version omits the awkward reference to Sprague.

The differences between the two passages for August 8 are less conspicuous but equally significant.

Beinecke Partial Copy

We were told also that a few minutes after our departure the roarings of Buffaloe had been heard across the river, and that Owen & two men with one cart had been despatched to kill 3 fat cows, and no more. Harris was sorry that he had not been here, and so was I, as both of us would have gone and joined & seen the fun.

AJ

We were told also that a few minutes after our departure the roarings and bellowings of Buffalo were heard across the river, and that Owen and two men had been despatched with a cart to kill three fat cows but no more; so my remonstrances about useless slaughter have not been wholly unheeded. Harris was sorry he had missed going, and so was I, as both of us could have done so.

On this Tuesday morning, Harris and Audubon had traveled about two miles north of the fort to seek fossils. On their return to the fort, they were informed that bison had been heard from across the river shortly after they left on their fossil hunt. The Beinecke copy then reports that three men had been "despatched to kill 3 fat cows, and no more." Maria's version interprets that order as a new restraint on the usual hunting practices and attributes the change to Audubon: "so my remonstrances about useless slaughter have not been wholly unheeded." In both versions, Harris and Audubon regret having missed the chance to accompany the hunting party, but the reasons given for that regret are strikingly different. Whereas Maria supplies an utterly innocuous reason, "as both of us could have done so," the Beinecke copy that Audubon asked his scribe to make provides a much more substantial and believable reason, "as both of us would have gone and joined & seen the fun," which is not at all consistent with the purported August 5 likening of the fates of the American bison and the great auk. But it is consistent with what Audubon writes a few lines later: when he and Harris learned that the hunters sent out earlier had seen more than three hundred bison, "that made Harris & I regret the more our having gone fruitlessly after stones."

Looming up out of these very different versions of journal entries are two very different versions of John James Audubon. One foresees the extinction of the bison in 1843, compares it to that of the great auk, and admonishes his colleagues to kill fewer bison, while the other Audubon does not utter a concern about extinction, expresses rather his excitement that "so many are yet every day to be found!," and regrets having missed the fun of joining in on yet another hunt. Determining which of these is accurate and which is false is important to our understanding of Audubon at this time in his life and in the history of American conservationist thought. Let me now carefully present the evidence and reasoning that lead me to conclude that the

Beinecke copy is faithful to the original journal and that Maria's version is faithful to her plan to produce a more refined and visionary Audubon in 1897.

I begin by assuming that both versions are based upon the same original manuscript journal that Audubon wrote. It seems highly unlikely that Audubon would have arranged for a scribe, probably Lewis M. Squires, to make a copy of his original journals but would have had him copy from some other hypothetical version of the journals that had already been drastically altered from the originals. Thus, both the scribe and the granddaughter worked from the same original document.

One of the most important considerations when evaluating Maria's entry for August 5 is that the great auk was not extinct in 1843. It was indeed on the verge of extinction, but that fact was not known at the time (Fuller 60–85). When Audubon published his brief essay on the great auk in 1838, he included several reports that the "'Penguin,' as they name this bird" was still breeding. One man reported catching a great auk on a fishing line in recent years while on a "passage from New York to England" and having kept the bird alive on board for several days. While Audubon was on his Labrador expedition in 1833, he further notes, "many of the fishermen assured" him that the great auk "breeds on a low rocky island to the south-east of Newfoundland, where they destroy great numbers of the young for bait." He was unable to verify this report himself because "this intelligence came to me when the season was too far advanced" (OB 4:316). That he continued to believe the species to be breeding as late as 1844—the year after Maria's "Great Auk Speech" states he thought the species extinct—is clear from a note added to the Octavo version of the essay: "Rare and accidental on the Banks of Newfoundland; said to breed on a rock near that island" (RO 7:245). The fact that Audubon owned a stuffed specimen of the great auk, having purchased it in London in 1836, may also have contributed to his sense of the species' continued existence (Fuller 156). When Maria Audubon was editing her grandfather's journals in the mid-1890s, the "disappearance" of the great auk was one of the most famous examples of human-caused extinctions. Symington Grieve had published his comprehensive study

The Great Auk, or Garefowl in 1885, and Elliot Coues, who was assisting Maria in the preparation of her edition, had made specific inquiries about the possible existence of great auks on his trip to Labrador in 1860. While Coues had not been able to find anyone who had actually seen a great auk, he was told they still existed on one island off the coast of Newfoundland (Cutright and Broadhead 11). Thus, in 1897, the extinction of the great auk would have been one of the most timely illustrations of what humans were on the verge of doing to the American bison, whose population had fallen from the tens of millions at the close of the Civil War to near extinction in 1884 (Lott 168–86).

To accept Maria's versions as faithful to the originals requires accepting some highly unlikely, even absurd, scenarios. Everyone who has ever tried to make an accurate copy of a text knows that it is the easiest thing in the world to skip over a few lines and inadvertently omit them from the copy. It is certainly possible that Squires simply omitted the "Great Auk Speech" from the August 5 entry. It is such a dramatic and eloquent piece of conservationist rhetoric, however, that it would draw attention to itself as something Squires should be especially careful to include. Nevertheless, its inadvertent omission is possible. However, the scribe's addition of the tone-setting exclamation point—"that so many are yet every day to be found!"—is another consideration. To believe that a paid scribe would take it upon himself to alter the tone of a particular passage requires believing that he had some agenda, some purpose causing him to make changes along the way. To accept Maria's version as the accurate one in this place, one must assume that Squires had such an agenda and added the exclamation point to change his employer's tone of lamentation (as in Maria's version) to one of giddy excitement that more bison "are yet every day to be found!" That would constitute a form of scribal sabotage, which is not supported or suggested by any evidence. Similarly, to accept Maria's version of the August 8 entry where Audubon and Harris regret having missed going with the hunting party because they "could have done so," one must believe that Squires was impish enough to substitute a reason—"as both of us would have gone and joined & seen the fun"—that directly contradicted the ethic asserted in the "Great Auk Speech." On the other hand, if Audubon did not write the "Great Auk Speech," and if on August 5 he was genuinely excited that "yet every day" there would be more hunting action for him and his companions who had traveled so far for precisely that, and if on August 8 he and his friend Harris genuinely would have enjoyed yet another chance to chase or watch bison being chased, the unlikelihoods and absurdities vanish.

Further support for the primacy of the Beinecke copy comes from Harris's journals. If Harris's good friend had experienced such a conversion on August 5 as Maria's "Great Auk Speech" suggests, Harris either did not know about it or chose not to comment upon it in his own journal. After their return that day from the previous evening's prairie camp in the rain, Harris turns only to his personal comfort: "A hearty dinner and a nap [. . .] completely restored me. The weather for the past few days has been very cool and bracing and I now feel my strength much recruited" (EH 168). Harris's journal entry for August 8 clarifies the meaning of Audubon's report that "the roarings of Buffaloe had been heard across the river" after he and Harris had departed that morning; it was a missed opportunity for a hunt: "On our return from the quarry, Mr Culbertson told us that he had sent Owen across the river to run a band of Buffalo whose bellowing had been heard soon after we started. I now recollected that I had heard the noise in the morning while watching for rabbits and took it for the bellowing of the Bull of the Fort. Had it not been for this unfortunate mistake I should have had another fine run after Buffalo cows" (170). His attitude toward missing this opportunity is certainly consistent with the Beinecke copy's explanation that he and Audubon "would have gone and joined & seen the fun."

Harris's journal also sheds light on Maria's claim that it was Audubon's "remonstrances" that caused the hunters to kill no more than three bison this day: "Owen returned soon after dinner having killed two cows and would have killed three, the number he was commissioned to bring back, but his stirrup broke" (EH 170). From Harris's perspective, at least, the issue is an equipment failure, not a "remonstrance" of his friend Audubon that fewer bison be killed. Culbertson, the superintendent of Fort Union, instructed his hunters in how many to kill based on the fort's needs. If Culbertson lowered the number to three on this occasion because Audubon advised him to restraint, there is no suggestion of that outside of Maria's "remonstrances" passage. Harris's focus purely on the sport is unmixed with conservationist concerns, at least partly because the bison seem to be *increasing* in numbers in the region of the fort: "It is now evident that the Buffalo are moving towards this quarter where they have been very scarce for the last two years. We are promised one more hunt before we leave, and I am in hopes that some such chance may be thrown in our way to give us still another opportunity. Both Bell & I have been remarkably successful for beginners having killed all the cattle we have started after" (170).

Only once do the journals evince something like Maria's "remonstrances." On either July 20 or 21, each of the three journalists—Audubon, Harris, and Bell—seems to have been affected similarly by a particularly wasteful hunt that day. According to Maria's edition, Audubon wrote, "What a *terrible destruction of life*, as it were, for nothing, or next to it, as the tongues only were brought in, and the flesh of these fine animals was left to beasts and birds of prey, or to rot on the spots where they fell" (AJ 2:107). The journals of Harris and Bell corroborate this emotional response (neither of the partial copies includes an entry for July 20 or 21). Bell is brief: "[R]eturned to camp very much regretting what we had done when they told us they could not carry any of the flesh home." Harris expostulates more fully and philosophically: "We now regretted having destroyed these noble beasts for no earthly reason but to gratify a sanguinary disposition which appears to be inherent in our natures. We had no means of carrying home the meat and after cutting out the tongues we wended our way back to camp, completely disgusted with ourselves and with the conduct of all white men who come to this country. In this way year after year thousands of these animals are slaughtered for mere sport and the carcasses left for the wolves. The skins are worth nothing at this season" (EH 149). Other than on this single day, however, none of the journals—except in Maria's edition—reflects a feeling among the men that they ought to curb their killing. Even on this day, the issue the three men address is the waste of the slain animals, not the diminishment of the size of the herds. And Harris provides a kind of confession that "all white men who come to this country" kill "for mere sport" and without restraint.

It was possible in 1843 to imagine the extinction of the bison (as George Catlin did before 1841 [1:247]), but given the great numbers Audubon and his companions witnessed and Audubon's evident lack of worry that humans might threaten the species' survival, it seems unlikely that he would have been moved to predict their extinction on this expedition. On July 23, for example, Bell reports that Audubon was placing bets in a sporting fashion on how long it would take Bell, Harris, and Culbertson to chase and kill four bison bulls, suggesting that he did not yet believe that such activity would threaten the species: "Mr Audubon bet we would kill all four in less than one hour from the time of leaving the fort." Then on July 28, just eight days before Maria's August 5 "Great Auk Speech," Maria reports that Audubon had been told that two hundred miles from Fort Union "there are eight hundred carts in one gang, and four hundred in another, with an adequate number of half-breeds and Indians, killing Buffalo and drying their meat for winter provisions, and that the animals are there in millions" (AJ 2:122). And on the very day that Maria claims Audubon expressed his fear that humans would eradicate the bison, Bell is relating his stories of the thousands of bison he and Owen have seen on a hunting excursion over the preceding four days. On the previous day, August 4, Bell reports in his journal, "Saw at least 5000 on our route." Bell's journal entry for August 5 includes this vivid scene of bison abundance:

> We then stopped at half past eleven to eat our breakfast and let our horses rest and eat, hundreds of buffalo within a few hundred yards of us. We then started again at half past one being anxious to reach the fort before night, so we trotted our horses nearly two thirds of the time, saw several antelopes, and a very large band of buffalo, and as we came opposite to them they got the wind of us, and started to pass in front of us. We then

rode on until we parted them and passed through. After about 300 had passed us, the others then passed behind us, in a line, and they kept comeing and comeing from the River and running to the hills. They formed a complete line, one, two, and sometimes three deep, for more than two miles long with scarcely an opening between them.

According to Maria's edition, Audubon is expressing his fear that the bison will follow the great auk into extinction at the very moment that Bell is telling of having seen bison by the thousands earlier that same day: "But this cannot last; even now there is a perceptible difference in the size of the herds, and before many years the Buffalo, like the Great Auk, will have disappeared; surely this should not be permitted. Bell has been relating his adventures." Maria's entry for August 5 certainly creates an anomalous juxtaposition. Consider also that on the descent of the Missouri River, neither Audubon nor anyone else showed any scruples about killing more bison. Indeed, if Bell was correct, on August 17 Audubon was inclined to take a bit too much credit for a bison kill. (See my discussion about the events of August 17 in part 2.)

A further piece of evidence I offer against Maria's claim that Audubon feared the extinction of the bison on August 5, 1843, comes from a letter he wrote to Spencer Fullerton Baird shortly after he returned home from the expedition. Baird's friends and family had dissuaded him from accepting Audubon's invitation to accompany him on the expedition, and Audubon writes now to say that his young friend should regret not having come along because of how utterly enjoyable the trip was:

Why, only think that I saw not one Rattlesnake and heard not a Word of bilious fever, or of anything more troublesome than Muschietoes and of those by no means many! No, our Trip was a pleasant one. Abundance of the largest Game was killed, and much more could have been procured had we wished for it; but when a fat buffalo weighing some 1500 pounds or upwards is dead and the camp is prepared and the beast is roasting by large Juicy pieces, who could have the heart to kill more for the sake of the Tongue, or for that of the Wolves? Why, not I, I assure you. (qtd. in Dall 93).[10]

It's really difficult to believe that a naturalist who three months earlier had feared that humans would drive the bison to extinction could now so gleefully celebrate the image of "large Juicy pieces" of the animal roasting on a campfire. But if the "Great Auk Speech" is a forgery and we remove it from the body of Audubon's writings, the number of such anomalies diminishes dramatically.

Finally, when Audubon composed the original essay on the American bison ("Buffalo") for his coauthor Bachman to rework for publication in their *Quadrupeds*, he opened with an overview of how the range of the bison had been diminished over the preceding half century: "In times not longer passed away than the days of our youth, Buffaloes roamed over the small and beautiful prairies of Indiana, and Illinois, and kinds of them stalked through the grand open Woods of Kentucky, and Tennessee, but towards the years 1808, and 1809, they had dwindled down to a few stragglers" (Beinecke). The herds, however, still existed in the West: "Their range has since that period gradually tended westward, and now you must take to the Indian country, and travel many a hundred miles beyond the fair vallies of the Ohio, towards the great Rocky chain which forms the back bone of North America, before you can see the Buffalo feeding in his sturdy independence upon the vast elevated plains." And there, he writes, "we may yet find thousands of Buffaloes." As Audubon presents the distribution of bison in this manuscript, which internal evidence shows that he composed after returning from the 1843 expedition, they are remote from human settlements but enjoying a "sturdy independence": "The range of the Bison is still very extended, although this animal (which was once met with on the Atlantic coast) has like many other species gradually receded, and

---

10 The date Dall gives this letter from Minnie's Land is November 3, 1843, and must be in error since Audubon arrived there on November 6. November 3 is also inconsistent with Audubon's reference in the letter's first line to a letter he received from Baird "of the 24th Inst.," meaning literally November 24. I have not seen the original.

gone west, and south before the spread of civilization, and the advance of the axe, and the plough." This essay is not colored by a fear of the possible extinction of the animal he and his companions hunted on the northern prairies: "At the present time the Buffalo is found in vast herds in some of the great prairies, and are more sparsely scattered over the whole length and breadth almost of the barrens east and west that adjoin the Rocky Mountain chain." In fact, he is so unconcerned about the human threat to the bison that he opens the next section of the essay with a promise, not of natural history or behavior, but of good hunting stories: "As various accounts of Buffalo hunts have been written already, we will pass over our earliest adventures in that way [...] and give you a sketch of the mode in which we killed them during our journey to the west in 1843."

It is truly unfortunate that, given what we know, we cannot rely upon the edition of Audubon's journals prepared by the last person to have had them all in hand—that is, without corroboration from other sources. The examples of the entries for August 5 and August 8 make me suspicious of many other passages in Maria's edition that may very well be faithful to the original manuscript journal, but in her edition you can't tell the gold from the brass. For example, on April 26 Maria reports that Audubon expresses regret when Captain Sire does not stop to retrieve a goose shot by Bell: "We saw a Wild Goose running on the shore, and it was killed by Bell; but our captain did not stop to pick it up, and I was sorry to see the poor bird dead, uselessly" (AJ 1:457). Bell's diary confirms that he "Shot one wild goose," but neither Sire's log nor any other evidence suggests that Audubon expressed pity for this bird. Similarly, on May 16 Maria shows him penitent once again: "I started a Woodcock, and caught one of her young, and I am now sorry for this evil deed." That last phrase certainly surprises the reader. He has left the boat with two others in order to "hunt to our hearts' content" (1:494), yet a few lines later he brands a simple act of hunting as an "evil deed." Other scholars have been suspicious of Maria's edition because of such glaring inconsistencies and anomalies, and I can't avoid suspecting that Maria rather ineptly sprinkled occasional phrases into her grandfather's prose in order to make him a bit more presentable for her sense of audience in 1897.[11]

All of this leads me to conclude that the "Great Auk Speech" is an eloquent piece of rhetoric that belongs in the canon of the American literature of conservation, but it was not written by a visionary conservationist in 1843. It was written by Maria Rebecca Audubon in the mid-1890s in response to the destruction of the American bison herds after the Civil War. The likely extinction of the species loomed large in the imaginations of Euro-Americans in the 1870s and 1880s. They "did not slaughter millions of bison [...] believing that nature provided an inexhaustible supply. Rather, they anticipated the extinction of the species. They regarded the disappearance of the herds as a triumph of civilization over savagery, because the extermination of the bison removed the nomads' primary resource and cleared the plains for Euroamericans" (Isenberg 162). Amid all her other motives for editing and rewriting much of her grandfather's journals, Maria Audubon also wanted to create an Audubon patriarch who, as early as 1843, the year of Maria's birth, could have foreseen and warned against the great destruction that occurred during her lifetime.

In the following chapters, I will explore—without the distraction of the conversion narrative—all the other evidence of Audubon's thought about how humans should relate to the natural world with the intent of establishing a more thorough and accurate understanding of the creator of *The Birds of America*. Because what he did and what he thought and wrote were not always perfectly in sync, a more complex and subtle Audubon will emerge. He did not live the life of a conservationist, but in his written word, we find an awakening vision of a more humble human posture on the land.

---

11 For a few expressions of doubt about Maria's edition, see Durant and Harwood 286–88; Knott 203n6; and Streshinsky xiii–xv.

PART II

AUDUBON'S MISSOURI RIVER
EXPEDITION OF 1843

His Eminence

From his seat in the front of the meat cart, he could see low, rounded hilltops scattered in all directions. No forest blocked his view of the northern prairie landscape, a rolling sea of mixed grasses, wildflowers, and the occasional ragged green cone of a juniper. The western breeze was cool in his hair, but the sun was already warming his grizzled beard. He had not shaved since leaving their new home on the Hudson River three months earlier, a full month earlier than he really ought to have departed. And he would sport this unkempt shagginess in order to return with the prairie wildness still about him.

Upwind four dark bison bulls, about a mile out, grazed in a hollow at the base of a rock outcropping. Culbertson on his hunter and Harris, Bell, and Squires on theirs galloped due north to approach their prey from behind a range of hills. How Harris and Bell had excelled out here. No one had expected to meet so quickly the physical demands of "the chase." This prairie hunting was flight and fury, a pounding, jarring race of frenzied, heavy mammals over dangerous unpredictable terrain until your horse bolts suddenly to the side and you shoot ass over teakettle ten feet from saddle horn to prickly pear, your rifle leaning cockeyed like a derelict scare crow, butt in the air, unusable until you clear out the soil.

He could smell them—beastly mammals, noble veterans of the prairies, frightful when pursued. He straightened up on the wagon seat when the bison began to lope away from the four figures that crested the ridge behind them. Two of the bulls pursued by two of the hunters, all at a dead run now, veered off to the west. He heard two gunshots. The remaining two continued their flight along the valley floor toward him. He heard and saw the smoke from two more gunshots, and a hunter shouted. His bull halted, stood still, waiting. The hunter stopped his horse, dismounted with pistol in hand. The final bull still steered a course for the base of his knoll. He saw the hunter toss gunpowder from his powder horn into his left hand and that into the barrel of his rifle, then take a ball from his mouth and push it in on top of the powder, tamp it down with one strike on his saddle, lean forward in his stirrups, aim, and fire. The bull's knees buckled under his massive bulk, and he crashed forward in a cloud of dirt and dust onto his muzzle, then horns. In his death, he fell over on his side—not sixty yards away. Perfect.

Mr. Audubon, the impresario of this expedition, drew the small calf-bound notebook from a pocket and began to pencil, in brief phrases to be developed later, the names of the men and what each did and when, the role each played in this perfect hunt.

The story of Audubon's Missouri River expedition is the story of a restless, gifted man who cared passionately, purely, even innocently about the world about him that his benevolent creator had made but then left his own troubled species alone to figure out how to live in it. He loved this world. Out in it, he felt free because of his strong legs, his long stride, his sure aim, the constantly renewed prospect of a new prize, a new bird brought down and possessed, and the stolen freedom of several months away from an awareness of money, "the needful" (*Journal of 1826* 99).

It can also be seen as the story of a people, a new nation, that spoiled everything they loved and touched, at first obliviously, but in subsequent decades—as the evidence mounted and the Euro-American population increased and spread ever farther westward—with increasing compunction and regret. The 1840s in America is the decade in which it ceased to be possible to deny that Americans were altering the natural environment by ways and means that were unsustainable over the long run. Seen in this light, Audubon's 1843 expedition takes on the appearance of a last act of such denial, the end of our innocence, so to speak.

Of course, the first and nominal purpose of this expedition was to advance the natural history of the mammals of North America that Audubon would coauthor with the minister-naturalist John Bachman of Charleston, South Carolina, to be entitled *The Viviparous Quadrupeds of North America*. The journals show, however, that—while some of the natural history work was done, and while a considerable number of drawings were completed—as Audubon and his companions acquired more experience on the northern prairies, the sporting aspect of the expedition held sway.

John James Audubon was a sportsman. Throughout his life, he was handsome and charismatic and women were attracted to and charmed by him. He himself once wrote, "Without female society I am like a Herring on a Gridle" (*Journal of 1826* 162). Still, he enjoyed the company and conversation of intelligent, informed sportsmen above that of any other type of person. He was an artist and a naturalist as well, of course. The basis of both those identities, however, was his ability with the gun and the considerable pleasures he derived from gunning. There is much to learn about the expedition from reading the journals, and the evidence will bear much analysis and interpretation, but above all, it seems to me, in their day-to-day lives, Audubon and his men focused increasingly on adventure and sport rather than on natural history and art.

Preparations

Audubon's westward momentum was born in 1806 when he returned to his eastern Pennsylvania estate, Mill Grove, from visiting his father in France to ask permission to marry his British neighbor Lucy Bakewell and he began to feel the pull of commerce and a livelihood down along the Ohio River Valley. The young couple's wedding journey down the Ohio River in 1808 propelled them along a powerful current on which floated thousands of new settlers ever and increasingly farther westward. Lewis and Clark had passed this way just five years before. The Corps of Discovery was known to have survived and to have returned to St. Louis, and, while no results of that expedition had yet been published, news of this successful venture into the West and back was in the air. After the young couple sojourned in Louisville and then settled at Henderson, Audubon and his business partner quickly dreamed up a venture to sell goods even farther to the west, in Sainte Genevieve, Missouri, and risked much to make the attempt. The energy of the Ohio River Valley set the imaginations of Anglo-Americans in motion—westward motion.

At least as early as 1820, Audubon developed a practical yearning to accompany a major expedition to the west and northwest. When the expedition of Major Stephen H. Long stopped in Cincinnati in 1820, where Audubon was somewhat precariously employed by Dr. Daniel Drake at the Western Museum, he met and heard stories from Long and several who accompanied him. Among these was an impressive group of some of Philadelphia's most highly trained and accomplished naturalists and artists: Thomas Say, co-founder of Philadelphia's Academy of Natural Sciences; Augustus Jessup, geologist; Samuel Seymour, painter of landscapes; and Titian Peale, scientific illustrator and son of Charles Willson Peale, one of the country's most famous naturalists and the founder of America's first natural history museum (Goetzmann 182–83). For a nature illustrator in need of a more constructive context, the idea of a well-commanded scientific expedition would have stirred the imagination. Later that year, when Audubon set out on a flatboat down the Mississippi River on a desperate mission to make money drawing people's portraits while also advancing his bird drawings, his desire for a lengthy expedition was inspired by an invitation to join a merchant on his next trip "up to the Osage Nation about 900 Miles": "so *Strong* is My Anthusiast to Enlarge the Ornithological Knowledge of My Country that I felt as if I wish Myself *Rich again* and thereby able to leave My Familly for a Couple of Years" (December 10, 1820, manuscript journal, Harvard). Subordinating his desire to be "*Rich again*" to his enthusiasm for ornithology, he reveals that he viewed wealth as necessary primarily to provide for his family while he would be traveling. His dream of an expedition was next excited three months later when he learned that a surveying expedition would explore the Red River Valley. His journal entry dated March 21, 1821, shows that he quickly rounded up recommendations that he be appointed artist to this expedition. When he was met with encouragement, he is exultant: "Walk^d out in the afternoon seeing Nothing but hundreds of New Birds, in Imagination and supposed Myself often on the Journey." This prospect of painting hundreds of new birds roused him to walking about the streets of New Orleans, "[n]ot unlike (I dare say) a Wild Man thinking too much to think at all." The power of this early dream of an expedition is evident in this vision from later in the same journal entry: "My Life has been strewed with Many thorns but could I see Myself & the fruits of my Labour safe, with My Beloved familly *all Well* after a return from Such an expedition, how grateful Would I feel to My Country and full of the Greatness of My Author." Although he was not appointed to the surveying expedition, by this time the felt need for a western expedition had

become foundational to his conception of his work, and it would continue
to mature through the very busy 1830s.

During the 1820s, however, there was no possibility for such travel. He
and his wife were forced to eke out an economic existence through various
kinds of teaching and drawing. Following his stupendously successful travels
in England, Scotland, and France from 1826 to 1829, he dedicated himself
to acquiring and drawing birds in northeastern states that he needed for
the ongoing publication in London of the plates of *The Birds of America*.
Lucy then accompanied him back to England, where they worked at what
had become the new family publishing business for a little over a year, but
when they returned to the United States, on September 4, 1831, Audubon's
dream of a western expedition was revived. He lost no time at all planning
an itinerary that would take him down the east coast of Florida and from
there, potentially, along the Gulf States and on across the continent to the
Pacific Ocean. He expected to be gone about two years. He characterized
his ambitious travel plan as follows in a letter he wrote for publication
from St. Augustine, dated December 7, 1831: "[T]he Floridas, Red river,
the Arkansas, that almost unknown country California, and the Pacific
ocean. I felt myself drawn to the untried scenes of those countries, and it
was necessary to tear myself away from the kindest friends" ("Letter from
Audubon to the Editor" 361). But the timing wasn't right for such a trip.
Lucy was staying with her brother's family in Louisville and sensing that
she was a burden. Some money was coming in from subscriptions for *The
Birds of America*, but even more was going out to pay the costs of producing
the plates and their travel expenses. Lucy was as supportive as a reasonable
person could be, but she was compelled to explain, in a letter dated March
19, 1832, all these and other difficulties that would arise from his prolonged
absence at this time:

> As to the trip across the Rocky mountains I have made every possible
> inquiry about from many persons who have travelled with the Indian trad-
> ers [...]. They all concur in the danger at this period to be encountered

from the disturbed indians. The Company leave Franklin in Missoury the last of April and expect to be absent a year, that you cannot join, and if you did, they keep the beaten track stopping only at night not for the benefit of Natural history. And my Dear it has been more than once suggested that your absence from London so long will be injurious to your work—I hope and trust this may reach you and that you will see the utility and force of what I say. (Beinecke)

Throughout the 1830s, while the engraved folio plates were being produced in London, the dream of a western expedition remained alive, but he would not be able to get away long enough until the plates were done and his wife and sons and their families were installed in the new home he would establish in 1842, Minnie's Land.

Following the completion of the plates in 1838 and the publication of the final volume of *Ornithological Biography* in 1839, Audubon and his family worked to secure their financial future for a good life at Minnie's Land, which was far from easy or guaranteed. Audubon traveled apparently tirelessly to woo more subscribers for, now, three projects: *The Birds of America*, an octavo edition of the *Birds* that would publish the prose bird biographies alongside the reduced (i.e., "octavo") lithographic plates, and the projected *Viviparous Quadrupeds of North America*. We see the *Quadrupeds* at an early stage in a letter to Professor Thomas McCullough of Nova Scotia dated June 26, 1841: "I received your kind letter of the 26th April in due course, but have not answered to it, positively because of my having been constantly engaged in the drawing of Quadrupeds (Viviparous) for my contemplated Work of the animals of that Familly which are to [be] found in *North America*, and I have made 25 Drawings containing 40 figures all the size of Nature within the last Two Months" (Princeton).[1] With the *Quadrupeds*

---

1   The earliest reference to a plan to produce a volume about quadrupeds that I am aware of occurs in a letter from John Woodhouse Audubon to his brother Victor dated April 6, 1834. John writes, "[B]ut the plan is to publish with you the, 'Quadrupeds of America' but I am not able to come to any conclusions for want of your advice, and that will be had when we meet" (Beinecke).

in development, Audubon's desire for a western expedition began to look more like a pragmatic need. He would have to travel to where the western mammals could be observed, hunted, and drawn. During this roughly three years of travel and overseeing the construction of Minnie's Land, Audubon combed through most of the eastern states and into Canada as far as Quebec. His need for income, of course, was a powerful motivation: income to balance expenses. As he wrote to his friend Richard Harlan of Philadelphia on June 30, 1839, "I find myself very little the better in point of recompense for the vast amount of expedition I have been at to accomplish the task. I find that unless I do labor more—or as Madame G. would say, *'Je me tue pour vivre,'* why, I will in fact die perhaps still poorer than I was when I began, which, God knows, was in all respects poor enough" (qtd. in R. Tyler, "Publication" 121–22).

Intensifying the drama and turmoil of this period were the sad, prolonged illnesses and deaths of the young wives of both sons, John's wife, Maria Rebecca, on September 15, 1840, and Victor's wife, Mary Eliza, on May 25, 1841. They were the twenty-three- and twenty-two-year-old daughters of the Charleston pastor and naturalist John Bachman, who would share the title page of *Quadrupeds* as coauthor with Audubon. Both suffered from tuberculosis. Maria Rebecca gave birth to two daughters before she died. One after the other, each son took his ailing wife to Charleston in the hope that she would recover. Late in 1840, Victor took Eliza to Havana for a while, but then they went back to Charleston, where he wrote home on May 5, 1841, that there was "no hope of Eliza's recovery" (Beinecke). She died shortly after their return to New York. On the day of Eliza's death, Audubon painted a family group of cottontail rabbits, depicting two parents overlooking their vulnerable young one. On the back of this painting Audubon wrote, "I drew this Hare during one of the days of deepest sorrow I have felt in my life, and my only solace was derived from my Labour. This morning our beloved Daughter Eliza died at 2 o'clock. She is now in Heaven, and May our God for ever bless her Soul!" (qtd. in Ford, *Audubon's Animals* 49).

During this period, John was charged with reducing the *Birds of America* double elephant folio plates to octavo size, using a camera lucida,[2] but in Charleston with his ailing wife, his productivity understandably slowed, stopped, then slowly resumed after she died. Victor managed to keep up his sales efforts and attention to business matters even while traveling with Eliza to Mobile, New Orleans, and Cuba. Audubon himself was regularly working fourteen-hour days.[3]

When the Audubon family moved into the new residence Minnie's Land in April 1842, in addition to the ongoing publication projects, there were increasing demands on everyone's attention in the form of John's two children and his second wife, Caroline Hall. And aside from an occasional reference to his "next coming Great Western Journey" (qtd. in Dall 78), Audubon did not yet have a specific route or plan in mind for his western expedition. A plan began to materialize, however, after significant conversations in Washington with Colonel Charles Pickering, the naturalist on the Wilkes expedition of 1838–42, who had just been appointed superintendent of that expedition's collections (Viola and Margolis 59), and Pierre Chouteau Jr., head of the American Fur Company. In order to find more subscribers, Audubon left home on July 11 to travel to Philadelphia, Baltimore, Washington, and Richmond, carrying with him the first four lithographic plates of the *Quadrupeds*. In Washington he found both Pickering and Chouteau staying at Fuller's Hotel and eager to help him decide where to go and how to get there. In his letter to Lucy of July 17, 1842, Audubon excitedly reports that he has received two tremendous offers of assistance from very influential sources. Pickering advocated for a land and river route that would take him along "the Saskatchewan River, where Boats were ready to Convey me to head Waters and horses afterwards through unto Corvil [probably Fort Colville, near Kettle Falls on the upper Columbia River] &ͨ &ͨ.—Pickering says that I am recommended in the very highest

2   A camera lucida is a device that employs a glass prism through which the image of an object can be projected onto paper to be traced.

3   Ron Tyler gives a full account of this period in "Publication."

terms at all the posts of the Fur Company down to Vancouver and up to Latitude 66 along the coast"—that is, as far north as the Bering Strait. The commandant at Fort Colville, Pickering assured Audubon, "has received orders from the British Government and the President of the London Fur Company to receive me with all Kindness and do all that I might desire." Chouteau's suggestion was geographically more modest, and possibly more appealing for that reason: "M͟r Chouteau on the other hand would grant me passage to the Yellow Stone and bring me back the following Year! There's News for you all" (APS). On the nineteenth Audubon "took a long walk with Mr. Chouteau," when he no doubt learned much more about what a trip up the Missouri River would entail (*Journal Made While Obtaining Subscriptions* 75).

Another invitation he received was from the wealthy Scottish nobleman and sportsman Sir William Drummond Stewart. He invited Audubon to join his party for an overland expedition into the Rockies, but ultimately, after hearing and considering Stewart's proposal for some time, Audubon did not wish to be a subordinate part of any expedition. He wanted an expedition focused sharply on the work he needed to accomplish.[4] About Stewart's party, he wrote, "He is most anxious that We should Join his party and offered us every Kind of promises &͟c but it wont do for us. he has first too many people of too many sorts [. . .]. he speaks of the Dangers of his trip, and I have no doubt on that a/c would like to add our 5 bodies, Guns &͟c to augment his Cavalcade" (Princeton).

For several years Audubon had been tentatively recruiting men of various capabilities and expertise to join him whenever he got his western expedition together. His first choices would have been his longtime friend Edward Harris and his son John Woodhouse, the latter of whom had accompanied him on the Labrador expedition of 1833 and proven himself quite capable. He approached the young naturalist and Harvard graduate Thomas Mayo Brewer in a letter dated May 26, 1838, about such a possibility: "How would

4   For Audubon's determination on this point, see his letters to Baird, November 29, 1842 (Dall 85) and to his family, April 2, 1843 (Princeton); see also AW 47–48.

you like to trip it over the Rocky Mountains next spring in company with
Ed. Harris, Townsend, and about forty others?" (qtd. in Brewer 674). As
it became clear that his son John's presence at Minnie's Land could not
be spared, his uncle William Bakewell wrote jocularly from Louisville, "If
John does not go with you he ought to be *shot with hot mush*, as he used
to say." Bakewell also reveals a shared concern that finding truly capable
expedition members will be challenging, especially in the East: "I hope at
all events you will find some clever thorough going Fellows to go with you
who know some thing about what they are undertaking & who are good
hunters, able to bring in fresh meat & do not expect feather Beds to sleep on.
Some might be found in the West. Those you find East are all *Shot Gunners*
and not Woodsmen, so that if they are not too timid to venture out there
will be more leather finding them, than finding the Game or Specimens you
are in pursuit of" (Beinecke). Even though Audubon liked his brother-in-
law quite well and knew him to be an experienced hunter, Bakewell could
not get away from his domestic duties for the projected months. John
Bachman, an accomplished zoologist and Audubon's coauthor, could not
be away from his parishioners so long. For a while there was a chance that
the French naturalist Joseph Nicolas Nicollet would join them, but he did
not (AW 45–46). Thus the recruiting was difficult. As Audubon wrote to
Spencer Fullerton Baird on November 29, 1842: "I have an abundance of
applications from different sections of the country, from Young Gents who
proffer much efficiency, etc., but I do not know them as I know you" (qtd.
in Dall 85).

Since Audubon's discussion with Baird about joining the expedition
occurred over two years, a look at their correspondence gives insights into
the terms of membership. A professional friendship first developed between
Audubon and the young admirer and burgeoning naturalist Baird beginning
in 1840. The seventeen-year-old Baird initiated their correspondence with
a letter dated June 4: "You see Sir that I have taken (after much hesitation),
the liberty of writing to you. I am but a boy, and very inexperienced, as
you no doubt will observe from my description of the Flycatcher" (qtd. in

Dall 44). Thereafter Audubon encouraged Baird to send him the skins of rare birds and mammals. Over time, America's most famous naturalist was convinced that Baird would prove a helpful companion and assistant on his "Great Western Journey" and invited him to join the expedition. Baird was honored but immediately intimidated by the prospect, largely because of his inexperience but also because his family discouraged him from so dangerous an undertaking. In his first letter to Baird following the invitation, July 20, 1842, the great American naturalist seems aware of his power to mesmerize the young man: "I have no very particular desire to embark as deep in the Cause of Science as the great Humboldt has done, and that, simply because I am both too poor in pecuniary means and too incompetent; but I wish nevertheless *to attempt* to open the Eyes of naturalists to *Riches untold,* and *facts hitherto untold.* The portions of the country through which it is my intention to pass, never having been trodden by white Man previously" (qtd. in Dall 78).

A few months later Audubon's terms became much more specific, if still somewhat malleable. In a letter dated November 29, 1842, Audubon urges Baird to commit to accompany him: "Would you like to go with me at any rate? [...] and undertake, besides acting toward me as a friend, to prepare whatever skins of Birds or Quadrupeds I may think fit for us to bring home." Baird could keep half the bird skins and one quarter of the mammal skins. "I will procure and furnish *all the materials* for skinning, preparing, and saving whatever we may find in ornithology and in Mammalia, and in all probability (if you think it absolutely necessary) pay one half your expenses from the time we leave Saint Louis until our return to that city. You will have to work hard, of course, but then I trust that the knowledge alone which you must acquire would prove a sufficient compensation, and as you already know me pretty well, I may freely say to you that I am not 'hard on the trigger'" (qtd. in Dall 85–86). Baird did not join the expedition, but he did become a professor of natural history at Dickinson College in 1846 and then the assistant secretary of the Smithsonian Institution in 1850, where he labored for the rest of his career, becoming the second secretary of the Smithsonian upon the death of the first in 1878 (Dall 140, 211, 393).

The four men who finally did accompany Audubon on the expedition were Edward Harris (1799–1863) of Moorestown, New Jersey; John Graham Bell (1812–89) of Sparkill, New York; Isaac Sprague (1811–95) of Hingham, Massachusetts; and one Lewis M. Squires from somewhere in the vicinity of Minnie's Land.

Edward Harris (see fig. 6) was a wealthy, successful farmer and husbandman with strong interests in natural history, especially ornithology. His friendship with Audubon went back to the summer of 1824 when the then-unknown Audubon traveled to Philadelphia to seek support for publishing his *Birds*. While Audubon's reception among the most influential naturalists in the City of Brotherly Love was cool, Harris—whom Lucy Audubon describes as "a young ornithologist of refinement, wealth, and education"—liked the tall traveler from the wilds of Louisiana and recognized the original genius of his bird paintings, some of which he purchased. Just before Audubon left Philadelphia, Harris handed him $100 and biographers a truly choice moment on which to build a legend, saying, "Mr. Audubon, accept this from me; men like you ought not to want for money" (L. Audubon 103).

Harris was still a good and true friend in 1843, but he was also a naturalist fully expecting to enhance his natural history collections significantly from so expensive and demanding a trip. He had to navigate difficult personal matters, rent his farm, and make special arrangements for expenses he would incur while traveling, but when Audubon announced in a letter that he had made an agreement with John G. Bell to skin animals for him, Harris had to pause, and he wrote, on January 31, 1843,

> You have given me no information as to the agreement you have made with Bell. Have you thought of me in making that arrangement? I feel satisfied that Bell would not go without full liberty to make collections of birds for himself, and as there would be no satisfaction to me to be of the party while he would be appropriating to himself all that would

be most valuable to me, (for I *cannot* skin). I will await your reply to this letter before I make any further arrangements. [. . .] I did not perhaps explain to you as I should have done my views on the subject, but I always spoke of *our* taking a skinner with us, and I fully intended to bear my share of the expense, and that the skinner should be to our sole use and benefit. (MH)

But Audubon had not forgotten his friend while working out his arrangement to hire Bell to work for him, and the misunderstanding was quickly cleared up. On February 7 Harris wrote back, "I want only the Bird Skins— Suppose that I pay one half the Salary to Bell and take half the Skins, leaving you the entire privilege of paying Bell out of your portion of the Skins. Then you would get your Skins of Quadrupeds for the payment of Bell's expences" (MH). Even though he writes later in this letter, "I now feel myself completely *in* for the expedition, come better come worse," unspecified "unforeseen circumstances" suddenly arose that caused him to fear he would have to break his promise to accompany Audubon on the trip, as he wrote on February 10: "Rest assured my dear friend that I had fully entered into the spirit of the enterprise and that I shall feel most keenly any disappointment that may occur" (MH). Torn between his need to be present to handle the new problems and his intense desire to take part in this expedition, his stomach plagued him with indigestion, but in the final days before departure, he somehow managed to stabilize the still-mounting crisis. On March 2 he wrote to Audubon, "I assure you I shall be delighted to be off and bid adieu to all care. My attack in New York left me in a poor state to undertake the preparations necessary for such an expedition, the consequence has been a severe attack of dyspepsia which is augmented by the cares which are increasing upon me as the day of departure draws near." His desire for the promised travel and adventure seems to have been curative: "I shall be as hearty as a buck as soon as I am once more free" (MH).

John G. Bell was a taxidermist and dealer in bird and mammal skins. He enjoyed a good reputation among naturalists, curators, and collectors. John

Bachman referred to him as "a first rate fellow & not slow as a naturalist" (337). Audubon concurred, recommending him for a position in 1842 in the following terms: "This will be presented to you by M.ʳ John G. Bell of this City, who goes to Washington with the view to offer his services as a preserver and mounter of Birds quadrupeds &ᶜ—and having known M.ʳ Bell for many years, I take pleasure in recommending him to you as an excellent Workman and an Industrious & sober person" (Audubon to John James Abert, July 4, 1842, Princeton). His reputation was clearly enhanced by his association with Audubon, but he continued to improve his techniques as a taxidermist and his excellence was widely acknowledged. In 1872, when the young Theodore Roosevelt wanted lessons in taxidermy, his father selected Bell as his instructor (McCullough 118).

Audubon hired the soft-spoken, rather introspective Isaac Sprague as an assistant artist for the expedition. As he wrote to Baird on January 31, 1843, "Besides John G. Bell, the stuffer, etc., I expect a young man from Hingham, Mass.ᵗᵗˢ., some of whose drawings I believe you have seen, and who will assist me when wanted, but who will especially draw plants and Views for backgrounds to our present work" (qtd. in Dall 89–90). He first encountered Sprague's watercolor drawings of birds and plants on August 19, 1840, in Hingham, Massachusetts, while traveling to solicit new subscribers to his various publications. He was powerfully impressed: "Saw some very remarkable Drawings of birds (far better than any ever made by the Immortal Alex.ʳ Wilson) by a young man named *Sprague*. Truly wonderful Drawings my Dearest friends. but this person was out shooting and I did not see him.—I however wrote a few lines on several of them the purports of which, I trust, will not displease him" (*Journal Made While Obtaining Subscriptions* 10). On Sprague's drawing of a Palm Warbler (*Dendroica palmarum*), known at that time as the "yellow redpoll warbler," Audubon wrote in pencil "first rate altogether, J. J. A."[5] Sprague had been apprenticed to a carriage painter for a time in his teens, but his pronounced interest in

---

5   This drawing and seventy-three others by Sprague are held at the Boston Athenaeum.

botany and ornithology led him to develop his artistic talents toward natural history. Even though he would produce many fine sketches of plants and animals along with detailed landscapes while on the expedition, he was probably the least suited of all the five traveling companions for the rough, rugged ways of life on the prairie. His journal tells us that in his life before he had never slept outside in the open air. After the expedition, his major work was as a botanical illustrator for Harvard's Asa Gray (see Randolph for a biographical sketch). (See fig. 7.)

Little is known about Lewis Squires other than what is reported in the journals of his companions. On fairly short notice, Audubon recruited Squires shortly after he knew that Baird would not be going, as he explained in a letter of February 23, 1843: "I have concluded to take a Young Gentleman in your stead who is a Neighbor of ours, but who alas is no Naturalist, though a tough, active, and very willing person" (qtd. in Dall 92). Tough, active, and willing he certainly was according to the accounts of Audubon, Harris, and Bell, if also slow to acquire some needed hunting skills. When they all returned to their respective homes, Audubon reported to Harris, November 30, 1843, "Squires is about going to China, where he may make money" (qtd. in McDermott 8n17). If Squires left journals and letters, they have yet to turn up.

Down in Charleston, John Bachman's contribution to the preparations for the expedition came in the form of a letter of instructions dated March 12, 1843 (Bachman 337–40). When he and Audubon first met in Charleston in 1831, Audubon latched on to the friendship Bachman offered for several reasons, but one was that Bachman was one of the most informed zoologists in the country and would no doubt prove useful to various projects. This letter shows that Audubon had much to learn from Bachman and that Bachman, perhaps naively, believed that the members of the expedition would attend to their work as naturalists first and foremost. After all, he was deeply invested in the success of the *Quadruped* project, having lost two daughters and now having two grandchildren by one of them dependent upon the future Audubon fortunes.

According to Bachman, Audubon and his companions should, in addition to observing and gathering specimens of the mammals, be alert for new species of birds, reptiles, and plants, gathering seeds, preserving snakes and lizards in alcohol, and preserving bird skins using "plenty of arsenic." He intended them to be thorough naturalists: "Mark the sexes & localities—date when killed—dimensions & weight—save all skulls. Write up your notes every evening [. . .]. Whenever you can sketch your animals in the flesh—note habits time of breeding—number of young—Holes and nests. Be careful to keep a specimen of every species you may procure although you may think them similar to those in our Atlantic states. You know my theory—that every species—except a few Bats—& a single species each of Bear & Deer is new in the far west" (338). If they followed his instructions, they would spend a great deal of time trapping marmots, ground squirrels, gophers, other squirrels, hares, and rabbits—animals less charismatic than elk, bison, and pronghorns. They also should interview men they meet with experience in the West: "Write down the accounts of trappers & Hunters when you are sure you can depend on their veracity." Perhaps his expectations were so high that some disappointment was inevitable: "In a word, give a true history of every species that inhabits the plains & mountains—the earth, rocks & trees. This is the duty of the Naturalist" (339–40).

As the time for departure drew near, there was a clear understanding that Audubon's family near and far would keep the popular press informed of his movements and deeds in order to maintain a general expectation of his return and of the new knowledge, images, and stories he would publish. The five members of the expedition were also aware that they were taking their place among the recent history of western travelers, Lewis and Clark and the Corps of Discovery primary among them. But the exotic, difficult, and romantic travels of others were also already reported in newspapers, magazines, and books. One of the earliest informed accounts of the western prairies and the Rocky Mountains is Edwin James's *Account of an Expedition from Pittsburgh to the Rocky Mountains under the Command of Major Stephen H. Long*, published in 1823. The travels and adventures of the naturalists

Thomas Nuttall and John Kirk Townsend were well known. Washington Irving's narrative of his travels on the western prairies, *A Tour on the Prairies*, was published in 1835. Irving's account of John Jacob Astor's travels, *Astoria; or, Anecdotes of an Enterprise beyond the Rocky Mountains* was published the following year. The charismatic Scottish nobleman Captain William Drummond Stewart first traveled in the American West in 1832. The artists Karl Bodmer and Alfred Jacob Miller also made long trips into the West in the mid-1830s. But awareness of the travels and publication of the artist and ethnographer George Catlin would have loomed most powerfully in their minds. His *Letters and Notes on the Manners, Customs, and Conditions of the North American Indians* was first published in 1841, and all four journalists of the expedition show some familiarity with this book of his travels, beginning in 1831, to many of the same places Audubon's party anticipated visiting in the coming months. Audubon and his four companions understood that they would be seen as joining the tradition of travelers in the American West.

On March 2, 1843, nine days before Audubon and company would begin their journey to St. Louis, Victor would marry Georgianna Mallory, a friend of Caroline Hall, his brother's second wife, giving both Audubon and Lucy a sense that their family was wholly patched together again on the verge of the patriarch's long absence.

Isaac Sprague had arrived in New York City on February 17, about three weeks before the planned departure. While his journal is disappointingly silent about how he spent this time, we do know that he produced a truly compelling and sensitive pencil portrait of Audubon. The pensiveness of the handsome fifty-seven-year-old suggests that Sprague perceived that aspect of his subject's personality, a trait he shared and understood. (See fig. 8.) When John G. Bell and Lewis M. Squires joined them, the four departed from New York City on March 11 on "the cars," riding the rails to Philadelphia (described by Sprague as "the finest city I have seen" ⌊March 12⌋), then Baltimore, where Edward Harris and his pointer Brag joined them upon their arrival there on March 12. Harris begins his journal upon their departure from Philadelphia the next day and shows that they all thought they were going as far as the Rocky Mountains: "Left Philadelphia for the Rocky Mountains via Baltimore, Wheeling, & St. Louis."

The rough stagecoach ride through the mountains did not begin until they reached Cumberland, Maryland, on March 14, after some eleven hours by rail from Baltimore. They were surprised that the coach fare allowed only 50 pounds of baggage per person, a limit they exceeded by 870 pounds, for which they were charged an additional $34.80, bargained down to $30: "a first-rate piece of robbery," wrote Audubon (AJ 1:454). According to Harris's list, they carried "9 Trunks / 1 Box / 2 Gun Cases / 4 Bags / 1 Carpet Bag / 1 Cloak / 4 Guns / 1 Dog." The coach set out into the dark around seven thirty that evening with two feet of snow on the mountains. They reached the highest point on this route, Laurel Hill, at ten the next morning, where they began their descent, as Sprague observed, "into the great valley of the Mississippi." But he was not simply celebrating his entrance into his country's largest watershed; the likelihood of an accident had just increased as well: "The first descent is quite steep for 5 miles and when covered with snow & ice rather dangerous, winding as it does along the brow of the hill, with a wall of rocks several hundred feet high on one side, and a precipice of great depth on the other." The five companions in the coach, nonetheless, were getting to know one another and developing already a spirit of shared adventure. As Audubon noted upon arriving at Wheeling, "My young folks are full of spirits" (AW 21). Sprague's account of the interior of the coach is illuminating and leavened with Shandyesque humor: "This is the hardest ride I ever have had having been now two nights with scarce any sleep. The interior of our coach presents a curious scene. Having exhausted all interesting topics of conversation, a dead silence ensues, each person fixes himself on his seat with a desperate resolution to wait the journeys end. But in a short time gets off his guard and commences a suspicious nodding and bowing to his neighbors, until some heavier plunge of the coach, completely upsets his gravity, and he is obliged to right himself and commence again." By Bell's count, the crossing to Wheeling took "36 hours, 9 hours behind the usual time" because of the snow. Arriving at Wheeling at 4:00 a.m., they had only about ten hours before their sternwheel steamboat *Eveline* (or *Evaline* or *Emily*, depending on which journal you consult) would cast off for the first leg of the river trip, to Cincinnati. For the season, it was "excessively cold" and snow continued to fall (EH 44).

Wheeling offered the first opportunity to send a letter home since their departure from Baltimore; thus Audubon's first letter from this trip is dated Wheeling, March 16, 1843 (AW). Throughout his life, Audubon had been a tireless correspondent, and he always expected Lucy and his sons to write often to him when he traveled. When he did not receive letters when he thought he should have, he expressed varying degrees of anxiety in his journals and letters. While I do believe that his need for news from his family was somewhat extreme for his day, and a little tiresome for his family members, I also know that we forget today what distance from loved ones meant in those days. Ships sank, houses burned, water supplies and infections from small cuts killed at much greater rates than at any time since then. On what might to us seem a relatively short trip—Baltimore to Wheeling and back, for example—your spouse could be dead and buried before you knew he or she was sick. On this day, however, the entire company was pretty clearly excited—despite the sleepless, jostling coach ride through the mountains—to be on the verge of a fast-moving river voyage to the distant city of St. Louis, where new adventures would begin.

If he was feeling the new distance from his family on reaching Wheeling, Audubon must also have sighed with relief from the flurry of busyness and the family saga and have looked westward with renewed pleasure of adventure and novelty. The sigh was also one of satisfaction: by 1843 he had fulfilled the promise he made to Lucy when they married in 1808 of establishing a permanent home together, but always any desire to celebrate was tempered by death, those of their two daughters in Kentucky and their sons' two young wives. Still, Audubon was emotionally buoyant and on his good days at this time would experience some satisfaction that he had left his family secure and all together in a new home, the result of his twenty-five-year-long effort to succeed with his publications. And as he descended the Ohio River on this trip, familiar places would remind him of scenes from his life from two and three decades earlier.

At Cincinnati, where he had worked at the Western Museum before setting out (on October 12, 1820) on his daunting river trip to New Orleans in an attempt to restore his family's fortune, the company transferred to the steamer *Pike*, for Louisville, where they arrived late on Saturday night, March 18, and spent the night on board before disembarking on Sunday morning. Since the steamboat to St. Louis, the *Gallant*, would not be ready for a couple of days, Audubon went to see his brother-in-law William Bakewell, and the others walked about the city awhile before taking their guns to hunt birds of potential value to their natural history purposes. A Mr. Fellows, whom they had met and befriended on the stagecoach, was pleased to throw a party at his home for the famous Audubon and his companions on Monday evening, March 20. The next day, Bell paid six dollars for an accordion (which is sadly not heard of again in any of the journals or letters). When William Bakewell wrote to Victor on March 24 about Audubon's visit in Louisville, he showed Audubon "out dancing the whole of us & Kissing all the Ladies in the range of his acquaintance" at Fellows's party as well as at one given by a Mr. Hite (Beinecke). Ever vivacious, Audubon would turn fifty-eight in just a few weeks, yet other, younger men and women were impressed by his energy, and he knew it.

In the same letter, Bakewell also expressed concerns about the coming travels: "[N]o doubt he will be detained by Ice in the Mississippi, the weather having turned very cold in fact no one ever saw the like at this time of the year. The ink is frozen & my hands in nearly the same fix & the River at this place full of floating Ice. [. . .] I think it will be impossible for your Father to leave St. Louis for 2 or 3 weeks" (Beinecke). Lucy wrote her first letter to her husband during this trip at about this time, March 19, and it is largely reassuring and optimistic: "The dear ones around me, both great and small are well and we all join in our good wishes for your health and safe return, tell Mr Harris I hope he will have as much satisfaction from the trip as I have in the knowledge that he is near you." Yet by April 2, when she wrote again, she had heard of the extreme cold and river ice and regretted that Audubon had not stayed home longer; she capped off a note about his shirts at *home* with "where you might as well have been a month longer." Victor revealed in the same letter that the household generally had been lamenting that Audubon left earlier than needed: he had learned that "you will not get off from S<u>t</u> Louis *before* the 15<u>th</u>—what a pity we did not know this sooner!" (Beinecke). As time passed, Lucy seems to have become ever more annoyed, writing on April 9, "I assure you I have all the time been quite vexed to think you had left us a month sooner that you need have done." But she was also becoming more worried about her husband's safety. Probably in response to Audubon's report of "an equinoxial Gale that amounted almost to a Hurricane and lasted nearly all night" in his letter of March 23 (AW), Lucy wrote, "I hope you will have no more hair breadth escapes or frights, in boats or else where but trust the same Providence which has hither to guarded you will still be with you, and all your companions and return you safely to us" (Beinecke).

In Louisville, Audubon also saw many old friends and several of Lucy's relations, including his brother-in-law Nicholas Berthoud, married to Lucy's sister Eliza. Berthoud had been developing a new business interest in St. Louis, so a number of these relations planned to visit Audubon and his party in St. Louis on a sort of farewell frolic.

After five days in Louisville, the party boarded the *Gallant*, "and such a motly of 100 passengers man never saw or can form an Idea of," Audubon wrote on March 23. But a hurricane-force snowstorm blew the day before, and the temperatures were so low that open ponds were freezing in a short time. There was a canal lock at Shippingport to allow easier passage by the Falls of the Ohio, but the captain of the *Gallant* took the less expensive option of having his boat guided through the falls. Many of the passengers, including Bell and Squires, chose to walk the distance round the falls, but "Harris and Sprague went *over the Falls*, and were pleased with the sight of these mighty rapids," Audubon wrote (AW). Harris no doubt understated the experience, noting that they "got safely over the falls, rubbing pretty hard on entering them."

The weather continued cold; Audubon wrote the next day, "Last night was the coldest of this Winter excepting one in Feby. The river was this morning covered with Ice, but we have proceeded on tolerably well." The boat passed "several villages now called cities and so improved in size that I with difficulty recognized them." But he did recognize his old sawmill in Henderson, which for him represented his largest business failure, leading to his bankruptcy in 1819. Now, however, he was America's most successful ornithologist and a great celebrity, recognized everywhere he went. He would not have gloated, however, for he knew how quickly anyone's fortune can tumble down. Still, he did disapprove of conditions on board the *Gallant*:

Our steamer is one of the filthiest I ever saw.—No towells, no soap, and every one this morning dipt the water to wash from the stream! Many of the passengers slept on the floor last night, some rolled up and tied in surcingles We have about 15 Horses on board, waggons, carts, carriages and furniture of all description belonging to new settlers going to the Missouri frontiers.—Harris and I have one *State room*. he sleeps upon 3 Boards and I upon 4. Our meals are none of the best, all is greasy and nasty. A first rate innitiation for the trip to the Yellow Stone; for we eat famously and sleep soundly. (AW)

Their final approach to St. Louis was inglorious. On Saturday morning, March 25, the *Gallant* made the turn northward from the Ohio into the Mississippi, but the water being low, they ran aground several times and once hit a snag that so jarred the boat that "a rush was made for the cabin door." The next day they continued to run aground and at one point passed a group of two hundred German Mormons "left behind us for want of water in the channel." At one point, another steamer, the *Cicero*, endangered both boats and all aboard by foolishly challenging the *Gallant* to a race and in so doing locking their bows together for the distance of half a mile. On the twenty-seventh the *Gallant*'s progress was slowed by ice, and Bell shot at some ducks between spells of hard-pouring rain. Audubon was glad Lucy was not with them: "Oh such fare, deck dirty &ᶜ &ᶜ. Why my Sweet heart, thou could not live among us.—We have no bread but Biscuit as hard as flint and scanty indeed are our meals, but we hope to reach St Louis this coming night. [...] Our boat leaks so from the roof that we have scarcely a dry spot to stand our feet upon! We stopped for the night about 12 miles below Sᵗ Louis in a dreadful snow storm and very cold" (AW).

On the final morning before reaching St. Louis, March 28, ice had practically closed the river and a snowstorm blew "hard enough to bear a man." The captain seems to have stocked no more food aboard for passengers than absolutely necessary, as Audubon reports, "No provisions on board, therefore no breakfast here" (AW). The hundred "motly" passengers were probably quite grumpy in such conditions and with no breakfast by the time they were landed at St. Louis at ten that morning (or eleven, or noon, depending upon which journal or letter you consult), but the Audubon party wasted no time finding lodging and beginning preparations for the next stage of their work.

St. Louis, March 28–April 24

St. Louis in 1843 was a vibrant, inchoate, noisy, muddy intersection for anyone traveling west by river or over land. The few thousand residents in the 1820s had grown to well over thirty thousand by the time Audubon arrived, and more were arriving every day. If you wanted to do anything in the West, Southwest, or Northwest, the first leg of your trip was to St. Louis. Rivers determined this, and the steamboat (which first reached St. Louis in 1817) accelerated the growth and activities of the human population, as well as the dramatic reductions of many animal populations, beaver and bison primary among them.

After unloading and securing their trunks and other gear and boxes, the Audubon party proceeded directly to the newest and most commodious hotel in the city, the Glasgow House, which had just opened and was expensive. Even though the cost was "9 dollars per week each," Bell was impressed: "It is one of the best regulated houses I have ever seen, & the best table" (March 28). Soon they would find cheaper lodgings; in the meantime, though, Audubon quickly found his brother-in-law Nicholas Berthoud and introduced himself at the main office of his friend Pierre Chouteau Jr.'s American Fur Company, where he met and immediately liked the French-speaking Joseph A. Sire, who would captain the steamship *Omega* up the Missouri to Fort Union. Audubon would continue to admire Sire's integrity and skills throughout the voyage. The bad news this day, however, was that the Missouri was still frozen over and they would not be able to get underway until the third or fourth week of April. They had entered a city at the end of its winter, poised for the breakup of the river ice. Berthoud's "large store filled with goods" was waiting, and "great numbers of Steamers at the *Levee*" waited to begin their voyages "for all the portions of the Western World" (AW 36–39). (See fig. 9.)

Three days later, March 31, three letters from home and one from John Bachman put Audubon in high spirits, and he assures his family that once this trip was over, "I do not think that I shall leave you or Minnie's Land for some time, if ever!" (AW). He also assures them that he and his companions have been "wonderfully kindly received" by everyone they've met or been introduced to, and all believe that the prospects for a successful and enjoyable trip are high. He was also pleased to know that newspaper notices of his passage through Louisville had appeared.

He wondered, though, why no one mentioned a poem entitled "To Audubon," just published on March 15 in the *Philadelphia Saturday Courier*, in which the poet and fellow naturalist Dr. J. K. Mitchell attempts to beatify the "brave wanderer" for his self-sacrifice, courage, and vision as he journeys to the northern prairies: "A blessing on thine Enterprise, For thee & thine, oh take!" Although he refers to the verse as "effusions," he half-credibly claims that it "is not bad" (AW).[6]

St. Louis was becoming the stage of Audubon's transition from "back East" to "out West." While encouraging his family to write him "extremely long letters on the 1st of every month," he was also anticipating the arrival of Lucy's sister and brother, Eliza Berthoud and William Bakewell, in the company of several other relations. But he had been disappointed as he

6   Dr. J. K. Mitchell was a well-known Philadelphia physician, naturalist, and occasional poet. His *Indecision, a Tale of the Far West; and Other Poems* had been published in 1839. One of Mitchell's aesthetic goals was to create a wild, natural pastoral to replace the traditional tame European Arcadian myth. The "bold Prairie-hunter" for his "sweet maiden" will chase the bison, for "our rifles afford / The joy of the chase and the food for the board." "Let England exult in her dogs and her chase, / Oh what's a king's park to this limitless space" ("The Song of the Prairie," in Mitchell 21–23). Rufus Griswold, in *The Poets and Poetry of America* (1842), had recently grouped him among the "under crust" of American poets. Squires copied the entire poem under the entry for August 10 in the Beinecke copy.

began to assess how to start his work as a naturalist: "Natural History has few or no advocates here, I have not heard a word about subscriptions from any one, and to my utter astonishment none of the men who have actually spent years and years in the Mountains, and Upper Missouri, know anything of the Quadrupeds there, beyond the Beaver, the Otter, the Coons, and such Foxes as pay in the skin. The same with the Buffaloes." Nevertheless, he has gathered information about the gophers, or "Pouched Rats," that are abundant in St. Louis: "I now know much of their Habits, and mayhap will draw some before we leave as they come out the moment the frost is out of the ground" (AW 42–43). Even though he and his companions had been receiving many visitors every evening since arriving in St. Louis, his mental focus was shifting ever more to the work at hand. (See fig. 10.)

To reduce their overall costs, and to begin gathering specimens, Harris, Bell, Squires, and Sprague crossed the river and made their way to the region near Edwardsville, Illinois, some twenty miles distant, on April 4. Audubon remained in the city in order to continue overseeing the preparations for their departure. When Captain Sire took him to see the progress made on the *Omega* up on the dry dock, Audubon was very pleased: "My State room is so large that I will keep me a good Bed Sted and bed.—We will have a large table to Draw and Write on and we will eat with the Captain who I must repeat [is] as fine a Man as I ever saw" (AW 45). He also looked about for well-recommended, experienced boatmen and hunters to take with him. He was especially hopeful that Etienne Provost would accompany him. Provost (pronounced "Provo") was just Audubon's age and had some thirty years of experience on the prairies and in the Rockies in the fur trade and other forms of employment. Through the 1830s, he had traveled, in various capacities, with the French naturalist Joseph Nicolas Nicollet, the U.S. government explorer John C. Frémont, the Scottish nobleman William Stewart, and the artist Alfred Jacob Miller. The "Man of the Mountains," as Provost was known, was well respected for his knowledge and skills, and no doubt also for stories he could tell. He had a home, wife, and daughter at Second and Lombard Streets in St. Louis. On the evening of April 1, Audubon spent an hour or so with Stewart, who had just survived shipwreck, from whom he may have heard praise for Provost as someone he could rely on.[7] Stewart talked, Audubon wrote, "with the lisping humbug of some of the English nobles" (AW 47).

Although a hard snowstorm struck on Sunday, April 2, causing some dismay among those eager to get underway, on Tuesday, when Harris and the others crossed the Mississippi for Edwardsville, the weather was fair, and the ice and snow in the bottomlands gave way to mud. Still, they got right to "work," hunting and trapping every species of bird and mammal: grouse, cranes, geese, rabbits, squirrels, snipes, turkeys, muskrats. On the first two full days, however, they learned that new degrees of caution and skill would be needed in these wilderness conditions. On Wednesday Sprague lost his balance while trying to cross a creek on a log and fell in, dropping his borrowed gun in ten feet of frigid water. The next day, after Sprague, Squires, and Bell managed to fish out the gun (with "a sort of oyster catcher" that Bell contrived), Bell's horse "threw [him] 10 feet from him against a tree" (April 6) when he fired at some ducks. Even though his injury was not serious, these new companions must have begun reconsidering the potential seriousness of even small accidents so far from help.

Otherwise, for the seventeen days they were away from St. Louis, April 4–20, these four were deliberately hunting and preserving (and eating) specimens and trying to be productive naturalists. Sprague, though, while a good companion and a willing gunner, began to reveal a sensitivity not apparent in the others' journals. He noted, briefly, on April 8 that he made a sojourn apart from the others: "One of my days of pleasure, rambling in the still forests alone. Saw wild Turkeys for the first time to day." On the thirteenth, when the group moved north "18 or 20 miles to a small town called Bunker Hill," Sprague was impressed by the beauty of the unfamiliar landscape: "This is the pleasantest place I have seen. The village is situated on a swell in the centre of a large prairie and commands a view of the

7   For Provost, see Hafen.

country in every direction." By contrast, Bell's accounts of his days focus strictly on killing animals and skinning and preserving their skins, which is what he was hired to do.

In the city, Berthoud invited his brother-in-law to stay in his home, a welcome invitation, as Audubon wrote April 8, "for money was melting saddly too fast" (AW 52). Audubon paid several visits to the elder Pierre Chouteau, who lived just west of St. Louis and was then eighty-five years old, and gathered specimens of the plains pocket gopher, which he referred to as "Pouched Rats." Several of these he drew, and he gathered information about the behavior of the species. He made a few other drawings or outlines as well. In the evenings, however, many people wanted to see him, some "to see simply what sort of a looking fellow I am" (AW 58). He was probably the most famous person in St. Louis at this time. A few days before their departure he wrote, "I have had an enormous number of visits from almost every body, and with dining out, and parties of evenings I am quite fatigued" (AW 60). Even though he lamented that "only 2 Men of Science belong to this City" (AW 58), there was an Academy of Natural Sciences, from which he accepted honorary membership. Another honor was the gift of one of William Clark's original journals from the famed expedition of the Corps of Discovery of 1803–6. D. D. Mitchell, superintendent of Indian affairs at St. Louis, presented this much-traveled volume bound in elk skin to him on April 19.[8]

Just at sunset on Friday, April 21, Audubon's companions returned to St. Louis to begin final preparations for departure. All four bought new hats the next day, the same day on which Audubon reported losing his "last upper tooth" ("now I must soak my biscuits &c" [AW 65]). On April 24, Harris reported a major purchase of "Sundries," amounting to $368.30. A packing list among the Harris papers at the Alabama Department of Archives and History includes quantities of blue, white, and vermillion beads; finger

---

8 This journal is now in the Eleanor Glasgow Voorhis Memorial Collection, Missouri Historical Society, St. Louis. It covers the dates September 11–December 31, 1805. See Moulton 2:557.

rings; Indian awls; five hundred gun flints, one hundred pounds bar lead, two kegs gunpowder, and thirty bags of shot; fifteen "boxes" of claret wine and five gallons of brandy; as well as tobacco, tea, sugar, "Pilot Bread," pork, rice, coffee, bacon, olive oil, ginger, and Epsom salts. Audubon had all the passports they would need as well as letters of introduction and requests for accommodations at the American Fur Company forts along the Missouri River. With the temperature rising to seventy degrees now, Bell set out the bird skins he had prepared in the recent weeks to dry in the sun, and he prepared new skins from the gophers recently captured.

Because his journals from this trip would be one of the most important resources for future publication, specifically in the *Quadrupeds*, Audubon assures his family in his April 23 letter, "My Journal will contain many curious facts, anecdotes &c and it will be kept duplicate, one copy will be sent to you, and the other will remain with us" (AW). Squires, as scribe, also had a job to do.

At about this time, an unknown person reported the following "personal interview" with Audubon in St. Louis, which was reprinted in several newspapers in 1843:

> Mr. Audubon is a man about the middle stature; his hair is white with age, and somewhat thin; he combs it back from an ample forehead, his face being sharp at the chin; has grey whiskers, an aquiline nose, and a hazle eye, small, keen and indicative of great tranquility, and sweetness of temper, cheerfulness and genius. He is a man of robust constitution though not of a stout frame. He told me he had not taken a particle of medicine for twenty years. He is capable of any fatigue; can walk thirty-five miles a day with ease, for months; can sleep any-where in the open air; endure all climates; his principal food being soaked sea biscuit and molasses. He cannot well masticate meat on account of having lost his teeth, from which he suffers, and is obliged to boil his meat to rags.

> He wore a dark frock coat, velvet vest and blue hunting shirt; is very pleasing and agreeable in conversation, and makes one perfectly at ease

in his presence. He says a man can live one hundred years with temperate habits, regularity, and attention to diet.

He was about starting up the Missouri—said he was entirely done with ornithology; his object now being to classify the American quadrupeds. He was severe on Buffon, whose book he regarded of no authority; said Buffon was a man of wealth, resided in Paris, and wrote his descriptions from dried skins, and drew largely upon his fancy. Mr. Audubon anticipated a good deal of pleasure, and much hard trapping, shooting, drawing and writing. He takes all his drafts from the animal as soon after it is taken as circumstances will admit. (Sage).

The scene of departure was not what Audubon and his companions might have wished for, but all five were in good spirits and had enjoyed good health, with the exception of Harris, who was recovering from some illness. Any excitement about beginning their long-anticipated ascent of the Missouri River was drowned out by the noise of the guns of the hundred or so trappers, "[s]ome drunk, some half so, and a very few sober" (AW 72). Nevertheless, on Tuesday morning, April 25, around ten thirty (or eleven thirty, or noon, depending upon which journal you consult), Captain Sire powered the *Omega* away from the wharf and headed upriver. The trappers were "French Creoles and Canadians" hired by the American Fur Company to work out of various forts. Following their uproarious embarkation, though, as Audubon explained, and after they had "dined, those that are sober are all singing on the top Deck, and Hallowing as if Mad Indians. The Drunken ones are all below sound asleep" (AW 72). Harris was considerably more disdainful about them: "Yesterday one who had enlisted for the trip was arrested for counterfeiting and murder. How many more of like character we have among us would be difficult to know, all we do know is that they appear to be the very offscouring of the earth, worse than any crew of sailors I ever met with." Bell's observation of the same moment is interesting, for he seemed to dismiss the shouting and firing as simply unimportant while he responded to the separation of friends at wharf side: "Started at ½ past 11 oclock amid shouting, firing & waving of handkerchiefs & friends bidding each other adieu, many no doubt forever."

The *Omega* made slow progress against the powerful spring current, but Audubon was pleased and surprised by how steadily the boat rode on the river. He would "be able to draw during *the day* instead of the night as I once thought would be my Lot." So he continued to write in his letter of April 25 until they reached St. Charles that evening, some forty river miles above St. Louis. He was excited at long last to begin their "labours," and he was eager for their first stop to cut wood, for "there will be our Traps &ᶜ and our Guns agoing.—May it please our God that I may procure a good round number of animals."

St. Louis to the Yellowstone River and Fort Union, April 25–June 12

When crew members tossed off the lines and the *Omega* dropped away from the wharf, its upstream progress was slow because of the powerful spring current and the weight of its full load of cargo and people. High winds from the west and northwest, sometimes of "hurricane" strength, were another force to contend with, together with snags, sawyers, and sandbars that impeded navigation. Audubon vividly distinguishes these hazards from any he had seen before in his letter of May 8, while also making the best use ever of the rarely seen adjective "brittly": "No one going down to New Orleans even 20 years ago can have an Idea of the Snags, Sawyers and Planters that are found in this 'Upper Missouri' they show their brittly prongs as if some thousands of mammouth Elk Horns had been planted every where for the purpose of impeding the navigation." This was truly a wilderness river, and he had never seen the like: "I am told that it will grow worse as we proceed further, but how can this be?" (AW 88). (See fig. 11.)

The other major hindrances to speed were the frequent stops to gather and cut wood for fuel. Captain Sire, assisted by his much-experienced pilot Joseph LaBarge, nevertheless managed on this voyage to complete the ascent of the Missouri to Fort Union in forty-nine days, the fastest ascent since the first one in 1832. In the preceding two years, the upstream passage had taken eighty and seventy-six days, respectively. Sire's log shows that some of his success this year was due to running all night when conditions were clear and visibility was good enough, despite his early decision not to run at night because of the risk: "[W]e have too much to lose to risk our cargo for the sake of gaining a little time" (3:985). Still, since this was Sire's third annual run up the Missouri to Fort Union, his knowledge of the river no doubt was a major reason they were able to average thirty-five miles a day.

On May 2 the *Omega* passed the village of Independence, as Sprague notes, "the last town of any size—that we shall probably meet with for some time." This was an important crossing for wagon trains at the eastern end of the Santa Fe Trail. Audubon mentions seeing "their Waggons and Carts awaiting their arrival at Independence which is the usual point of departure" (AW 88). Sprague registers the party's general awareness of this: "From this place the Santa Fe traders take their departure. Their goods are conveyed in large wagons drawn by mules." Fifteen days after the *Omega* passed this place, the explorer John Charles Frémont made his way westward to the Kansas River and reported, "Trains of wagons were almost constantly in sight; giving the road a populous and animated appearance, although the greater portion of the emigrants were collected at the crossing, or already on their march beyond the Kansas river" (107). Audubon had been as far north on the Mississippi as Ste. Genevieve in 1811 with a business partner when the region was still at the beginning stages of white settlement, but now he was heading into the wilds of the northern prairies at a time of transition from the former days—when only the more or less qualified and capable, or desperate, took on the challenges of wilderness travel—to the 1840s, when "gentlemen" began using influence to attach themselves to various exploring parties. Frémont's complaint about an incident caused by one such gentleman in his party, which endangered his entire expedition, is strategically understated: "This accident, which occassioned delay and trouble, and threatened danger and loss, and broke down some good horses at the start, and actually endangered the expedition, was a first fruit of having gentlemen in company—very estimable, to be sure, but who are not trained to the care and vigilance and self-dependence which such an expedition required, and who are not subject to the orders which enforce attention and exertion" (174).

Since for this ascent we have only Maria's edition of Audubon's journals, we must rely on his surviving letters, which are full and interesting, but the original journals would be more so. Nevertheless, when his letters corroborate material reported in *Audubon and His Journals*, as often happens, we can place some cautious degree of confidence on at least the general reliability of a given passage. In his letter of May 2, for instance, he reports to his family, "I Spend nearly all my time at my Journal Philosophising as it were on all that I see and that occurs during our slow days of progress" (AW 80). And in the entry in Maria's edition for three days earlier, April 29, we see him "Philosophising" on the mighty power of the Missouri River currents to alter and destroy:

> On looking along the banks of the river, one cannot help observing the half-drowned young willows, and cotton trees of the same age, trembling and shaking sideways against the current; and methought, as I gazed upon them, of the danger they were in of being immersed over their very tops and thus dying, not through the influence of fire, the natural enemy of wood, but from the force of the mighty stream on the margin of which they grew, and which appeared as if in its wrath it was determined to overwhelm, and undo all that the Creator in His bountifulness had granted us to enjoy. The banks themselves, along with perhaps millions of trees, are ever tumbling, falling, and washing away from the spots where they may have stood and grown for centuries past. If this be not an awful exemplification of the real course of Nature's intention, that all should and must live and die, then, indeed, the philosophy of our learned men cannot be much relied upon!" (AJ 1:460)

Each day took them farther from civilization and deeper into a primal wildness where the raw forces of a watershed could be observed largely unaltered by humanity, and Audubon was glad for the privilege, which conjured for him the transience of life and the imminence of death amid the sublime forces of untamable nature.

Audubon knew, however, that this wild context was also conducive to his own species' potential for unnatural vileness. In the May 2 letter, after noting his recent philosophizing, he writes that they had just heard from

people aboard a passing steamer "of an awfull Murder of one Man belonging to the Santa fee expedition for the sake of 32,000$." The accused men,
"[t]he rascals that Shot and plundered him," were on their way to St. Louis
for trial. Audubon's wish is quite blunt: "I Wish I was in the Jury, I would
have both and all of them Shot at the door of the Court House as soon as
the verdict of Willful Murder is past against them." By contrast to nature,
Audubon was thinking as he left civilization behind, humankind is a moral
mess: "Strange to say those very men Stood in their own portion of the
country as Men Honourable and of good Standing. So much for the love
of plunder and of Money" (AW 80).

Even though Audubon had much experience traveling in several variet-
ies of wilderness in the South, East, and North, he was especially enthused
to be entering the western wilderness. In his letter of May 8, he celebrates
the fulfillment of his dream of western travel long deferred: "We are now
positively out of the *United States* boundaries Westward, and now all the
people we will see will be Indians, Indians, and nothing but Indians," since
he believed they had crossed "the line dividing the State of the Missouri from
the wilderness as I am now pleased to call the Country ahead of us" (AW
85). If his geography was slightly mistaken (the *Omega* had not quite reached
the boundary at Council Bluffs), his excitement for expected adventures
is clear. Although the larger game animals had not yet appeared, he wrote,
"We are told however that next week the Wolves, the Foxes, Black-Tailed
Deer &c will show themselves and that as we will have opportunities of
crossing the bend we will have some fun! I hope so I am sure" (AW 86–87).

From the beginning, though, it is evident that Audubon and his men
intended to take advantage of their time by shooting birds and mammals
in the hope of finding new species and gaining valuable preserved skins.
Bell had a work space set up and was "making" skins regularly. Every stop
for wood was seen as a chance to take to shore to hunt the new species they
fully expected to be inhabiting these distant regions. Near the Black Snake
Hills, May 4 was a particularly good day. Both Harris and Bell shot new
birds that Audubon later named for them: Harris's finch (RO plate 484)

and Bell's vireo (plate 485). In Audubon's letter home on May 5, their collective enthusiasm is evident: "We are all perfectly well, and when we ever go ashore, our Guns tell wonderful tales" (AW 83).

What Sire does not mention in his log is the gunning, which began in earnest once the *Omega* entered the Missouri River. Especially the journals of Bell and Harris reveal that the dominant behavior of many on board was shooting. Hardly an animal was seen within firing range that was not fired upon by multiple guns. The image of the *Omega* that emerges from the journals on this leg of the expedition is that of a floating gunship firing constantly and indiscriminately at all that lived along the riverbanks and in the air above. There are several general reasons for the ceaseless killing— food, sport, science—but the reader today is stricken by the pervasive and then-common disregard for the wildlife, an anthropocentric presumption that humans are under no obligation to extend ethical considerability to animals. On the wild prairies there was no call or perceived need for restraint. Besides, most of these men were paid hunters of one sort or another. Harris's entry for May 15 reveals the pervasiveness of this ethical disregard for the animals in their sights. He explains that he and his companions had walked after dinner to a "Heronry" he had discovered the day before and "shot four of them and a Raven which came to feast on their eggs when they found the herons absent." He then complains that the trappers continually interfere with their hunts and that they are poor marksmen: "The trappers who are very much in our way on our shooting excursions had been shooting all the morning at them and had only killed one. We find these men (who are scarcely any of them old trappers), to be the worst shots we have ever met, with either a rifle or shot gun, and how they are to subsist in the woods after they are turned loose to shift for themselves is not easy to conceive." For Harris, a thoroughly informed ornithologist, shooting nesting herons is not objectionable: Bell will remove and preserve the skins, which will enhance his or someone else's natural history collection. The only issue, in Harris's view, is the interference they endure from the trappers and their shoddy gunning. The "work" of Audubon's expedition members was to

bring back "a good round number of animals," and the gunning reported
in the journals reflects their uninhibited enthusiasm for this work.

Despite the unwelcome presence of the trappers aboard the *Omega*, an
atmosphere of cordial sociability continued among Audubon, his companions, and Captain Sire. Sprague worked on some of the first drawings of
birds and mammals whenever the *Omega* stopped for a while, usually in the
evenings and mornings. He also hand-colored a lithographic print for Sire.
The others took advantage of stops to hunt onshore, and Bell continued
to preserve skins. Several miles above Bellevue, at the mouth of the Platte
River, on May 10, while the Indian agent and his dragoons inspected Sire's
cargo for whiskey (which at the time was not permitted to be imported
into the Upper Missouri), Harris, Bell, Squires, and Sprague traipsed off
with guns in pursuit of birds. Audubon chose to visit with several officers
for cordial conversation, potential stories about quadrupeds, and probably
also the pleasure of being recognized and congratulated far from home.

An engaging development in the journals on this leg of the expedition
occurs in the voice of Isaac Sprague, which is more expressive on this leg
than on any other, his tone varying with the situation. An early landscape
description, May 3, is tempered with an understated awe: "Passed Fort
Leavenworth. A frontier post beautifully situated on a high bluff, 431 miles
above St Louis. From this fort there is a fine view of the country for many
miles around, consisting of immense forests and boundless prairies as far as
the eye can reach." When describing his response to the strange appearance
of a group of Indians, he uses a light ironic humor: "Some of them were
fine looking fellows, others looked like the Devil, many of them having
their faces painted with red, black or yellow, which did not add much to
their beauty as I could see" (May 6). With regard to the captain and crew,
however, when the boat ran aground (May 23, for example), he could be
quite supercilious, evidently considering himself in a class apart from them:
"This afternoon they managed to run the boat on to a sandbar—and so
admirably have they stuck her into the mud, that all efforts to start her have
proved ineffectual." In some moments, he expresses sheer delight, such as

the evening they crossed the Great Bend on foot and camped on the prairie, May 26: "In a very short time we had arranged our camp, kindled a splendid fire—and despatched two hunters in quest of game. They returned in less than an hour bringing a fine black tailed deer—This was soon dressed and sundry portions thereof in the shape of joints & steaks after being roasted on sharpened sticks before the fire—were despatched with a keen relish and pronounced *delicious*." (See fig. 12.)

On May 17 a plate on one of the *Omega*'s boilers burned out because of the extreme muddiness of the river water, despite their nightly cleaning. While they were eager to reach their destination, Fort Union, Audubon's party were also glad for this chance to hunt over the next two or three days. In the meantime, dead bison were seen floating down the river, the first reported on May 16 and others almost daily thereafter. The cause was a snowfall of two feet several days north of them that "killed thousands of Buffalo calves which now strew the Praries" (EH 67). Audubon explains more closely, "They swim the River in vast multitudes, are drowned and left on either shores for hundreds of miles above; and many have already passed us floating, bloated and putrid!" (AW 95). They soon had all the more reason to be on the lookout for the first live bison. On Thursday, May 18, the supervisor of Fort Pierre, William Laidlaw, on his way downriver, stopped his fleet of four flat-bottomed "Mackinaw boats" to visit and to offer to convey letters to St. Louis, where he was bound with his cargo of 250 "Buffalo hides" (EH 67). Laidlaw explained that the severity and length of this winter had driven bison farther south than usual and that they would begin to see live bison soon after the *Omega* got underway again. He was right. At three the next morning, the steamer began pushing upstream again, and before noon the following day, Saturday, May 20, the first bison appeared, "at a great distance passing the crown of a hill" (EH 68). In Sprague's first impression, he seems somewhat surprised by the peacefulness of the scene: "These noble animals when they are quietly feeding, or reposing on the grass, so nearly resemble our domestic species that I could hardly realize their being wild" (May 21). Audubon, however, recalled something else that Laidlaw had told them: "that we would kill more of all these animals than we would know what to do with" (AW 95).

Laidlaw's prophetic—if also confessional—statement inaugurates important questions about this expedition: How much killing on the western prairies in 1843 was too much? And on what basis might Audubon and his companions decide to restrain themselves? On May 23, when the first bison were killed by hunters from the *Omega*, it is clear that there was some compunction among the men about wasting the meat from slain bison. Harris explains that shallow water compelled the hunting party to leave behind the bulk of their kill: "[U]nfortunately for us and for them there was not enough water for us to land where they were, and it was about 2 miles farther before we came to shore, consequently they were obliged to leave most of their meat. They killed 4 Buffalo and had to leave nearly all for the wolves, a tongue and about 40 lbs. of meat were all they brought on board" (EH 70). Sprague, on the hottest day of the expedition thus far, was in a foul mood and less willing than Harris to excuse the waste: "These animals are thus wantonly destroyed—but a very small portion of the flesh being used, often times none—and the skin being of no value at this season—they are left to be devoured by wolves and buzzards." Similarly, Audubon was inclined to blame the hunters: "The Rascals were so lazy as [. . .] only to bring one tongue, and no portion of the Hump" (AW 101). Whether anyone was to blame or not, these three voices on the very first day of bison kills on this expedition show some concern with the ethics of the hunt, but it is a muted concern, showing up only rarely and not again for two months, until July 20.

Such abstractions become niceties, however, by contrast to the work the expedition members needed to accomplish and the mechanical and navigational challenges Captain Sire and his crew had to confront almost constantly. On May 24, for example, Harris and Bell went ashore in the *Omega*'s "barge" with crew members charged with finding suitable wood to replace several spars broken the day before. Now that they were on the "Upper Missouri," the two naturalists were motivated to procure the new bird species they anticipated finding. On this day's hunting, they shot several known species but also several specimens of what Harris correctly believed would prove a new species of meadowlark. The identification of the western meadowlark is an unusual case because Bell and Harris first suspected it was a new species based on the bird's song rather than its appearance. Audubon credits Bell with first noticing this difference (RO 7:340), showing that in addition to being an excellent hunter and preserver of skins, Bell had the discerning senses of a good naturalist. When Bell helped Harris hear how different this new meadowlark's song was from that of the familiar eastern meadowlark, Harris consulted the natural history books they had brought along to determine physical differences between the two species. The physical descriptions in the published books, however, did not help "establish any specific difference." Yet Harris became convinced that the song itself was diagnostic: "[I]t is utterly impossible that the same bird in different parts of the world can have notes so totally different." He was right, of course, but they were naturalists on the move. For the time being, he would simply make his field notes and hope to later identify the physical distinctiveness needed to publish a "good species": "*Mais nous verrons*" (EH 70–71).

Audubon and Harris had developed a fairly efficient method for finding and identifying new species: go to the new territory, shoot everything, determine in the books which specimens were new species, and sell or barter surplus skins later. It is clear also that each man worked with an expectation that Audubon would name a new bird species in his honor. Having found a new finch species, Harris began to refer to further specimens as "my finches" (entries for May 16, 18, EH).

While the party members enjoyed their relatively privileged status as the invited guests of Pierre Chouteau, Sire and his crew struggled against high winds, strong currents, low water, shifting sandbars, and snags. Sire's log makes plain how different were the concerns of the expedition members and those of the crew. On May 20 Sire wrote, "All day long the wind blows as it only can on the upper Missouri. Often we scarcely move at all." The next day the wind was "almost irresistible." On May 22 a group of natives

Sire believed to be a war party of "those rascally Santees" fired five shots at the *Omega* to make it stop; one ball passed through a man's pant leg, and all were surprised no one was hit. After running aground on May 23, on May 24 Sire wrote, "[W]e succeed in extricating ourselves, but we go aground again, again get off, and after having sounded again find only one passage and that a doubtful one. We lurch and break one of our rudders, but 10 minutes afterward we are afloat. We put to shore to mend the rudder, and meanwhile I have some wood cut from drift." Bell's observation about the weather that day is telling: "Windy & pleasant." Sprague reports he "[m]ade a sketch of the bank of the river opposite where we lay."

At this time, Audubon's attention was divided among various concerns. He was writing abundantly with potential publications in mind: "I keep my Journal very full of all I see, do, and hear" (AW 96–97). He was also preoccupied with matters at Minnie's Land, as he wrote of his concerns to his family: "That you are all quite well, and that you all think of me as often as I do of you All.—I am sometimes uneasy about Dearest Mother, but I hope that you and the Doctor will take the best care of her you can, and indeed of every one of yourselves my Dearest Friends!" (AW 98). Still, he did not long for the end of the trip, for he had business to conduct: "Would that I was there for a few hours and back here again tomorrow morning" (AW 97). While reminding his family to take care of themselves, he also reminded them that they had a role to play by publicizing his current travels. When they received the copy of his journal that he would send from Fort Pierre, he wrote, "You may then find some extracts of interest for publication; but I would be nice and circumspect in all such cases as I prefer keeping as much as possible for our Books on Quadrupeds.—To say something about our being here or there and a few other matters would probably be more prudent than to write at length" (AW 100). He was also hopeful, however, that his family could share some of his experience. On a somewhat comic or ridiculous note, he informs them that "[n]either Harris, Squires, or myself have shaved since we left St Louis, and I have not once pulled off my breeches when I have tumbled down at night to go to sleep" (AW 98).

There was the sublime as well, however: "Of Buffloes we have seen many, many thousands, for either from any part of the Boat or even from my own room door, I see them in large herds feeding quietly, whether on the great Prairies or on the crests or edges of the Hills and Ravines" (AW 106).

One of the most purely enjoyable experiences for all four of the journalists occurred at the Great Bend, where the Missouri forms a loop of some twenty-six miles before reaching a point only two or three miles from the beginning of the bend. Captain Sire put Audubon, Harris, Bell, and Sprague ashore to cross the isthmus on foot, camp on the far shore, and await the arrival of the *Omega*. Michaux and two other hunters accompanied them, according to Harris, "to carry our luggage, act as guides, &ᶜ" (EH 73). As always, they hoped for new species of birds and mammals, perhaps even fossils. Audubon found the hour-and-a-half crossing difficult, "for it is slow work to ascend Hills covered with greasy clay and covered with pebbles at every step" (AW 106–7). Harris agreed but saw the landscape as nonetheless satisfyingly wild and scenic: "We started on a beautiful and very extensive level prairie lying just above the highest freshets, and reaching about half way across, the rest was steep and rugged hills of several hundred feet in height" (EH 73). On this prairie also was the first prairie dog village they had seen. (See fig. 13.) With less than three hours of daylight left, they attempted to shoot some but procured none because they lacked "time to wait for them to come out of their holes to reconnoiter after their first alarm" (EH 73). Despite this failure, their distance from the noisy steamboat and the vista from the high crest were relieving and vivifying, and spirits were high among the travelers. After descending to the opposite bank and setting up camp under a clump of six cottonwoods, Michaux and another hunter, just before sundown, carried into camp a yearling mule deer buck, a species new to Audubon and company. Harris describes the evening as follows: "It was not long before some of the choice morceaux were roasting before the fire impaled upon sharp sticks stuck in the ground. We all agreed it was the best Venison we ever tasted, and none failed to do ample justice to the repast" (EH 73–74). Sprague, as mentioned previously, was also enjoying the evening and noted that the venison was "despatched with a keen relish and pronounced *delicious*" (May 26). Harris later commented that "he should not want any more food for 3 days" (AW 107).

Judging from how vividly Harris and Sprague rendered this evening in their journals, and Audubon in his letter—Bell is still rather stingy with words in his entries—this night probably remained one of their favorite memories for years to come. Audubon gazed at the fire and his companions through the eye of an artist, of course: "The sight of our camp in the Darkness of the night (and I looked at it several times) was almost sublime and would have made a grand and imposing Picture" (AW 107). Since Maria Audubon's edition of her grandfather's journals provides one of the most eloquent and vivid passages in all of the journals for this scene, I will quote it at length and hope that it is faithful to the original:

The darkness of the night, contrasting with the vivid glare of our fire, which threw a bright light on the skinning of the Deer, and was reflected on the trunks and branches of the cottonwood trees, six of them in one clump, almost arising from the same root, gave such superb effect that I retired some few steps to enjoy the truly fine picture. Some were arranging their rough couches, whilst others were engaged in carrying wood to support our fire through the night; some brought water from the great, muddy stream, and others were busily at work sharpening long sticks for skewers, from which large pieces of venison were soon seen dropping their rich juices upon the brightest of embers. The very sight of this sharpened our appetites, and it must have been laughable to see how all of us fell to, and ate of this first-killed Black-tailed Deer. After a hearty meal we went to sleep, one and all, under the protection of God, and not much afraid of Indians [...]. Our fires were mended several times by one or another of the party, and the short night passed on, refreshing us all as only men can be refreshed by sleep under the sky, breathing the purest of air, and happy as only a clear conscience can make one. (AJ 1:516)

This night was all the more memorable for the thirty-one-year-old Isaac Sprague because he had never before slept outside with nothing over his head: "To me the situation was one of novelty it being my first encampment in the open air. However, I slept quite as sound as usual—and experienced no inconvenience whatever." No other experience on this expedition provided such unqualified pleasure.

After resuming their places aboard the *Omega* the next day, they began to see increasing numbers of bison. Bell reported "at least 5000" the same afternoon. Audubon teased his family in his letter: "[W]e have seen so many Buffaloes on both sides of the River, and swimming that I dare not mention their numbers" (AW 108). When they passed Fort George on May 28, however, their attention turned to other matters. While pilots and crew struggled to navigate through shallow water and ran aground, Sire, Audubon, and others went ashore to walk to the fort. There they acquired a two-month-old bison calf, whose head Sprague began to draw "the size of life." On May 30, with the *Omega* making little progress each day, the superintendents of the next two fur company forts upstream—Honoré Picotte of Fort Pierre and Francis Chardon of Fort Clark—traveled down to greet the distinguished traveler and his companions and to offer their assistance. Picotte went back to Fort Pierre and returned with a scow and a barge to help lighten the load of the *Omega* over the remaining shallow river miles. Picotte also began the bestowing of gifts by presenting Audubon with "the finest pair of Elk Horns, that ever were seen in *this country*" (AW 109). Audubon was also already anticipating the work that Sprague would do at Fort Pierre: "Sprague will make several outlines of very young Calves with my camera [i.e., a camera lucida], and group them to form part of the Plate of the *Buffaloe!*" (AW 109). (See fig. 14.)

Even though this ascent to Fort Pierre was the fastest ever, with the *Omega* arriving at three in the afternoon of May 31, Captain Sire wasted no time. Audubon was impressed: "Our Captain who is one of the most pushing men, went to work at once to unload one half of his cargo, and put off 50 men with their baggage to the Fort" (AW 110). With the *Omega* considerably lightened now, everyone anticipated much faster travel up to Fort Union. In the meantime, however, Sprague recorded his admiration for the encampment of Sioux, "or as they call themselves *Dah cotah* Indians," near the fort: "Many of the men are really noble looking fellows—They are generally large, tall, and finely formed, and as they stood in groups, or pranced up and down the shore, on horseback—with their long hair and showy dresses streaming in the wind they presented a truly picturesque appearance." This refined young man from Hingham, Massachusetts, now 1,274 river miles above St. Louis, was observing the most exotic sights of his life: "They ride remarkably well—both sexes—dashing ahead at a furious rate—and clinging so closely to the horse as to appear a part of the same animal." He had work to do, but he was also somewhat dazzled to be there, as were they all. In his excitement, Audubon sought to span the distance between this western outpost and Minnie's Land: "Would to God that you could all of you see us at this moment. Full of health and high spirits" (AW 110). (See fig. 15.)

The next morning, June 1, Sprague set to work with the camera lucida, drawing outlines of several captive bison calves at the fort, while Bell and Harris walked with light guns to procure more birds. Audubon wrote letters and worked up his journal, anticipating Captain Sire's return from Fort Union to St. Louis and his delivery there of many packages and bundles of gifts, letters, souvenirs, elk antlers, and a copy of his journal made by his secretary, Squires. Later that afternoon, after the *Omega* was pushing upstream again, they were all reminded of the perpetual threat of human violence on this lawless frontier. Chardon, hitching a ride back to Fort Clark, discovered "an Indian [. . .] concealed on board whom he had flogged at one of the upper stations, and that he had no doubt his object was to kill him." Near the shore, the man's "pack was thrown on the bank, and with a push Chardon sent the Indian after it up to his waist in water" (Chittenden, *History* 1:373). The fur trade historian Hiram Chittenden characterized Chardon as "an able but unscrupulous man, and something of a desperate character when his evil nature was once aroused" (*History* 1:373). Harris foresaw a killing: "This affair will never be forgotten by the Indian, and Chardon is enough of an indian to bear it in mind also, so that it will probably sooner or later result in the death of one or the other of the parties" (EH 84).

The remaining passage to Fort Clark, near the Mandan village, was difficult because of a north wind and snags, but Captain Sire was not surprised, for this was "the worst part of the Missouri" (3:996). It was here, however, on June 6, that they learned that Audubon's goal of reaching the Rocky Mountains on this trip would not be met. Four Mackinaw boats heading for St. Louis brought word of renewed hostilities between Blackfeet Indians and fur company employees in the region of the new Fort Chardon. Harris recorded the change of plan: "This state of affairs will prevent our visiting that interesting region." Fort Union, then, would be their western-most residence.

Harris, Audubon, and the others had read and heard about the 1837 smallpox epidemic that wiped out the Mandans, so when the *Omega* arrived at Fort Clark and the Mandan village on June 7, they hoped to visit the remnant population, who had removed two or three miles farther up the river. Sprague put their surviving number at "about 150." The Mandans were particularly interesting to the travelers because of the earth lodges that distinguished them from the Sioux and other northern plains native cultures.

Sprague granted himself a considerable amount of time in his June 7 account of their visit to the Mandan earth lodges when they stopped at Fort Clark and crafted some of the best prose written on this expedition. By following the stages of the construction of these shelters in his carefully detailed description, he implicitly pays homage to an ingenious people wiped out by smallpox just six years before his visit:

> Instead of tents of skins, supported by poles, they build a permanent hut—commencing by digging a circular cellar about two feet deep, around the outside of this are set stakes or posts 5 or 6 inches in diameter quite close together, rising 4 or 5 feet above the ground, with a slight inclination inward—
>
> On the top of these are placed other poles which rise to one common centre at an angle of some 30 or 40 degrees. The whole is then covered

with a layer of willow twigs and small poles lashed across and over the whole a thick coat of mud or clay, which gives them the appearance of a hemispherical heap of dirt somewhat resembling in shape a bowl bottom up.

From the ground up he vivifies this image of a Mandan earth lodge that would have been home to fifteen or twenty people. When they visit the much larger "medicine lodge," the images Sprague selects are subtly haunting, understated: "On the floor laid an old man quite blind, who, as our guide informed us by signs, had come there to die—he shook hands with all of us—Another Indian sat wrapped in his robe on one side of the lodge but he neither moved nor spoke." No other journalist reports this dying man shaking their hands. This is Sprague's sensitivity to the loss and remembered horror they were witnessing. Sprague's inclusion of the silent, still man suggests the utter helplessness of a people decimated by a disease beyond their comprehension. Sprague and the others had all read Catlin's romanticized account of the noble Mandans and entertained varying degrees of skepticism about that representation, but Sprague's account of this visit seems to acknowledge that the Mandans Catlin knew in the early 1830s were nothing like the natives they were encountering in 1843 on the other side of the devastating smallpox epidemic of 1837: "A visit to their village such a day as this destroys all the romance of Indian life. It was stormy and cold, and they appeared anything but comfortable. The water dripped through the roof in many places so that the floor was quite wet." This writer's spare style shows an instinct for allowing precise images to speak for themselves.[9]

The final five days of the *Omega*'s navigation to Fort Union were more or less routine ones, which means grueling and challenging. They departed from Fort Clark early, around 2:45 a.m., on Thursday, June 8, and reached Fort Union about 7:00 Monday evening, June 12. Along this final passage they were entering a new biozone. On June 10 Harris wrote, "We now appear to have really got into a game country"; on this day they began to

9 For a firsthand account of the smallpox epidemic of 1837, see Chardon 123–39; see also Chardon's account reported in AJ 2:42–47.

see bighorn sheep for the first time and increasing numbers of elk. And they seem to have become all the more inured to the killing of the large mammals. On June 9, when the steamer stopped for the night, Bell and Harris—who had become regular hunting partners, both having excelled at the new skills required of hunters on the northern prairies—"went out with our Guns," and soon Harris was in pursuit of an elk and Bell in pursuit of a bison in different directions. After the ship's bell called them back aboard, Bell reported to his friend Harris that he had found a group of bison and "had been amusing himself with firing at this gang of Buffaloes, which then consisted of 7 and a calf, he fired 5 times, once with small shot at the calf which he wounded, and he also wounded badly a young bull, but the Bell ringing he had to leave them, they were no doubt killed" (EH 95). The phrase "amusing himself" suggests the degree of indifference to the killing and the waste. So too does Harris's entry for June 11, in which he reports without any evidence of compunction the killing of a bison from the steamer: "[A] round of three or four guns were fired first without sensible effect and he turned to run up the ravine, when another round brought him to and he tumbled headlong into the ravine, and we lost sight of him—having plenty of meat on board the Capt would not stop for him and we left him for the wolves and Ravens." The journals consistently show that the first impulse of most on board upon seeing an animal was to shoot first and ask questions later.

On Monday, June 12, the forty-ninth day after departing from St. Louis, Captain Sire and crew completed the fastest ascent yet of the Missouri River to Fort Union, according to Bell, "at 7 oclock amid roaring of cannon from fort and boat." The arrival of a steamboat from St. Louis was always the occasion of celebration and revelry, even amid the heavy labor of unloading the upstream cargo and reloading with skins and furs. Sire's pilot on this trip, Joseph LaBarge, later described the event as follows:

Among the important events of every voyage were the arrivals at the various trading posts. To the occupants of these remote stations, buried in the depths of the wilderness, shut out for months from any glimpse of the world outside, the coming of the annual boat was an event of even greater interest than to the passengers themselves. Generally the person in charge of the post, with some of the employees, would drop down the river two or three days' ride and meet the boat. When she drew near the post, salutes would be exchanged, the colors displayed, and the passengers would throng the deck to greet the crowd which lined the bank. The exigencies of navigation never left much time for celebration and conviviality. The exchange of cargo was carried on with the utmost dispatch, and the moment the business was completed the boat proceeded on her way (Chittenden, *History* 1:132).

Audubon noted that he and his companions were "all well and in good spirits" and that the supervisor of the fort, Alexander Culbertson (1809–79), "received us all very kindly, and I trust that he will make our stay here comfortable" (AW 114). Culbertson was eager to please so distinguished a guest and his party. According to Harris, he "offered us his services and the command of all the means of the establishment to further the accomplishments of our views." He had prepared a room of twelve by fourteen feet in the fort for their lodging and would offer other space for the various types of work they would be conducting. Culbertson also introduced Audubon to experienced hunters at the fort who would go out after any and all species of animals Audubon desired. From these first hours, Audubon already anticipated that his hunters would keep him and Sprague constantly supplied with animals to draw: "I hope to be kept very busy drawing along with Sprague until our departure" (AW 116).

This first night, however, they all spent on board, writing letters and preparing packages that Sire would deliver in St. Louis after his rapid downstream return voyage. (It took only fifteen days.) Squires was busy finishing a copy of Audubon's journal, which he would complete the next day, June 13. In Audubon's letter home of the same date, he informed his family, "I send the Copy of my Journal by Squires, to the Care of the House at St Louis to be forwarded to you by a safe private Friend of theirs, and

I hope you will receive it safely."[10] He also advised them "to let *the World* know where we are"—ever the impresario. This night, though, he was also assessing all he had seen thus far on this upriver journey into the past, to a condition of wildness no longer available in the East: "We have seen an immensity of Game of all description. Yesterday and within only 3 miles of this place, we saw 22 Mountain Rams in one flock, and saw them for nearly 10 minutes running up and down Hills as if so many unaccountably active Sheep.—Grisley Bears are abundant too, and Wolves are not to be compted, so numerous are these Beasts every where." If only his two sons were with him: "Now if Johny was here, he would delight in shooting Townsend's Hare, Bighorns, Elks, Antelopes, Black tailed Deer &c and Victor would paint some very strange scenery, though not very fine for Pictures." Shooting and drawing were the diastole and systole of this man's heart, and he saw the same rhythm played out in his two sons. But amid his excitement this night, he dedicated himself to the work that would begin in earnest the next morning: "I am going to collect all possible information about Quadrupeds &c during my stay here and from good sources. My head is actually swimming with excitement and I cannot write any more" (AW 116).

10 About this copy of the journal, Eliza Berthoud wrote to Lucy, from St. Louis, June 28, 1843: "We have this morning received a letter from Mr Audubon dated Fort Union June 13*th* 1843 in which he requests we will write to you, & send by the first good private opportunity his journal:[...] opportunely our Friend Mr Fishel expects to leave tomorrow, & as he resides in your neighborhood will I am sure take all care of any dispatches" (Beinecke). Lucy wrote to her husband on August 7, 1843, "Your Journal we received safely at last a day or two ago, I have only had time yet to look at a page or two, the contents I dare say will enrich the Quadrupeds" (Beinecke). This is likely the copy of the journal sent to Bachman in 1846, which he complains "terminates abruptly," adding that "[y]ou had only got to the mouth of the Yellow Stone." He characterizes the content of the journal: "To me it has been a very interesting one although there is less in it of quadrupeds than I had expected to find. The narratives however are particularly spirited—often amusing & instructive. On what you write on the spot I can depend." He is trying to be positive, but he's clearly disappointed by the dearth of information useful to their future publication: "I cannot find that you ever set a trap or looked at the smaller rodentia. This was a terrible error." And "I think you all spent too much time among the Buffaloes & hunting to feed other people" (Bachman 346).

Fort Union and the Prairies,
June 13–August 15

Everyone felt the excitement. Not only would the Audubon party be free now to commence the work they had come to the northern prairies to do, but they would also begin to hunt from horseback and acquire the skills needed to successfully pursue and kill the legendary bears, elk, pronghorns, and bison. (See fig. 16.)

Culbertson began their instruction on June 14, shortly after they all wished Captain Sire "a speedy passage home" (EH 99). After the six members of the party were all squared away in their new quarters (Provost had joined them and agreed to pilot their boat down to St. Louis on their return), Culbertson demonstrated "how wolves were run down on horseback." Mounted on his "beautiful Blackfoot Pied mare," he spotted "Mr. Wolf," as Harris referred to the animal, and pursued at full speed until he disappeared from view over a small hill, but he quickly returned, still at full speed, with the slain wolf dramatically slung over the horse's neck in front of the saddle. Harris noted that Culbertson could hold the wolf in place because he was not using either hand to guide the horse; he guided it rather by changing "the inclination of the body, the reins being thrown upon the neck, as they necessarily were during the chase for the convenience of loading and firing." None of these easterners had ever seen such an amazing performance of man on horse. And Culbertson sustained their amazement by having several other hunters from the fort join him in a competition to determine which one could reload and fire his "flint guns" the greatest number of times while on horseback at a dead run. All four journalists were impressed by the skills displayed, but Harris's description is the most detailed and shows that he fully intended to master this method of hunting:

> They generally put five or six bullets in their mouth, and when they fire they pour a charge of powder into the left hand from the powder horn which hangs over the right shoulder, throw it into the barrel, which is hastily struck on the saddle to shake down the powder so as to pass into the pan to prime it, then throw in a bullet wet with saliva of the mouth which causes it to adhere to the powder and prevents it falling out when the muzzle is depressed to fire. In this manner these gentlemen fired from 12 to 14 times each in riding about a mile. (EH 100–101)

Culbertson and his hunters were no doubt showing off, but they were also challenging the new would-be "buffalo hunters" to learn the necessary skills and introducing them to the dangers of this kind of hunt. Of Audubon's party, Bell and Harris would become the most successful prairie hunters. Squires would excel at the challenging prairie horsemanship but generally fail at the hunt itself. There is no evidence that either Sprague or Audubon ever aspired to become a bona fide prairie hunter. But all of them had now been introduced to the thrilling danger of the chase. The prevailing spirit among the men was enthusiasm for the pursuit of adventure, which Culbertson more or less guaranteed them. Most eagerly anticipated was the full-speed horseback pursuit and killing of wild bison.

The journals also reveal that hunters and artists alike would not waste their precious time. With Audubon as the organizational hub of operations, a routine was quickly established. Harris, Bell, Provost, Owen McKenzie, and other hunters regularly crossed the river to bring back animals of all sorts to preserve, or eat, and for Audubon and Sprague to draw. On the second day of their residence at the fort, Audubon engaged a hunter to go out specifically for pronghorns, generally referred to as "antelopes." Because the elusive bighorn sheep are particularly difficult to hunt, he offered a skilled hunter ten dollars for each and every one he could bring in. Audubon's intent was full-time employment for his entire crew. As intense as his appetite for adventure was, when it came to advancing his publications, he could work twelve- and fourteen-hour days, and his expectations of others were high. He noted a typical day of work at the fort: "Bell was busy all day with skins, and Sprague with flowers, which he delineates finely" (AJ 2:52). On a day when Bell failed to bring in a wolf he killed, Audubon lectured him: "I felt vexed that he had carelessly suffered the Gray Wolf to be thrown into the river. I spoke to him on the subject of never losing a specimen till we were quite sure it would not be needed; and I feel well assured he is so honest a man and so good a worker that what I said will last for all time" (AJ 2:40). When Audubon was trying to recruit Spencer Fullerton Baird for the expedition the previous November, he warned him fairly, "You will have to work hard, of course." But he also reminded him, "[A]s you already know me pretty well, I may freely say to you that I am not 'hard on the trigger'" (qtd. in Dall 86). As the five companions settled into a work schedule, a spirit of camaraderie continued to unite them. Sprague, however noticeably industrious, was also noted for his "usual reticence" (AJ 2:82).

One of the principle pleasures at the fort was the shooting of wolves at dawn and at dusk from the ramparts. They were attracted to the scraps, bones, and offal of animals killed by hunters and thrown outside the fort walls. The journals show that this sport shooting was practically a daily activity. Audubon arranged for himself and his colleagues to be called at sunrise before the fort gates were opened if wolves were present and within rifle range: "and I hope that to-morrow morning we may shoot one or more of these bold marauders" (AJ 2:41). Harris particularly enjoyed this gallery shooting, according to Audubon: "We are having some music this evening, and Harris alone is absent, being at his favorite evening occupation, namely, shooting at Wolves from the ramparts" (AJ 2:85). The journals make abundantly clear that a wolf was a creature that needed to be killed.

Like practically all naturalists collecting items of natural history at this time, Audubon also intended to procure skulls of Native Americans.[11] Today, of course, we find this practice disturbing, but Audubon apparently saw nothing untoward about it, although he certainly knew that other Natives would be offended and probably attack anyone caught pilfering one of the coffins placed in trees near the fort. The following passage from a June 18 journal entry makes clear that he decided to take the skulls as soon as he learned that dead bodies were in coffins nearby: "I took a walk with Mr. Culbertson and Mr. Chardon, to look at some old, decaying, and simply constructed coffins, placed on trees about ten feet above ground, for the purpose of finding out in what manner, and when it would be best for us to take away the skulls, some six or seven in number, all Assiniboin Indians. It was decided that we would do so at dusk, or nearly at dark" (AJ 2:38). On June 22 he notes matter-of-factly, "As I was walking over the prairie, I found an Indian's skull (an Assiniboin) and put it in my game pouch" (AJ 2:51). He adds no comment. On Sunday, July 2, Audubon violated the burial of a Native man well known to residents of the fort and destroyed the coffin. Because it reflects the utter disregard of Anglo-Americans for Native Americans widespread at the time, I quote from this graphic passage at length:

11   The Philadelphia physician and naturalist Samuel G. Morton (1799–1851) will suffice to illustrate the prevailing attitude in Audubon's day toward collecting the skulls of Native Americans, ancient and modern. Morton published two works based on his collection of nearly a thousand human skulls, *Crania Americana* (1839) and *Crania Aegyptiaca* (1844). Morton's statement of purpose in the earlier volume also indicates the breadth and geographic range of his collection: "[T]o give accurate delineations of the crania of more than forty Indian nations, Peruvian, Brazilian and Mexican, together with a particularly extended series from North America, from the Pacific Ocean to the Atlantic, and from Florida to the region of the Polar tribes" (iii). His purpose was also to encourage "throughout this country, a greater interest to this important and attractive study" (iv). To further develop his own collection, Morton "respectfully solicits the further aid of gentlemen interested in the cause of science, in procuring the *skulls of all nations*, and forwarding them to his address in this city" (v). Practically every natural history collection or museum included artifacts and skeletons from Native American burial sites. Audubon's young friend Spencer Fullerton Baird would oversee one of the most extensive in his work at the Smithsonian Institution beginning in 1850.

Mr. Denig and I walked off with a bag and instruments, to take off the head of a three-years-dead Indian chief, called the White Cow. Mr. Denig got upon my shoulders and into the branches near the coffin, which stood about ten feet above ground. The coffin was lowered, or rather tumbled, down, and the cover was soon hammered off; to my surprise, the feet were placed on the pillow, instead of the head, which lay at the foot of the coffin—if a long box may so be called. Worms innumerable were all about it; the feet were naked, shrunk, and dried up. The head had still the hair on, but was twisted off in a moment, under jaw and all. The body had been first wrapped up in a Buffalo skin without hair, and then in another robe with the hair on, as usual; after this the dead man had been enveloped in an American flag, and over this a superb scarlet blanket. We left all on the ground but the head. Squires, Mr. Denig and young Owen McKenzie went afterwards to try to replace the coffin and contents in the tree, but in vain; the whole affair fell to the ground, and there it lies; but I intend to-morrow to have it covered with earth." (AJ 2:72–73)

Because this passage exists only in the bowdlerized edition of Audubon's granddaughter, we can't know whether Audubon's original language was as plain and unemotional as this; the passage nonetheless shows the indifference of most white Americans to the humanity of Native Americans in Audubon's day. The fact that Maria—who is known to have omitted objectionable material from her edition of her grandfather's journals, including anything inconsistent with her image of him as "a refined and cultured gentleman" (AJ 1:ix–x)—kept this passage in shows that continuing indifference through the 1890s.

This indifference to the humanity of Natives is perhaps even more surprising in this case since Audubon was told the man's name, White Cow, and some of his story: "He was a good friend to the whites, and knew how to procure many Buffalo robes for them; he was also a famous orator, and never failed to harangue his people on all occasions. He was, however, consumptive, and finding himself about to die, he sent his squaw for water, took an arrow from his quiver, and thrusting it into his heart, expired, and was found dead when his squaw returned to the lodge" (AJ 2:73). Even the pathetic story of White Cow's illness and suicide also apparently failed to elicit an emotional response from the man who destroyed his coffin and now possessed his head, "under jaw and all."

The great event that all the men anticipated was the first bison hunt on horseback, and Culbertson was gradually preparing them for a day that could be quite dangerous for the novice prairie hunters. Since Harris and Bell showed the most promise, Culbertson told them they would go first. On June 28 Harris revealed the general plan: "When Buffalo are found Bell and I are to make our maiden effort at running them, we are to be furnished with the two best Buffalo horses belonging to the Fort, and Mr. Audubon and Mr. Culbertson will follow as spectators drawn in the Carryall by old Peter [a reliable mule]. We are to take an extra cart to bring home any animals which may be procured for our use." While Harris, Bell, and Squires continued to go out hunting on their own practically every day, and Squires occasionally accompanied a hunter from the fort on an overnight hunt some distance away, they were shooting mainly birds and smaller mammals but still gaining experience on the rugged terrain. On July 1 and 2 Culbertson added to the other day's activities "a sham Buffalo hunt." On the first one, Audubon reported, Culbertson, "Harris, and Squires started on good horses, went about a mile, and returned full tilt, firing and cracking. Squires fired four times and missed once. Harris did not shoot at all; but Mr. Culbertson fired eleven times, starting at the onset with an empty gun, snapped three times, and reached the fort with his gun loaded. A more wonderful rider I never saw" (AJ 2:71). The next day, only Squires of the Audubon party participated, and Audubon enjoyed seeing his secretary's great improvement: "I was glad and proud to see that Squires, though so inexperienced a hunter, managed to shoot five shots within the mile, McKenzie eleven, and Mr. Culbertson eight" (AJ 2:73). These brief but vigorous exercises were no doubt competitions between prairie hunters, and an occasional tourist, but they also served to sharpen skills.

From late June into mid-July, the journals of Harris, Bell, and Audubon report much puzzlement over two problems: the possibility of one or two new species of northern flicker and their collective failure to capture or shoot an especially elusive small hare. In entry after entry, Harris reveals his mounting fascination with the flicker problem that has arisen with the capture of young birds with markings inconsistent with those of the two known species: the eastern yellow-shafted (golden-winged) flicker and the western red-shafted flicker: "We are all in a heap with these Golden wings & Red shafts—We have the *Red-shafted Bird* with *Red cheeks*, and *Newly-fledged birds* with *red shafts* and *black cheeks, Yellow-winged birds* with *Red cheeks* and with *black cheeks*, both adults" (EH 122).[12] Harris's intellectual tenacity as a naturalist shows through in these entries. He knew that any new "good species" of bird would be published in the final volume of the Octavo edition of Audubon's *The Birds of America*, which was then being prepared, and he and the others devoted considerable physical and mental labor to gathering specimens and determining how many actual species made up this family of woodpeckers on the Upper Missouri. Finally, while on this trip, they were unable to decide: "We must wait further developments before we can pretend to decide this curious question" (EH 123).

Harris's lengthy lamentations over their failure to capture the hare (which Elliott Coues identified as the wormwood hare [*Lepus artemesia*], AJ 2:49n1) have a comic appeal for us—and for Harris—but they also open a window onto the simple but powerful desire of a naturalist, rather than that of a sportsman, that motivated Harris, Audubon, and the others to persist from day to day. I quote here from Harris's July 10 entry:

In the afternoon I took a horse and rode up the river to the spot visited on Friday to renew the search for the little Hare which is evidently in considerable numbers. And yet so timid are these little animals that we can discover no signs of them at a distance of more than two or three

12 The yellow-shafted and red-shafted flickers are now considered as different populations of one species, the northern flicker (*Colaptes auratus*). Audubon and his companions were witnessing the results of their free interbreeding where their ranges overlap.

yards from the thick bushes where they take shelter during the day, and where they are perfectly secure from their enemies. About half an hour before sunset I placed myself on the bank of a deep ravine opposite some bushes on the other side, where the signs were numerous and fresh. I remained with the utmost patience until 9 of the clock when no longer being able to see I gave it up for a bad job. Before leaving I approached the bushes quietly and could hear the little rascals moving about in them. There is nothing in the country which has puzzled us so much to procure as this little rabbit, said by everyone to be abundant. If we go into the bushes, principally rose bushes, they are so thick that we cannot see them when they start, and if we lie in wait for them in the morning and evening they wont come out to be shot, and there is no dog to be had who will follow their tracks.

Nearly three years later, when Bachman was developing the natural history essays for the *Quadrupeds* in Charleston using Audubon's Missouri River journals, he berated Audubon for this failure to bring home specimens of the small hare: "Why man what poor trappers you proved yourselves to be. Do you know what I would have done? Why make a brush fence a foot or 18 inches high in the thick brush wood 100 yards long. Set a snare at every gap ten feet apart & I would have had a couple every night. I never have failed—rabbits—Northern Hare—marsh hare—I care not—in they go & by the neck they hang" (347). Bachman always suspected that this expedition had been more about sport and art than science.

On July 14, the day before their departure on a three-day excursion "to look for Elk, Bighorns and Beaver" (EH 135) (see fig. 17), Culbertson and his Blackfoot wife, Natawista, orchestrated a rousing display of riding and shooting skills that included elements of pageantry. Since Audubon's journal entry in *Audubon and His Journals* is so vivid and coherent and generally corroborated by the other journalists, I quote from it at length:

After dinner we had a curious sight. Squires put on my Indian dress. McKenzie put on one of Mr. Culbertson's, Mrs. Culbertson put on her own *superb* dress, and the cook's wife put on the one Mrs. Culbertson had given me. Squires and Owen were painted in an awful manner by Mrs. Culbertson, the *Ladies* had their hair loose, and flying in the breeze, and then all mounted on horses with Indian saddles and trappings. Mrs. Culbertson and her maid rode astride like men, and all rode a furious race, under whip the whole way, for more than one mile on the prairie; and how amazed would have been any European lady, or some of our modern belles who boast their equestrian skill, at seeing the magnificent riding of this Indian princess—for that is Mrs. Culbertson's rank—and her servant. Mr. Culbertson rode with them, the horses running as if wild, with these extraordinary Indian riders, Mrs. Culbertson's magnificent black hair floating like a banner behind her. As to the men (for two others had joined Squires and McKenzie), I cannot compare them to anything in the whole creation. They ran like wild creatures of unearthly compound. Hither and thither they dashed, and when the whole party had crossed the ravine below, they saw a fine Wolf and gave the whip to their horses, and though the Wolf cut to right and left Owen shot at him with an arrow and missed, but Mr. Culbertson gave it chase, overtook it, his gun flashed, and the Wolf lay dead. They then ascended the hills and away they went, with our princess and her faithful attendant in the van, and by and by the group returned to the camp, running full speed till they entered the fort, and all this in the intense heat of this July afternoon. Mrs. Culbertson, herself a wonderful rider, possessed of both strength and grace in a marked degree, assured me that Squires was equal to any man in the country as a rider, and I saw for myself that he managed his horse as well as any of the party, and I was pleased to see him in his dress, ornaments, etc., looking, however, I must confess, after Mrs. Culbertson's painting his face, like a being from the infernal regions. (AJ 2:88–89)

Following these initial "fine evolutions," as Harris referred to the riding displays, he mounted a horse in order to try reloading his gun at a full gallop for the first time, a skill he knew would be helpful on a bison hunt. While he managed to load his gun rather awkwardly once—"I got the load in after a fashion"—his second attempt failed. He nevertheless did some challenging high-speed riding while he and Culbertson chased two more wolves that afternoon. Like Squires and Bell, Harris was becoming an experienced prairie hunter.

The journal entries of Harris, Bell, and Audubon for the next two days, July 15 and 16, tell a remarkable story of fabrication and stage direction for the purpose of cultivating and preserving Audubon's celebrity. The evidence shows that either Bell or Harris largely copied the other's entries for these two days and that Audubon copied Harris's entry for the sixteenth into his journal but substituted himself for Harris in a crucial moment of thrilling danger and last-second salvation.

On Saturday, the fifteenth, as Harris explains, "Mʳ Audubon, Mʳ Culbertson, Bell & I in the carryall with two horses in tandem, Squires and Owen & Provost on horseback and three men with a mule to the cart on which was placed our skiff and a tent and our baggage, started this morning after breakfast on an expedition to the Yellowstone which we have been talking about for some time." (Only Sprague of the party chose not to go on this trip.) After crossing the river at the fort, they traversed the rugged terrain eastward to the confluence with the Yellowstone, which they followed upstream to a region about twenty miles from the fort, observing many elk tracks but seeing no game. However, the next morning, July 16, following a severe thunderstorm in the night, Culbertson spotted a bison bull some three or four miles distant. He and McKenzie mounted their horses and set out in pursuit, with the men in camp watching. After McKenzie, Culbertson, and the bison disappeared from view, Bell and Harris, assuming the bison was killed, drove the carryall, with another horse in tow, to retrieve the meat and skin. At the site of the slain bull, McKenzie noticed another bull walking slowly toward them.

At this point, the respective journal entries of Harris, Bell, and Audubon begin to tell a very different sort of story. July 16 had been the most exciting day of hunting they had had so far, and it is clear that Harris and Bell cooperated in the writing of a complete, exciting narrative of the day; much of their respective entries for this day is identical, reporting the same sequence of events in the same phrasing. Because each entry includes some lines or details not included in the other, I cannot say definitively who copied from whom. In either case, however, Harris and Bell agree that the following exciting series of events occurred, although they disagree about an opening detail: When McKenzie pointed out the second bull approaching, Harris and Bell both claim that they were the only one of the three men present (Culbertson had returned to camp with the tongue of the first bull) with lead balls for his gun, which he then handed over to McKenzie, the more experienced hunter, because the meat was much needed at the fort. Then Harris and Bell placed themselves "on an eminence to view the chace." The bison broke into a full run when McKenzie came within seventy or eighty yards of him. Because his mare was already winded from two previous hard runs that morning, McKenzie applied "the whip pretty freely" and soon came "within shooting distance and fired a shot which sensibly checked the progress of the animal and enabled him quickly to be alongside of him when he discharged the second barrel into his lungs, passing through the shoulder blade, which brought him to a stand." Harris and Bell then ran as fast as they could until they were close enough for McKenzie to hear them shout "not to shoot again." They wanted in on the kill. I quote the conclusion of the hunt from Harris's journal at length:

> The Bull did not appear to be much exhausted, but he was so stiffened by the shot in the shoulder that he could not turn quickly, taking advantage of which we approached him, as we came near he would work himself slowly around to face us and then make a pitch at us—We then stepped to one side and commenced discharging our six-barrelled pistols at him with little more effect than increasing his fury at every shot. His appearance was now one to inspire terror, had we not felt satisfied of our ability to avoid him. I came however very near being overtaken by

him through imprudence. I placed myself directly in front of him and as he advanced I fired at his head and then ran directly ahead of him, not supposing he was able to overtake me, but casting my head over my shoulder I saw Mʳ Bull within three feet of me, prepared to give me a taste of his horns. The next moment I was off the track and the poor beast was unable to turn quickly enough to avenge the insult. Bell now took the gun from Owen and shot him directly behind the shoulder blade, he stood tottering for a few moments with an increased gush of blood from the mouth and nostrils, fell easily on his knees and rolled over on his side and was soon dead.

It is an exciting hunting story, and both Harris and Bell agree that Bell fired the last shot to save his friend Edward Harris and that Audubon had remained in camp during the entire episode. In a letter Harris wrote to his brother-in-law shortly after the expedition, he explained that Audubon had been fishing the while: "I returned to Camp, and found that Mr. A. had been successful in catching fish" (qtd. in McDermott 31).

When Audubon heard this story, however, and later read Harris's journal account, he liked it, and then he had an idea. Since the entire expedition was mounted in order to advance his work on the *Quadrupeds of North America*, and since he was making his companions famous by naming new bird species after them and by writing about their adventures in the future publication, Audubon—most likely with Harris's permission—appropriated the story and substituted himself for Harris. And he seems to have done so that very night when he worked up his journal for that day. As the grand-daughter's edition of Audubon's destroyed original journal reports, Bell's final shot saved the life of America's most famous naturalist, not that of the unknown wealthy farmer Edward Harris:

> Through my own imprudence, I placed myself directly in front of him, and as he advanced I fired at his head, and then ran *ahead* of him, instead of veering to one side, not supposing that he was able to overtake me; but turning my head over my shoulder, I saw to my horror, Mr. Bull

within three feet of me, prepared to give me a taste of his horns. The next instant I turned sharply off, and the Buffalo being unable to turn quickly enough to follow me, Bell took the gun from Owen and shot him directly behind the shoulder blade. He tottered for a moment, with an increased jet of blood from the mouth and nostrils, fell forward on his horns, then rolled over on his side, and was dead. (AJ 2:94–95)

We know that this substitution of characters is not Audubon's granddaughter's fabrication because the original manuscript essay entitled "Buffalo," which Audubon wrote following the 1843 expedition for inclusion in *Quadrupeds*, tells the same story (Beinecke). And when Audubon published the story in the 1851 second volume of *The Quadrupeds of North America* (44), Bell once again saved Audubon, not Harris. Since it is unlikely that Maria Audubon ever saw the diaries of Harris or Bell, it seems that she published her edition without knowing that her grandfather manufactured this thrilling episode to enhance his public image as one who faced dangers in his promotion of American natural history.

The relationships existing among these journal entries (and other, subsequent texts) also suggest an understanding among these companions that journal entries could be shared because time was short and good stories retold in detail were as valuable as physical artifacts and souvenirs. (Sprague remained outside this league, and the relative brevity and shallowness of his entries may have resulted from his fear that others might ask to read them on the expedition. With this thought in mind, it is interesting that on the day of their departure from Fort Union, August 16, Sprague's journal entries immediately grow longer and more reflective, as if with everybody packed up and moving rapidly downriver, it was much less likely that someone might have a chance to read his small, tight notebook.) But, as Audubon is the impresario and will produce the book that will make them all famous, they seem happy to share with him. There was a spirit of cooperation in order to produce a great hunting story to take home, retell, and, in Audubon's case, publish. And it is the best kind of hunting story, full of action, danger, physical courage, and last-second rescue. We would call the fabrication a "PR stunt," but from their perspective, it simply enhanced a lavishly illustrated work of natural history and adventure. In all the surviving journals from this expedition, however, only the entries for July 16 show this extent of collusion, if that is not too harsh a term.

The historian David McCullough and the natural history writer Scott Weidensaul have demonstrated that the fabrication has had the desired effect. In his study of the young Theodore Roosevelt (1981), when McCullough recounts the time when John G. Bell was selected to instruct Roosevelt in the art of taxidermy, he notes, "Once on the Yellowstone he [i.e., Bell] had actually saved Audubon's life by grabbing a rifle and shooting a wounded bull buffalo that was within feet of trampling Audubon to death" (118). If Bell had saved the life of Audubon's good friend Edward Harris—which he did—McCullough would not have mentioned the event, which Audubon well understood—as did Harris and probably Bell.[13] Weidensaul, more recently (2008), accepts Audubon's version of the story and from it concludes that Audubon named Bell's vireo for Bell because he "saved his life on the upper Missouri" (98). Audubon liked his friend Bell quite well, but he named the new bird for him because Bell was the first to shoot a specimen of the new bird (see Bell's diary entry for May 6, in part 5, and AJ 1:473), not because Bell saved his life, which he gladly would have done if Audubon had not been back in camp catching catfish.

On Thursday, July 20 (Bell has the date wrong, as the twenty-first), the long-promised bison hunt was staged. Culbertson led Audubon, his companions, and several hunters with an array of chase horses, mules, and two meat carts to cross the river and climb the rugged hills into a day of extreme prairie heat. Bell recorded a high that day of ninety-nine degrees.

13 Phillips B. Street first reported the discrepancies between the journals of Harris and Audubon for July 16. However, because he did not know about Bell's corroboration of Harris's version of the story or about the other corroborating documents, he speculated, "Is it possible that faulty copying of the Audubon manuscript in preparation for its publication [that is, in AJ] was responsible for the omission of credit to Harris?" (180). Peck reiterates that possibility (185n42).

Especially Squires, Harris, and Bell had been looking forward to this event for weeks. Audubon was excited too, but he always knew he would be mainly a spectator, and his job would be to capture the sights and sounds of the bison chase in words for possible publication.

About noon four bison bulls came into view, and Harris makes it clear that everyone knew exactly what they would do:

> We quickly saddled our hunters and arranged our dress and accoutre-ments for the chace and started with the whole of our equipage in a direct line for the herd covered by a small eminence behind which we left the carriages with Mr Audubon so that he could ascend the mound and have a fine view of the chace while we filed off to the right, keeping out of sight behind the hill so as to come upon them in such a way as to force them to run in the fine level valley [. . .], thereby ensuring a fine run and at the same time give Mr Audubon a good view of the chace.

Bell expressed his expectation from Audubon the writer: "[N]o doubt he will give a glowing description of the whole scene."[14]

Culbertson had instructed the men that there was an ethic to follow in hunting bison on horseback. Once a man has fired a ball into a bison, that is *his* bison to hunt and no one else should attempt to kill it. In this case, since there were four hunters (Culbertson, Harris, Bell, and Squires) in pursuit of four bison, each man had a right to claim one. When the chase began, the bison separated into two groups; Culbertson and Bell pursued two; Harris and Squires, the other two. Bell writes that the bison he and Culbertson chased "ran round and round a small hill like circus horses four times, and much faster than I had any idea of, and so close together that we could not part them." When Culbertson managed to shoot his, which died within sixty yards of where Audubon watched, he then turned to help Bell pursue his, but he violated the established order by firing into Bell's bull "about 3 inches below where I first shot him," to which Bell objected vociferously: "I then told him to leave him, that I could easily kill him myself and which

14 For the account published in the Royal Octavo *Quadrupeds*, see QU 2:44.

was contrary to our agreement at the start." The journals clearly show that Bell, Harris, and Squires were sportsmen acutely aware that their time on the prairies to accomplish great hunting feats was running out.

Harris allowed Squires first choice of the bison they were chasing, but when Squires first fired, his bison turned on him and caused his horse to turn so abruptly that he was thrown hard to the ground. Harris delayed his pursuit long enough to see that Squires was on his feet again and that the bison was not going to attack him. Harris's vivid and vigorous narrative of the continued action shows that he was now a fully accomplished prairie hunter and that he gloried in that fact: "[I] pushed on after my game, which I soon overtook and fired striking him in the middle of the thigh, a few strides more brought me directly opposite him and I shot him in the lungs just behind the shoulder. I rode along side of him reloading my gun, which I had just accomplished when I saw blood gushing in a stream from his mouth and nostrils in such a way that I knew he must soon fall, I therefore turned away and galloped after Squire's horse which I caught and brought back to him."

Harris and Squires then resumed the chase of the remaining bull, but Squires had difficulty reloading on the run and, because he was feeling pain from his fall, gave up this chase. When Harris continued, the bison ran down a steep clay bank and charged at Harris on his horse when he followed. The horse wheeled to one side, throwing Harris to the ground between the horse and the bison. Luckily for Harris, when his gun failed to fire at this vulnerable moment, the bison continued his flight. Harris returned to the bull he had killed, "happy to find that I was not much bruised and had escaped the horns of a vicious old Buffalo Bull, and purchased besides another good lesson in running Buffalo." After cutting off the bison's tail as the customary trophy (see fig. 18), he returned to the others and learned that Culbertson and Bell "had each killed their Bulls and I was happy to find that Mr Audubon had had a splendid view of the whole chase." Even though he had not pursued the bison himself, in his hat Audubon was sporting the tail of the bull Culbertson had killed. When Culbertson and

Audubon had walked out to see the dead bull, "Mr. Culbertson insisted on calling it my bull, so I cut off the brush of the tail and placed it in my hat-band" (AJ 2:104).

With three of the four bulls down and the hunters' respective trophies in place, the entourage began making its way toward the evening's new encampment with all but Squires content with the results of their first bison hunt. When Bell spotted eight bison not far off, however, and Squires wanted a chance to redeem the day in the last hour before sunset, his friends agreed to a second charge. Audubon left the men from the fort to set up camp and walked up an embankment from which he could observe this second chase of the day. Culbertson shot *his* bull after a short chase, and Squires, suffering from his injuries, pursued his bull long enough to fire into his body three times but then gave up the chase without knowing whether he had killed the animal. Bell and Harris pursued two bulls in a different direction for some two miles before Bell killed his bull and Harris mortally wounded his, which was still standing, "with the blood gushing from his mouth and nostrils, at the same time showing a strong disposition for battle," when Bell, as Harris expresses it, "rode up to see the death of mine" (EH 149). Thus we learn that watching the death of these large mammals was an attraction for the newly accomplished prairie hunters. Harris chose not to fire again to end the bull's suffering because he "saw that he was failing fast." Bell corroborates that he and Harris watched the slow death: "We stood and watched him some ten or fifteen minutes when he laid down to rise no more."

As far as Audubon, Harris, Bell, and even Squires were concerned, the pursuit and shooting of bison from horseback was the culminating experience for a man on the prairies, but this, their first bison hunt, caused expressions of compunction from the three journalists who participated in or witnessed it. And because all three expressions of regret were written at about the same time (likely the next day, July 21, after their return to the fort), it seems that they discussed among themselves their response to this first day of bison killing. The fact that no similar regrets about any of the earlier killing appear in the journals suggests that it was their first direct involvement in the slaying of such large mammals that caused the regret.[15]

Harris's lament is the most fully developed and principled of the three. After his account of watching the death of the second bison he killed that day, he wrote, "We now regretted having destroyed these noble beasts for no earthly reason but to gratify a sanguinary disposition which appears to be inherent in our natures. We had no means of carrying home the meat and after cutting out the tongues we wended our way back to camp, completely disgusted with ourselves and with the conduct of all white men who come to this country. In this way year after year thousands of these animals are slaughtered for mere sport and the carcasses left for the wolves. The skins are worth nothing at this season" (EH 149). Audubon's narrative of the events of that day makes an abrupt transition from his evaluation of whose set of bison horns was largest ("Bell's horns were the handsomest and largest, mine next best, and Harris's the smallest") to his statement of regret: "What a terrible destruction of life, as it were for nothing, or next to it, as the tongues only were brought in, and the flesh of these fine animals was left to beasts and birds of prey, or to rot on the spots where they fell" (AJ 2:107). Bell's statement is the least fully developed of the three and reveals that he and Harris talked about their regret shortly after the killing: "After we had the tongues, we cut off the tails, which was all we took off these fine large fat animals and returned to camp very much regretting what we had done when they told us they could not carry any of the flesh home." Whatever pang of conscience he felt faded quickly, however, and he "slept very comfortable."

15 On his own first bison hunt in 1832, Washington Irving expressed a similar regret upon making his first solo kill: "I am nothing of a sportsman; I had been prompted to this unwonted exploit by the magnitude of the game, and the excitement of an adventurous chase. Now that the excitement was over, I could not but look with commiseration upon the poor animal that lay struggling and bleeding at my feet. His very size and importance, which had before inspired me with eagerness, now increased my compunction. It seemed as if I had inflicted pain in proportion to the bulk of my victim, and as if there were a hundred-fold greater waste of life than there would have been in the destruction of an animal of inferior size" (178).

Nor did Harris lose any sleep. The rest of his journal entry simply explains what a good hunter he had become. It reads like an experienced hunter's instructions to a novice:

In approaching the herd, [the horse] needs no guiding, you throw the reins on his neck and he runs to within 15 or 20 feet on the right side as soon as you fire he shies off gently to the right to avoid a rush from the animal, if you have not killed you load your gun at once while he continues to run along the side of the animal with the reins on his neck. This is rather difficult for a beginner but I succeeded once very well, the other charge I was obliged to stop him [i.e., the horse], this makes the chace too long and is one of the great faults of a young beginner, indeed it is the main thing in Buffalo hunting to load your gun quickly while the horse continues at full speed. There is but little difficulty in hitting so large a mark and the faster your horse runs at the moment at taking aim, the more steady will your aim be. (EH 149–50)

Into their fifth week of residence at Fort Union, Audubon and his companions (save Sprague) were more engaged in the hunt and the camaraderie of hunters than they were in the scientific and artistic purposes of the Missouri River expedition. The closing image of Harris's entry for this day is of the stories told among friends: "[W]e were soon among our friends recounting our exploits over a good cup of coffee, some hard buiscuit and the boss or hump of one of the Bulls killed in the morning." At this time, Culbertson told Audubon, "Harris and Bell have done wonders, for persons who have never shot at Buffaloes from on horseback" (AJ 2:107). These easterners were now playing at the sporting life on the wild western prairies. They had read others' accounts of similar experiences and saw themselves as the next wave of those who would be read about, admired, and marveled at. And Audubon's fame and planned publication would see to that.

All celebrities, of course, are attractive prey for satirists, and about this time, a hoax was perpetrated on Audubon in the form of a fictional letter supposedly written by Audubon from "110 miles above Fort Union" to his friend John Bachman in Charleston. In the letter, "Audubon" informs Bachman that he has discovered a mammal with the hind legs of a kangaroo, a sheep's tail, short forelimbs ending in sharp claws, a head and antlers like a deer, and a thick midriff productive of oil. Its meat is "very white and tender, tasted very similar to veal." It grows to six hundred pounds and measures nine feet, four inches from head to tail. Indians in the region call it "Ke-ko-ka-ki, or jumper" and have successfully domesticated it. He had purchased two live jumpers, male and female, and would send them back to the fort. The satirist's target seems to be Audubon's vanity: "I think, without doubt, in point of usefulness and value, I may pride myself in surpassing most of my compeers, in thus bringing so great a discovery to light" ("Interesting Letter"). An ironic response to the hoax letter also circulated in the newspapers and showed that Audubon's enemies were still at work, trying to further diminish his credibility. The author of the response pretends to defend the veracity of the jumper letter against those who have charged that it is a hoax, but then comes the stab in the back:

> It is no doubt a mistake in supposing there are vast numbers of these animals—the probabilty is that these two are the *last* of the race, and like his bird of Washington, will never be seen again.[16] Let him not then be deprived of his honors prematurely by men who know not the difference between a *bat* and a bird—and as our country has the credit abroad, through his industry and wonderful discoveries, of some hundreds of birds more than were described by the *Pioneer* in Ornithology, Wilson! So let it be known that we have animals unthought of by the Naturalists of Europe. ("Mr. Audubon")

It is signed only "W." but, especially since it was published in a Philadelphia newspaper, has all the markings of one of Audubon's ornithological

16 Audubon claimed that his "Bird of Washington," plate 11 of the original folio edition of *Birds of America*, was a distinct species, but this claim has generally been dismissed as an error. Audubon wrote that he saw the species three times while living in Kentucky, but not since then. He reported its wingspan as three feet wider than that of the bald eagle; Audubon's bird is thought to have been an immature bald eagle.

enemies, only one of which, George Ord, is known to have been active in enmity so long.[17]

From the time of the first bison hunt until the departure from Fort Union by Audubon's party on August 16, the journals present an astonishing schedule of killing, especially those of Bell and Harris. Audubon and Sprague note that some of the work toward the *Quadrupeds* is being done, but mostly by Sprague, who tends to stay at the fort drawing and going on limited excursions to collect seeds and plants. Audubon, however, was frequently out with the hunters. On July 27, for example, he sketched a favorite campsite known as the Three Mammelles in his field notebook while on a hunt for bison cows and specimens of any other desired species. (See figs. 19–21.) As the end of their time at the fort drew nearer, Audubon was ever more concerned that he would not acquire sufficient skins and heads of grizzly bears and bighorn sheep, and he was still attempting to find hunters who would procure these for him. Of bison he had enough, having seen untold thousands over the past two and a half months, and having heard a report on July 28 of bison "in the millions" two hundred miles to the north (AJ 2:121).

For all the men except Audubon, this expedition was probably the greatest adventure of their lives, and each one told stories about it as long as he lived. As their time at Fort Union grew short, at least Squires, Audubon, and even Sprague regretted having to leave and head home. On July 31, Squires proposed to Audubon "to let him remain here this winter to procure birds and quadrupeds, and I would have said 'yes' at once, did he understand either or both these subjects, or could draw;

17 Alice Ford claimed that Audubon wrote the letter himself as a practical joke on Bachman (*John James Audubon* 407). She offers no evidence, however, and does not explain her reasoning. She does cite a letter from one Audubon enemy to another, George Ord to Charles Waterton, dated August 23, 1843, in which Ord quotes extensively from the fictional letter in case Waterton has not seen it. Ord concludes "that this letter is nothing but the production of some wag, who thought fit to amuse himself at the expense of the modern Mandeville: it is what is vulgarly termed *a hoax*" (APS). Audubon certainly was mischievous enough to try such a practical joke, but it seems unlikely that he would have risked his credibility as a naturalist in such a public way. He could simply have sent the letter to Bachman privately if he wanted to play a practical joke on him.

but as he does not, it would be useless" (AJ 2:125). Squires, of course, probably thought more of continued adventures than actual usefulness. Sprague expressed his own uncertain desire to stay through the winter, as Audubon reported on August 1: "I spoke to Sprague last night about remaining here next winter, as he had mentioned his wish to do so to Bell some time ago, but he was very undecided." Sprague, of course, could have been extraordinarily useful to Audubon's work on the *Quadrupeds*. Even Audubon himself wished he had more time: "My regrets that I promised you all so faithfully that I would return this fall are beyond description." Even though he was fifty-eight years old, he explained, "there is so much of interest here that I forget oftentimes that I am not as young as Owen" (who was born in 1826) (AJ 2:126). Even when he did not participate in the chase, his enthusiasm for its vast dramatic action—skilled, spirited men on well-trained, courageous horses in furious pursuit of the shaggy, monstrous emblem of the wild northern prairies—inspired him to observe every hunt he could.

Audubon's pure delight in the chase and the kill is apparent in his vivid and animated account of a bison hunt conducted by himself, Harris, Culbertson, and several hunters from the fort on August 4:

> We soon saw that the weather was becoming cloudy, and we were anxious to reach a camping-place; but we continued to cross ranges of hills, and hoped to see a large herd of Buffaloes. The weather was hot "out of mind," and we continued till, reaching a fine hill, we saw in a beautiful valley below us seventy to eighty head, feeding peacefully in groups and singly, as might happen. The bulls were mixed in with the cows, and we saw one or two calves. Many bulls were at various distances from the main group, but as we advanced towards them they galloped off and joined the others. When the chase began it was curious to see how much swifter the cows were than the bulls, and how soon they divided themselves into parties of seven or eight, exerting themselves to escape from their murderous pursuers. All in vain, however; off went the guns

and down went the cows, or stood bleeding through the nose, mouth, or bullet holes. (AJ 2:129)

His complete satisfaction with this place and his engagement in it is evident in the complementary presence in this passage of the pastoral—the bison "feeding peacefully" in the "beautiful valley below us"—the heroic—"the chase began" and "off went the guns"—and the comic—"and down went the cows." The last sentence above is nearly a found poem, with its parallel phrasing and the near rhyme of "nose" and "holes." John James Audubon was a happy sportsman on this expedition: there was "so much of interest here."

Much of the imagery of killing in the journals is unquestionably gruesome, and it seems to reflect the men's complete acceptance of this unavoidable fact or condition of their hunts, especially in the journal of John G. Bell. While we must remember that this is the journal of an experienced taxidermist, who tended to view living animals as so much skin and bones, still the relentlessness of his accounts of killing has an almost mesmerizing effect. A close look at his entry for August 3 will serve as a window onto the attitude toward the animals they were killing that the journals suggest was the prevalent one.

The purpose of this excursion of August 2–5 with Owen McKenzie was to hunt sage grouse, or cocks of the plain, for Audubon. The morning of Thursday, August 3, began when Bell's horse woke him and he saw a bison bull a few yards off. The two hunters killed the bull with four shots. While McKenzie went to retrieve their knives and Bell sat on the dead bison, another bull approached. At a distance of forty-five yards, Bell shot and killed the second bison from his perch atop the first. Since the second bull was "the best of the two," Bell and McKenzie "cut out some of the best pieces and returned to camp and made up the fire." Before they could cook their breakfast, however, they saw and pursued a grizzly bear, but it escaped into some bushes. "We then returned and roasted our meat and got our breakfast. There were hundreds of buffaloe near us this morning." By six thirty they were underway on horseback for a day of hunting. When they approached ten elk on a sandbar, the elk "got the wind of us" and fled. When Bell then thought the grunting he heard in some bushes was a bear, he fired without seeing the animal and out rushed a mortally wounded bison bull: "He was a remarkably fine one, but we left him as we had done the other two for the wolves. We then started again at 8 oclock." After a few miles, McKenzie's shot at an elk broke her back, causing her to struggle "very much" and make "a mournfull noise." They killed and skinned her and tied the skin around a tree to dry until their return: "We then started again at ten oclock and saw hundreds of buffaloe all along our route."

When they discovered back at their campsite that they "had lost our meat that we roasted this morning," McKenzie shot another bull "that was then crossing a branch near us." The description that follows is among the most grisly, even macabre, in all of the journals from this expedition and is remarkable for the complete absence of any emotional engagement in the life of the animal:

> I then gave him two shots and still he did not fall, but stood still. I then gave him three pistol shots, with the same effects. I then went to him and attempted to pull him down by taking hold of his tail. Owen then stuck him in the side with his knife. I then stuck him in the other side and still he stood still. Owen then cut a large hole in his side and run his hand in and cut him in the lungs. He then soon fell, but attempted to get up three or four times, and I then cut off his tail, and it was several minutes before he was dead. We then commenced taking off part of the skin, so as to cut out some ribs.

Before they could return to the campsite, however, they watched fifty or so bison crossing the stream and climbing the bank near them. Bell seems impressed by the strength and agility of the animals: "I was quite surprised to see with what ease they dashed through the water and mud and how quick they got up the bank, which was seven or eight feet high." The opportunity for yet another kill was so tempting that McKenzie "was for killing a cow and had I not prevented him would have done so, as we could have killed three or four had we been so disposed." While Bell does not explain the cause of his restraint in that moment, it may have been simple hunger: "We then made a fire and roasted some ribs and ate our dinner."

The feeling of Bell and McKenzie seems to be that a three-quarter-ton bison is a luncheon for two. Readers today, of course, view this scene through the lens of the near extinction of the American bison in the 1880s and the history of the partial recovery of the species. As a society today we are also much further removed from the conception in Audubon's day of humans as hunters who kill animals for food and clothing. Bell and McKenzie, on the other hand, made their livings and fed their bodies by killing bison, elk, hares, and grouse. And they did so in the context of the seemingly endless surrounding populations of all those species.

The hunters rested after their luncheon but started again at three thirty in the afternoon. As unfeeling as Bell might seem to be while his young friend cuts a hole in the bull's side large enough to reach his arm in to cut the lungs so that the blood will spill out faster, his accounts of the vast numbers of bison they watch at such proximity do suggest a kind of awe:

> [We] saw hundreds of buffalo. I counted 185 in a band near us, and another 250. We then arrived at our camping place at six oclock, saw hundreds of buffaloes a few hundred yards from us. We then unsaddled our horses and let them go, and went along a ravine untill we crept within 8 or 10 feet of the path they were traveling and laid ourselves down to see them. We remained there for some 7 or 8 minutes, when Owen raised his head a little above the wormwood, when a large bull saw him and came within 5 or 6 feet of him to see what it was, but as we remained perfectly quiet, he passed on. This occured two or three times. Once they were alarmed and ran up the bank, but as they saw nothing more, they passed on again in the path.

The hunters' wish not to be detected shows their desire to prolong the experience, a rare chance to observe wild bison going about their daily lives free from human interference—free, perhaps, from the preceding fifteen thousand years of human presence in North America. If Bell had such thoughts, he did not write them in his journal, but I have to believe that a New Yorker on those remote prairies in 1843 in a quiet moment with no other human presence would understand that he was witnessing the world as it was in ages long past.

For their evening meal, they mortally wounded a "fine fat cow": "She then got up and fell down a bank, rolling over and over some six or seven times before reaching the bottom. She had one leg broke and was quite dead when we got down to her. We then cut out some ribs and marrow bones and started for our camp." The entry for August 3 closes with a hunter's expression of utter satisfaction with a day: "It is now half past seven and our ribs are roasting finely as we have plenty of wood and a good supply of water. I am very sorry Mr Harris could not come with us as he would have enjoyed it very much. Owen is a very good man to camp out with and a very good hunter." When this day was done and Bell settled in to write his account, the first word he put on paper was "[p]leasant."

Bell's entries for the next two days relate events and experiences similar to those of August 3, but on the fifth, the day he and McKenzie were rushing to reach the fort before nightfall, Bell writes more expressively of the great number of bison they traveled among and watched. They saw

a very large band of buffalo, and as we came opposite to them they got the wind of us, and started to pass in front of us. We then rode on until we parted them and passed through. After about 300 had passed us, the others then passed behind us, in a line, and they kept comeing and comeing from the River and running to the hills. They formed a complete line, one, two, and sometimes three deep, for more than two miles long with scarcely an opening between them. They are like sheep, when one starts the others will all follow. To me it was a most beautiful sight, to see them passing within 150 yards of me running as fast as they could, with their tongues hanging out, in the distance as they went up the hill, which was quite gradual in its ascent. They would collect in large bodies and move slowly to the right and left. They looked like as many soldiers marching along. We stood still until they were nearly all past, and as they passed us we would halloo at them to make them run faster.

That evening in the fort, as Audubon was wrapping up his account of the day, he noted, "Bell related all his adventures to every one" (AB). Bell, the newly initiated New Yorker, and McKenzie, the native-born prairie hunter, had just spent four days at play on a primal landscape, the only two human beings among thousands of wild bison, elk, bears, deer, and wolves. Bell expected to sleep well in the fort this night, for on the prairie, "I could not sleep very well on account of the bulls bellowing all night long and a few wolves howling near our camp" (August 3).

That Saturday, August 5, was an especially busy day at the fort. Three separate hunting parties returned. Audubon had sent Squires out with Provost and La Fleur, another hunter, for bighorn sheep. They succeeded only in wounding one but lost it. Audubon and Harris had gone out with Culbertson on the fourth for bison. They returned with the meat and heads of four cows. Bell and McKenzie had been sent out for sage grouse but did not see one in the four days of their hunt. They did bring in, however, a porcupine and a live fledgling bald eagle.

What most struck Audubon that day was the apparent, if also incredible, resilience of the bison as a species when one considered all the causes of death they encounter. On the *Omega*'s ascent of the Missouri, he had witnessed multitudes of bison calves drowned by the rising waters and more powerful currents caused by the spring thaw in the Rockies. He also learned at the fort that "[b]uffaloes become so poor during hard winters when the snow covers the ground to the depth of 2 or 3 feet, that they lose their hair, become covered over with scales, on which the Magpies feed and the poor beasts die by hundreds." He certainly knew of predation by wolves and bears, but his own species was probably the greatest enemy of the bison: "One can hardly conceive how it happens, that notwithstanding their general deaths and the immence numbers that are murdered daily over the immence wastes, called prairies, that so many are yet every day to be found!"[18] Bell's account of the thousands of bison he and McKenzie had traveled through that morning and the day before could only have reinforced Audubon's amazement at this persecuted species' continued survival.

Audubon's August 5 entry also contains the first evidence of an unexplained tension that emerged between Sprague and Audubon: "Sprague never spoke to me on my return, but that is his way." No other journalist, including Sprague, ever refers to this difficulty or suggests a possible cause. The journals do show that Sprague spent much more time at the fort than any other member of the Audubon party. He may have felt that he did not quite fit in and have kept somewhat aloof from the others who were so eager for the hunt. We don't know, but the trouble recurs several times in Audubon's downstream journal entries.

As much as Audubon enjoyed being on the prairies, he definitely missed his family back at Minnie's Land. We see this in the journal entry for August 7, where he notes a bad dream: "I had such a horrible dream last night about our dear son Johnny. So horrible was what I saw in that dream, that I awoke & found my face covered with tears, and my whole body in a high state of perspiration. I slept badly the rest of the night & was up with the ringing of the bell at daylight." His reference to recurring bad dreams about Lucy three weeks later, on August 29, suggests that his thoughts were increasingly turning homeward after several months away. His thoughts were reciprocated at home, where Lucy, on the same day that Audubon reported his bad dream about John, wrote to her husband, "The little ones at home are quite well and grown, the rest are as well as can be expected and we shall all rejoice when you are safe at Minnies Land." She also announced that she expected John's wife, Caroline, to deliver her baby "about the 20th" (Beinecke). Their third grandchild, the future editor of Audubon's journals, Maria Rebecca, would be born on August 19.

18 All citations from Audubon's journal entries for August 5–12 are from the Beinecke partial copy (AB), which is reprinted in part 3. In part 1 I discuss important differences between AJ's version of the entries for August 5 and 8 and those in AB.

With possible publication in mind, Audubon was also collecting stories and information from living sources he would soon leave behind. Under August 7, he (or Squires) copied a long passage from Culbertson's journal about a violent battle in 1833 between men at Fort McKenzie at the Marias River in league with a group of Piegan Indians and some four hundred Assiniboines. Under the same date appear two men's stories of grizzly bears. Under August 9 he copied long passages from Bell's recent journal entries as well as another long passage from Culbertson's journal about a conflict with Crow Indians in 1834—all of which stories Audubon emphasizes are "facts" and not at all like "all the trashy stuff written & published by Catlin!"[19]

The continued failure to procure either the "little hare" or bighorn sheep becomes a leitmotif in the journals at this time. The flat-bottomed Mackinaw boat, named the *Union* ("in consequence of the United exertions of my companions to do all that they could for me in this costly expedition"), was nearly ready to launch, and Audubon daily sent out one or more hunters to shoot bighorns before he had to depart. On August 8 both Audubon and Harris expressed their concern that time was running out for more bison hunts. They regretted having gone out that morning in search of fossils because they missed thereby an opportunity to join a hunting party that Culbertson sent out to shoot "3 fat cows, and no more." As Audubon expressed it, "Harris was sorry that he had not been here, and so was I, as both of us would have gone and joined & seen the fun." Harris expressed his regret as well, for "I should have had another fine run after Buffalo cows" (which run faster than bulls). But he found consolation in the substantial progress he and Bell had made as prairie hunters: "Both Bell & I have been remarkably successful for beginners having killed all the cattle we have started after." The compunction about the killing that Harris, Audubon, and Bell all expressed on July 20 or 21 seems quite forgotten on August 8.

19 Audubon and his companions generally disparaged what they considered George Catlin's overly romantic representation of the Native Americans he had encountered on the plains ten and twelve years earlier. See, for example, AJ 1:497–98.

The final week or so before departure was a very busy time, of course, with outfitting the boat, packing, and going out on the last hunts. Audubon and Squires worked many hours on the journals. Having located a copy of the *Philadelphia Courier* at the fort in which J. K. Mitchell's adulatory poem "To Audubon" was printed, Audubon had Squires copy it. And Audubon developed lengthy descriptions of the method of skinning bison and of the challenging terrain of the bighorn sheep known as the "Mauvaise Terres," or badlands, most of which would later provide vivid images for his published essay on the "Rocky Mountain Sheep" (QU 2:163–72).

On the morning of the final bison hunt from the fort, August 12, Harris was pleased to see Culbertson's confidence in him and Bell as hunters: the supervisor sent out two meat carts with them rather than just one. Owen McKenzie was the third hunter. Culbertson ordered six cows. In the first "race," as Bell explained, each of the three hunters "killed a cow, and shot down several more but could not find them when we returned." Soon, however, when Bell's horse showed himself to be too young and inexperienced, Bell returned to the fort. After Harris killed his first cow, he wounded a second one, but she managed to run with the fleeing herd of fifty or sixty bison until Harris "gave up the chase, very much to my regret and mortification." On this, his last performance, as it were, he intensely wanted everything to go right. He would never have this chance again. He and McKenzie joined up and determined "to procure 3 more cows to fill another cart." After riding about two miles, they came upon a band of bison lying in a hollow, numbering between two and three hundred. When the chase began, Harris for the second time lost the cow he wounded in the herd, but he soon selected another cow, which he wounded and successfully pursued even though riding in the midst of many bison running at top speed. He was surprised, however, that the several more shots he fired at her had no apparent effect: "[A]t last I discovered that in my anxiety to kill her I had not paid sufficient attention to the level of the gun and the balls had rolled out." Eventually the wounded cow, "bleeding at the mouth and nostrils," slowed to a walk and Harris gave her "a death wound." This had been an unusually long chase, and since he "was out of sight of Owen and the cart, I merely cut off her tail and turned back," leaving her "for the Wolves."

Harris's history on the prairie—his progress from novice to accomplished hunter—strikes me as an epitome of the experience of the white hunter who came to the prairies between the 1820s and 1840s. A period of initiation and education precedes the first successful hunts. Over time, a hunter has seen untold thousands of bison, so the occasional lost wounded bull or cow does not seem significant, though there is an inevitable moment of compunction, such as Harris writes of in his entry for July 20, when he felt "completely disgusted with ourselves and with the conduct of all white men who come to this country." With more experience, however, the hunter becomes ever more inured to the killing and sees the waste as acceptable because it is unavoidable—and, with regard to the overall bison population, insignificant.

Sprague's history on the prairie provides an interesting contrast. He was the least gregarious of the five, but adequately sociable. He was never tempted to develop as a prairie hunter like Harris and Bell but was content to work on his drawings, to botanize, and to hunt birds, which he did effectively. But his journal does not reflect how he felt about his experience at Fort Union until the last day, August 16. On the ascent of the Missouri, he seemed somewhat aloof, even haughty at times. But on August 16, his journal shows that he had become quite attached to the remote outpost and the men there. Over the preceding ten weeks, the young artist from Hingham, Massachusetts, developed a romanticized appreciation for the lives the men at the fort lived: "Here far away from civilization, the traders pass the best of their days—some from a love of adventure, some for gain—and others for crime are driven from civilized society." The prairies, for Sprague, had a purifying, restorative power. And he was grateful for this new understanding and for the experience he had been allowed at the fort: "About 12 o clock we bade good bye to the inmates of Fort Union—and it is not probable we shall ever meet again—They have endeavoured as much as in their power to make us comfortable—and notwithstanding some disappointments—I have passed many pleasant hours here—It is with regret I take leave." What for Bell and Harris was a heroic sportsman's journey, for Sprague was more of a romantic quest.

Because Audubon was fifty-eight years old, he did not develop as a prairie hunter as his friends did, but his pleasures and satisfactions from this time on the northern prairies were tremendous. But he had to already know that his coauthor Bachman would be disappointed with their scientific results.

Fort Union to St. Louis,
August 16–October 19

The Newberry partial copy of the Missouri River journals shows that conditions onboard the *Union* were crowded: "We left at 12 in a Mackinaw Boat 41 feet long and 8 feet wide at the Bottom; and this proved sadly too small, for what with our effects and Barrels of pickled skins as well as ourselves, and M̱ Alex̱ Culbertson, Wife and Child[20] that accompanied us; we were *packed* in and under the roof as if so many herrings in a Box." According to Sprague's count, there were "14 persons in all." There were also live animals: a badger in a cage and a black-tailed deer (referred to by Bell as a fawn), whose means of confinement is not described.

With fourteen people to feed and no way to preserve meat, it was necessary to kill animals practically every day. Evidence of how serious a matter this was for the crew appears in Audubon's entries for Friday, September 1, through Monday, September 4. After having "Caught 14 cat fish" on August 31, "all most excellent," no new meat was brought on board over the next four days, a fact that Audubon mentions every day. On only the second day of this ordeal, Audubon saw the importance of fresh meat on this journey: "We had no fresh meat on board, and uneasily perceived the effect of this on the faces of the men of our Crew" (AN) The crew consisted of Provost and five others, and most of the hunting probably had to be done by Culbertson and Audubon's four companions. Still, the journals from this leg of the expedition show that the relentlessness of the killing reported at

20 Culbertson was transferring to Fort John on the Platte River.

Fort Union did not abate as they headed home. Everyone killed at every opportunity—even Sprague joined in briefly—without commentary on ethical considerations, or even mild regret, as had occurred briefly on July 20. I remark on this because it shows that this expedition took place at a time when it was still possible to kill the large game mammals of the prairies without much concern that humans might diminish their overall numbers and make disturbingly wasteful use of the animals' bodies.

That Audubon and his small group felt no need to limit their killing is clear from their downstream behavior. In fact, a spirit of competitiveness prevailed among the gunners. While Audubon had witnessed several bison hunts near Fort Union, he had seldom participated because, at fifty-eight, he was less fleet than formerly. On this descent of the Missouri, however, opportunities arose and he killed—or helped kill—four bulls between August 17 and 22, on which day he notes, "I shot the 4$\frac{th}{}$ Bull I have killed this trip." Indeed, he might even have claimed more credit for killing in his journal than was actually due to him. On August 17 he writes, "I Killed 2 Buffalo Bulls" (AN). His skinner, Bell, however, writes, "Mr Audubon shot at 2 bulls. I then fired and killed them." In Bell's brief entry, he could have simply recorded his two kills, but he clearly wanted to include the failure of Audubon to shoot as skillfully as he did. At the beginning of this return journey, some evidence of a new tension among fellow travelers ("herrings in a Box") begins to appear, especially when we read various journalists' entries alongside one another. When Audubon shot a bull (his fifth) crossing the river on August 25, he wrote, "Shot another Bull. Bell gives little credit to any person but himself" (AN). Bell's journal for that day suggests that Audubon might not deserve full credit for the kill: "Saw several wolves & swans, & killed a bull while crossing the river." Thus the comparison of entries tells us that after the famous American naturalist and the thirty-one-year-old taxidermist, of some reputation already, fired from the gunwale of the *Union* at the bull struggling to cross the river, they argued about who made the kill. From our perspective today, this can seem funny in a way, but it also reveals an unvarnished desire among the veteran hunters on this expedition to be credited for their kills.

By contrast, the thirty-two-year-old Isaac Sprague, while a veteran fowler, was rarely tempted to compete with the others. The journals suggest that he was generally friendly and sociable, but he was also rather introverted, naturally quiet, and of a distinctly romantic sensibility. On the second day aboard the *Union*, August 17, Sprague felt a degree and quality of delight from his surroundings that rarely appears in any of the journals from this expedition: "Fine day—very pleasant on calm still days floating along this wilderness—especially in those places where the banks of the river are overhung by the ancient forest. Occasionally the boatmen strike up one of their wild Canadian boat songs, keeping time as they row with their oars—These delightful days afford me an enjoyment much superior to anything I ever experienced in the bustle and turmoil of civilized life." If he had written so vividly and evocatively every day, Sprague's journal would have been published long ago, probably in his lifetime, and it would have been praised and valued—in the nineteenth century, at least—along with works by his contemporaries Henry David Thoreau, Susan Fenimore Cooper, Washington Irving, and John James Audubon. But he did not, and we are left simply wishing that he had. But on August 17, he was going home and beginning to understand that "home" for him now would always be lacking something.

In a rare moment for Sprague, on August 20, he joined the hunter Bell to go after a herd of bison "that were feeding on a level plain at a short distance." The event that ensued that day suggests that, of all the members of the expedition, Sprague may have had the greatest potential for compunction over the unrestrained killing. This is Sprague's narrative of the successful hunt:

> Seeing two buffaloe bulls leisurely approaching the river, Bell & myself concealed ourselves in some bushes that grew along the bank, and waited for them to come up—The first one—an animal of immense size and most formidable appearance—came directly toward me, and

when within a dozen yards I leveled and blazed away. He sprange off a few yards and fell, but not dead, the shot taking effect rather too high on the shoulder to be immediately fatal. The other, fully equal in size and quite as ferocious in appearance, being 2 or 3 hundred yards behind was not alarmed at the report of the gun, but came directly on unconscious of danger in the same track until quite near, when I discharged the contents of my second barrel at him he bounded ahead a few paces and fell dead! The first one, startled by the second report and the sight of us as we came from our cover, started and ran a short distance when he staggered, fell and died!

While the journals do not tell us precisely how many times before this Sprague had fired at bison, he did so rarely, and this killing of two bulls was certainly a unique event for him. His first response to the killing is that of a satisfied hunter: "This was good shooting, to kill two buffaloes weighing 2000 lbs each—with a single ball each." But, ultimately, as he records his hunt in his journal, he is not pleased with what he had done: "[B]ut the sight of them as they lay bleeding on the earth seemed to me too much like butchering—I immediately made a sketch of the nearest one as he lay, afterwards cut off the tails as trophys and left them for the wolves!" That a quiet artist from Hingham, Massachusetts, who innocently enjoyed shooting small birds could kill four thousand pounds of bison with two small lead balls astonished and disturbed him, and I suspect that he felt somewhat like Washington Irving did when he killed a bison in 1832, as if he "had inflicted pain in proportion to the bulk of my victim, and as if there were a hundred-fold greater waste of life than there would have been in the destruction of an animal of inferior size" (Irving 178).

Two days later, a crisis was ignited. While Sprague's journal mentions nothing about any difficulty, both Audubon's original field notebook and the Newberry partial copy report an emotional clash between the two artists. On August 22, after Sprague informed Audubon and others in a returning hunting party that he "thought he had seen a Grisley Bear," the party "went toward the spot. The fellow had taken under the high bank, and it was killed in a few seconds. Mr C. shot it first through the neck. Bell elsewhere and I in the belly." Sprague then apparently objected vociferously to their killing of the bear. It also appears that Audubon then reminded him that he was under some obligation to assist in spotting animals for the others to kill, to which "Sprague said that he never promised us *anything*. Said that if he saw another Bear he would not Mention it to us &ͨ &ͨ and was quite abusive in other words." Audubon then gave him an ultimatum: "I told him that he might go any direction he chose and he said he would depart tomorow morning, but of this I am doubtfull." Since this eruption occurred near the mouth of the Little Missouri River, some forty miles from the next downstream fort, Sprague's threat to leave seems an extreme and unreasonable reaction to the killing of the bear, several of which had been killed by the Audubon party earlier. If he did leave, he would be on foot, alone, and abandoning his belongings. There is no evidence in any of the journals that Sprague had the potential for such an emotional flare-up preceding this incident.

Sprague's baffling emotional state continued for a week. Three days later, August 25, the same day that Audubon killed his fifth bull and complained a bit about Bell and on which the *Union* reached the Mandan village, Audubon writes, "Sprague has not open his lips to me, and I hope he will remain with the Mandans!" Having not spoken to Audubon "since the death of the Bear," Sprague now tried to arrange to stay in the village when the *Union* moved on: "Sprague attempted to remain there but was defeated by Mͬ. C. contrivance." The next day Audubon notes "Sprague's surliness." On the twenty-seventh Sprague assisted the others in shooting at a bison in the river. Audubon describes the bison's behavior as inexplicable: "We saw a bull on a Sand Bar.—The fool took to the Water and swam so as to meet us. We shot at it about 12 times about the head.—I shot it *through the Nose*. He was shot by Sprague thrugh one eye.—Bell and Harris thrugh the head &ͨ and yet this Bull made for our boat and came until Mͬ C. touched it with a poll.—he turned round and went off across the River, but acted as if Wild and Crazy." When the *Union* stopped for the night, Sprague apparently went botanizing onshore and behaved oddly upon his return, as Audubon writes, "I was amused at Sprague bringing some plants on board and hiding them sudenly although close to me at the time.—he is in fact as strange a being as the Bull we have just now shot at!" After a week of such perplexing behavior on too small a boat for privacy, on August 29 Sprague tried to heal the breach: "Sprague apologising to me, we shook hands and now all will be forgotten, on my part at least."

This entire incident is all the more puzzling because Sprague's journal makes not the least suggestion that anything was wrong. In fact, he seems quite content. On August 24, for example, after he enjoyed some wild cherries, he was quite engaged by the "curious" features of that night's camp: "Our encampment presents a curious spectacle; composed as it is of wild looking figures in Indian-hunter's and civilized costume each engaged after his own fashion in cooking, eating, or preparing for the night—some talking English some French some Indian." And on the day he apologized, August 29, he seems to have been in a calm, contemplative state: "Time passes rather dull when tied up in such places—read some of Roger's poem Pleasures of Memory[21]—reminded me of home—I think of home and am there although 2000 miles away." (See fig. 22.) The whole incident remains a puzzle, anomalous, and it likely occurred simply because of travel fatigue and too much closeness on the *Union*. It is also possible that Audubon's perception of Sprague's behavior was skewed or exaggerated. Perhaps Sprague was temporarily super-sensitive to the killing because of his regret over killing the two bulls, and he then felt responsible for the death of the grizzly. And when he said he would not tell them about any others, Audubon snapped back disproportionately, and thus ensued the silence, some chagrin and embarrassment, and the final apology—with Audubon still a bit dismissive and haughty, perhaps feeling defensive about his own overreaction.

The high winds of the "Upper Missouri" often tossed the flat-bottomed

---

21 Samuel Rogers (1763–1855), "The Pleasures of Memory" (1792).

Union about violently, causing seasickness in some and occasional fear. The night of September 6 and 7 was particularly turbulent. The wind shifted from the south to the northwest, and when it blew so hard they thought the boat might be swamped and sink, they all went ashore, carrying as much of their guns and ammunition as they could. Culbertson's wife, Natawista, had her baby with her, about whom Audubon writes admiringly: "Mrs C. with her Child in her arms made for the willows and had a cover for her babe in a few minutes!" (AP). Like all storms, this one passed, but the high winds of the northern plains were a regular condition of navigation there.

The powerful currents of the river and rising water offered other dangers. On September 8 Audubon describes the violent breaking up of sandbars in these conditions. The river had risen over six feet since noon the day before: "The effect of sudden rises in this river is wonderful upon the Sand Bars, which no sooner covered by a foot or so of water, at once breake up, causing very high waves to run, through which no boat could run without eminent danger.— The *Swells* are felt for upwards of 50 Yards and are felt as if small waves at Sea!" (AP).

The travelers were also witnessing the autumn migration of many bird species along the Mississippi Flyway. On August 31 Audubon estimated that the pintail ducks passing overhead numbered in the millions. Frequently the various journals refer to great numbers, sometimes thousands, of passenger pigeons, Carolina parakeets, golden plovers, snipes, gulls, geese, and others. Such skies, of course, will never be seen again because of extinctions, habitat destruction, and the ever-greater human presence in North America, but one potential value of these journals is as a source of environmental history.

The *Union* arrived at Fort Pierre around 5:30 p.m. on September 8 and stayed until the fourteenth because Audubon decided to take advantage of a larger boat available to them there that had to be repaired and modified. The prevailing wind and rain did not encourage much outdoor activity, causing Audubon to write repeatedly, "nothing done." On Monday, September 11, however, severe weather threatened the still-loaded *Union*, as Sprague writes: "The wind which continues to blow from the south, as it has for two days past, increased this forenoon to a perfect hurricane and raised such a commotion in the river, that we were obliged to unload our boat to prevent her being filled and sunk." No one gives the dimensions or name of the new boat, but it was loaded and ready for departure on the fourteenth. There were only ten aboard now. Culbertson and his family stayed behind in order to continue on to Fort John on the Platte River. Among other items acquired at Fort Pierre, the cargo now included the live swift fox that Francis Chardon had promised to Audubon on the upstream passage, June 7. (See fig. 23.)

On September 16, as they approached the Great Bend, Sprague's picturesque contentment is evident in his journal: "Lovely morning—perfectly calm and beautiful—and when we float along with the current—the men resting on their oars—not a sound is heard to disturb the profound repose that reigns over this vast wilderness." At the Great Bend, the men were interested in black-tailed deer, which several of them had so enjoyed dining upon the previous May 26, and wood for oars. They were also interested in commemorating their two stops at "our old camp of the 6 trees" (AN). Squires nailed a board to the center tree, on which was painted or carved "JJ Audubon," and Sprague drew a sketch of the camp. (See fig. 24.)

Everyone was glad to see bison on September 21 since none had been seen since the departure from Fort Pierre a week earlier. Provost, Bell, and Harris managed to bring in fresh meat from one of the bulls they saw.

Two days later, the upstream steamer *New Haven* delivered to Audubon a letter from his sons dated July 22, in which Victor wrote, "News from your arrival at the Yellow-stone has reached us and been published" (Beinecke). Audubon was no doubt pleased to know that his sons were keeping the public interested in his whereabouts, but he also laments "that my Lucy had not also written" (AN).

After several days of cold wind and rain and general discomfort, all the journalists agree that September 28 opened with "[a] beautiful morn." A hunting party, including Audubon, went ashore to hunt, by Audubon's order, only elk bucks. At this late date for the expedition, Audubon was still disappointed by the elk specimens he had procured, and on the previous day he had hired another hunter at Fort Vermillion to increase the chances of killing more. Sprague notes that "Mr A. is very anxious to obtain good specimens" of elk. Audubon and the others returned to the boat that evening having failed in their hunt, even though Audubon had a very close encounter: "I saw a prodigious large one who stood before me within 5 or 6 steps, but my Gun snapped and away he ran and Jumped in the River and swam accross carrying his horns flat down and spread on each side of his back." His bad luck continued that day as he was reboarding the boat: "On getting in the boat this evening I had a severe fall which has bruised my Knee and my Elbow" (AP). This injury was severe enough to prevent him from walking the next day.

This day, however, that began with "[a] most superb morning" brought a completely different experience to Sprague: "[C]lear, calm, and delightful, the finest we have had for a long while—but to me these fine autumnal days bring a recollection of the pleasures I have enjoyed on similar days at home—and notwithstanding every thing is more pleasant, I feel an approach to homesickness that is not thought of when the weather is bad." Even though Sprague had tried to remain at Fort Clark and the Mandan village, his desire now drove him homeward. No doubt all five members of the Audubon party were experiencing mixed feelings about the imminent end of this adventure, regret mixed with relief and perhaps the beginnings of joy at the prospect of being home in a few weeks.

When they reentered the United States on October 6, Audubon's injury of September 28 still troubled him. On October 4 he notes, "My Knees very sore yet," and on the fifth he fell again: "I had a Slide on the bank as it had rained" (AP). But everyone enjoyed a most cordial evening at the dragoon post, Fort Croghan, commanded by Captain John Burgwin. It was with Lieutenant James Henry Carleton (1814–73), however, that Audubon struck up a friendship. Since the dragoons had been ordered to abandon the fort, they traveled along with the Audubon party, most on horseback overland, but Carleton and eighteen of his men (according to Bell) joined the men in the boat. The evenings included much conversation, possibly some whiskey, and games of whist. Audubon characterizes Carleton as a "fine companion and a perfect gentleman." The lieutenant must also have admired the *Quadruped* plates Audubon carried with him, for Audubon gave him one on October 8, for which Carleton gave Audubon a black bear skin and a set of elk antlers. Their regard for one another seems to have quickened over these few days of travel together, October 5–11, for on the ninth the two men "exchanged Knives," and Audubon gave Carleton three more *Quadruped* plates, which Carleton answered with the gift of "a damascus Double Barrelled Gun" (AP).

Carleton remained at Fort Leavenworth, which the boat reached on October 10, and Audubon and his new friend parted ways the next day when the boat continued on for St. Louis, but Carleton always remembered his brief encounter with Audubon with great fondness. In a letter from his home in Bangor, Maine, to Audubon at Minnie's Land dated April 2, 1848, Carleton reminds Audubon of "our delightful journey together down the Missouri" and of the *Quadruped* plates "you were so kind as to give me." He assures Audubon that the plates "are the choice ornaments of the parlors of my friends in Boston and here. They are, for you must know, admired by every one." While Carleton was suffering from ill health, he was glad to have heard of Audubon's "continued prosperity; and that you are still enjoying your green old age, a comfort to your friends, an honor to your country, and to the world" (Audubon Collection, Missouri History Museum). The professional soldier seems to have genuinely revered Audubon and valued his acquaintance with the celebrated artist-naturalist who had sought out the wild northern prairies.

The week of Carleton's company, all the journalists agree, was marked by lovely weather, making the memory of it all the more pleasant. And as the landmarks passed with increasing frequency the closer they came to St. Louis, the men's excitement must have risen. On October 9, "at the commencement of a bend in the river about 4 miles across and some 8 or 10 around," Sprague accompanied Harris and Bell to walk across and meet the boat on the far side. The weather and scenery were simply irresistible: "The point was covered with heavy growth of black walnut, oak &ᶜ—noble trees that have stood here for ages. I delight to roam through these old forests alone—on such fine, calm autumnal days as this—every thing is so quiet— and there is nothing to disturb the profound silence that reigns around save the occasional chatter of a squirrel—the tap of the woodpecker—or the scream of the parrots as they sport among the branches of these ancient trees." The profound silence, however, was disturbed: "But in the midst of this, and quite out of place, I thought, we found that curse of the red and white man—a distillery!"

The halcyon days ended, however, and on the fourteenth they were "[s]topped at 12 by a gale, which ran us ashore." The injury to his leg of September 28 must have healed, for even at fifty-eight years of age, Audubon undressed and slipped into the muddy river "to push off the Boat, after which I got so deep in the mud I had to dragg myself backwards" (AN). This image of the great American naturalist naked on his back in the mud at the end of the Missouri River expedition may lack dignity, but it manifests the reality of the trip: through main strength and awkwardness, Audubon and his companions had taken on and largely met the challenges of the Missouri River and the northern plains and were now returning slightly battered and partially successful.[22]

Four days later, at St. Charles, they lost their pilot, Provost, who, according to Audubon, "was drunk this day & went off by land to Sᵗ Louis" (AP). The Scotsman William Drummond Stewart, with whom Provost traveled for a while in 1837, once characterized him as "Old Provost the burly Bacchus" (qtd. in Hafen 91). Having successfully piloted the party and all their cargo from Fort Union to St. Charles, covering on average twenty-seven river miles on each of the sixty-five days of the descent despite the obstacles of wind,

---

22 Audubon's granddaughter's revision provides a more dignified image: "We ran ashore, and I undertook to push the boat afloat, and undressing for the purpose got so deep in the mud that I had to spend a much longer time than I desired in very cold water" (AJ 2:174).

snags, and sandbars, the already-legendary Etienne Provost, after seven months in the company of Audubon and his companions, seems to have considered that his job was done. He no doubt had friends in St. Charles, and if he arranged for some sort of transportation there, he may well have reached his wife and daughter at their home in St. Louis before Audubon arrived at three the next afternoon, October 19.

It is not surprising that the journals grow all but silent from this day onward. After taking lodging at the Glasgow House, which they had found so expensive the previous spring, the party had business and pleasure to attend to. All their gear, baggage, barrels, boxes, antlers, and cages were unloaded and taken to the warehouse of Audubon's brother-in-law Nicholas Berthoud. While Bell's journal is the only one that mentions a "warm bath," I suspect that much bathwater was heated up that day in the Glasgow House. Harris worked up the final account of the expedition's finances. Audubon wrote letters home on the nineteenth and the twentieth, presumably to send ahead of him by steamboat so his family would be able to anticipate his return as soon as possible.

St. Louis to Minnie's Land, October 22–November 7

After three busy days, all but Harris boarded the steamboat *Nautilus* for Cincinnati. Harris apparently wanted more time in St. Louis before returning to his home in Moorestown, New Jersey, and remained in the city until October 30, when he took passage for Louisville. He intended to take advantage of opportunities to visit various locations and people along the Ohio River Valley, including the then secretary of state Henry Clay at his Ashland estate in Lexington, Kentucky. However, because of an "attack of asthma" in Louisville on November 4, he changed his plans and took a steamboat to New Orleans.

The others made their way steadily and without incident up the Ohio. Only Bell and Sprague noted the progress of their travel. At Louisville on October 26, the *Nautilus* stopped for two hours and they enjoyed a brief visit with William Bakewell, Audubon's younger brother-in-law. At Pittsburgh on the thirty-first, after Audubon arranged the shipping of the bulk of his baggage and cargo separately, the four companions booked their passage by canal boat and inclined planes across the mountains of Pennsylvania to Harrisburg, where they would continue by rail.

A fateful encounter occurred on the first day aboard the canal boat. The young Charles Wilkins Webber (1819–56), a travel adventurer and an aspiring writer, was on his way home from a western trip when he booked passage on the same canal boat as Audubon and his party. Eight years later, Webber published a lengthy account of his five days of travel with Audubon across Pennsylvania to Philadelphia in *The Hunter-Naturalist*. There he boasted of his privilege of giving up his right to a more comfortable berth on the canal boat to the epitome of his hunter-naturalist ideals, Audubon.

Webber for years had been inspired by Audubon's example of the hunter-naturalist, and he had traveled widely in order to become worthy of the epithet himself. He arrived at Pittsburgh on October 31 and chose to continue on to Philadelphia by the "Canal Route" rather than the "Stage Route." After spending a dull day in Pittsburgh, "this dim Cyclopedian city," waiting for the nighttime departure of the canal boat, "we got underweigh, with such a cargo of pigs, poultry and humanity, as even canal boats are seldom blessed with. I stood upon tiptoe for the fresh air in the cabin, until the time had actually come when people *must* go to bed; when that awful personage, the Captain, summoned us all together, and informed us that every man, woman and child aboard, must stow his, her or itself away along the face of the narrow walls, in the succession of their registration during the day." Since Webber was first on the list, he got first choice of where to bunk: "As it happened that this right of choice was finally definitive for the route, and determined whether one should sleep upon a hammock, or the floor, or the tables, for several successive nights—it was a matter of no little moment" (92). He first learned that Audubon was on board when he overheard a gentleman say that Audubon was last on the list and might not get a bed. Webber felt his heart leap, for he thought Audubon was still on his "Rocky Mountain tour" and because Audubon was "[t]he man of all others in the world I wanted to see most." The gentleman—likely either Bell, Squires, or Sprague—laughed "as he pointed to a huge pile of green blankets and fur which I had before observed stretched upon one of the benches, and took to be the fat bale of some Western trader." Audubon had been taking a nap. Instantly, Webber waived his "right of choice in favor of Mr. Audubon" (93).

Webber pours forth pages of adulatory and purple prose in praise of his hero, but it reflects a contemporary professional author's belief that Audubon's celebrity was so great that the reading public would judge such lavish praise to be appropriate—or, at least, understandable, not surprising.

Webber stepped toward the waking Audubon: "Yes, it was Audubon in his wilderness garb, hale and alert, with sixty winters upon his shoulders [...]. He looked as I had dreamed the antique Plato must have looked, with that fine, classic head and lofty mein! He fully realized the hero of the ideal. With what eager and affectionate admiration I gazed upon him, the valorous and venereble Sage!" It gets worse when Webber characterizes Audubon's dedication to "the holy priesthood of nature! I felt that the very hem of his garments—of that rusty and faded green blanket, ought to be sacred to all devotees of science, and was so to me" (94). When we read that Webber found Audubon "not very talkative," it's easy to suspect that Audubon might have been put off a bit by so ardent an admirer, despite how eagerly he had sought fame in the past. Nevertheless, Audubon spent time with Webber, who had traveled extensively in the South and West. Webber reports the live wild animals Audubon was bringing home with him: "the swift Fox, the snarling Badger and the Rocky Mountain Deer." Audubon showed him "some of the original drawings of the splendid work on Zoology of the continent, which his sons are now engaged in bringing out." And Audubon permitted Webber to walk with him "for hours along the tow-path ahead of the boat," causing Webber to observe "with astonishment, that, though over sixty, he could walk me down with ease" (95).

Webber maintained his acquaintance with Audubon after this trip and got to know both of his sons, with whom he is said later to have planned "a monthly magazine of mammoth size, to be illustrated with copper-plate colored engravings by Audubon" ("Charles Wilkins Webber"). In three different issues of the *American Whig Review* in 1845, he wrote rhapsodic review essays about all things Audubon under the pseudonym Charles Winterfield. But when he published his *The Hunter-Naturalist* in 1851 after Audubon's death, he wanted to memorialize the first meeting with his idol on his return from the Missouri River expedition and give Americans a national hero: "Thus it was I came first to meet him, laurelled and grey, my highest ideal of the Hunter-Naturalist,—the old Audubon!" (97).

Audubon was indeed at the height of his fame when he returned from

this expedition. Victor and John Woodhouse had done their part by keeping the public informed of his whereabouts as well as of the ongoing and
future publications. But on the day of his return to Minnie's Land, his first
concerns were for his family: "Reached home at 3 o'clock P.M. 6th Nov.
1843 and thank God found all my familly quite well!" (AP).

Fortunately, Victor's wife, Georgianna, captured the moment of his
return vividly in a letter:

> He returned on the 6th of November, 1843. It was a bright day, and the
> whole family, with his old friend Captain Cummings, were on the piazza
> waiting for the carriage to come from Harlem. There were two roads,
> and hearing wheels, some ran one way and some another, each hoping
> to be the first to see him; but he had left the carriage at the top of the
> hill, and came on foot straight down the steepest part, so that those who
> remained on the piazza had his first kiss. He kissed his sons as well as the
> ladies of the party. He had on a green blanket coat with fur collar and
> cuffs; his hair and beard were very long, and he made a fine and striking
> appearance. In this dress his son John painted his portrait. (qtd. in AJ
> 2:175–76) (See fig. 25.)

PART III

The Three Forgotten
Manuscript Journals

THE BEINECKE PARTIAL COPY

This twenty-six-page partial copy of Audubon's original full journals from the Missouri River expedition, including the entries for August 5–13, 1843, is held at the Beinecke Rare Book and Manuscript Library, Yale University, in the Morris Tyler Family Collection of John James Audubon (Gen MS 85, box 8, folder 446). The hand is most likely that of Lewis M. Squires, Audubon's paid scribe on the trip.

The manuscript is brittle and stained in the upper and lower right-hand corners. Some words in the stain are no longer visible or so faint that they are illegible. Other words have been lost where part of a page corner has flaked away (see fig. 26).

Journal No 3

AUGUST 5, SATURDAY

The weather was cool. the wind at North wet & cloudy, [****] menacing rain. We made toward the road we had come upon [****] and on our way Harris shot a young of the "swift Fox" which we could have caught alive had we not been afraid of his running into some hole. We saw only a few bulls, antalopes & wolves. The White horse that had gone out as a *hunter*, returned as a *pack* horse loaded with the entire flesh of a buffaloe cow, and our 2 mules drew 3 more and the heads of 4. This morning at day light when we were called to drink our coffee, there was a buffaloe bull feeding within 20 steps of our tent, and it removed slowly towards the hills as we busied ourselves to make preparations for our departure. We reached the fort at

12 & had a good dinner. Squires, Provost, & La Fleur had returned. They had wounded a big horn & lost it. Owen & Bell returned this afternoon. They had seen no "Fuchs of the plain," but brought an Elkskin female, and a female porcupine as well as a young white headed Eagle. Buffaloes become so poor during hard winters when the snow covers the ground to the depth of 2 or 3 feet, that they lose their hair, become covered over with scales, on which the Magpies feed and the poor beasts die by hundreds. One can hardly conceive how it happens, that notwithstanding their general deaths and the immence numbers that are murdered daily over the immence wastes, called prairies, that so many are yet every day to be found!—Sprague never spoke to me on my return, but that is his way. Bell related all his adventures to every one [****] and I am off to bed pretty well fatigued. Our boat is going [****] and I wish I had a couple more big horns Males [****.]

[Measurements of swift fox given here.]

[****] skin of this anim[****] [****] [****] did not even look at [****] [****] took it outside [****] [****] its coat in pickle where I have no doubt it will keep well enough.

[Measurements of "Old Female Elk shot by Owen" given here.]

[Measurements of female porcupine killed by Bell given here.]

AUGUST 6, SUNDAY

Weather thick and looked for a thunder storm, which however we have not had. Sunday as usual was a dull lazy day. After dinner Bell came to me and told me all about his expedition and gave me the measurements of the Female Elk, and the porcupine given above. He finished the skin of the Porcupine, the Elk skin was put in pickle by Provost. The latter went off to Ft Mortimer to see Boucherville & others & try if they would go after big horns tomorrow morning. I had another male of the Sperm Hoodii[1] destitute of testes by the operation of {Carias} and this is the second time

---

1   According to Elliott Coues, "Sperm Hoodii" is "a synonym of *Spermophilus tridecem-lineatus*, the Thirteen-lined, or Federation Spermophile," or ground squirrel (AJ 2:37).

this has happened to us scence here. These animals are killed by the herd guards with their whips. This afternoon we had an arrival of Indians, who were here two weeks ago. They had been to Ft [****] reported that a battle had taken place betwen the Gros [****] in which the former had lost one man. [****] prairies, through the severity of the winter [****] shockingly poor even in the views of [****] are caught in pens in the manner [****] described when they are dispatched [****] [****.] In the winter of 1840 [****] [****] having drifted there [****] [****] following them on Horseback, and fencing them in these drifts which are in places as much [****] feet deep. They were brought home on a sleigh & let loose in [****] rooms. They were to appearances so very gentle that they suffered them to handle them, although the animals were loose. They were removed to the Carpenters shop. There one broke its neck by leaping over a turning lathe. The others all died in the same manner for they regained their former wildness & jumped kicked &c untill they were all dead. They had been caught with lassoos placed on the ends of long poles. Young buffaloes 1 & 2 years old have been caught in the same manner by the same gentleman, assisted by Le Brun & 4 Indians. He took down the river 13 of the latter. The antelope cannot be {tamed} when thus caught, & it is very rarely {indeed} that they can be raised when caught young—Mr Wm Sublette had one, however, a female, which grew to maturity and was so gentle, that it would go all over his house, mounting & descending steps, & even going to the roof of his house. It was alive when I first reached St Louis but I was not aware of it, and before I left it was killed by a Bull Elk belonging to the same Gentleman. Provost returned and said that Boucherville would meet him & La Fleur tomorrow morning early, *but I doubt it.* God bless you all, good night.

AUGUST 7, MONDAY

I had such a horrible dream last night about our dear son Johnny. So horrible was what I saw in that dream, that I awoke & found my face covered with tears, and my whole body in a high state of perspiration. I slept badly the rest of the night & was up with the ringing of the bell at daylight. Provost,

Bell, & La Fleur started after breakfast and after having waited for Boucherville nearly 4 hours, they left at 7 and the Indians were curious to know
where they were bound, and looked at them with more astonishment than
we liked. At 9 we saw Boucherville acc[****] all mounted, and they were
surprised [****] waited for them, or rather that he had [****] a bottle of
Whiskey & they started [****] must have overtaken the other party in about
[****] helped {but} I may have some big horns [****] hasten [****] than
any [****]. Sprague [****]ed out and collected seeds. Harris & I amused
ourselves in searching [****] sand stones, with impressions of leaves &c
and found a few. Mʳ Denig says that the band of Indians called the [*ra*]
de Ca[****] (Assinniboins) killed a black bear on White Earth river, about
60 miles from the Mouth, they are occassionally killed but it is of very,
rare occurrence. Mʳ D saw the skin of the bear in their camp last winter.
And a raccoon was also killed on the Big Cheyenne river, by the Sioux,
and they knew not what to make of it. The following a/c of a skirmish that
took place at Fᵗ McKenzie in the Black foot country is coppied from Mʳ
Culbertsons journal.

Aug 28 1833. At the break of day we were aroused from our beds by the
report of an enemy being in sight. This unexpected report created naturaly
a confusion among us all. Never was there a set of unfortunate beings taken
in such complete surprise, as we were. By the time that the alarm had spread
the fort, we were surrounded by the enemy, which proved to be Assinniboins
headed by their chief Gauche (or the Antelope) and the number as near as
we could estimate was about 400. Their first attack was upon a few lodges of
Paigans, who were encamped at the Fort. They also being surprised could
not escape. We exerted ourselves to save as many as we could by getting
them into the Fort. But the foolish squaws when they started from their
lodges took each a load of old saddles & skins which they threw in the door
and stopped it so completely that they could not enter. Here this enemy
massacred several. In the mean time our men were firing with muskets &
shot guns. Unfortunately for us we could not use our cannon, as there were
a great many Peigans [****] us and the enemy thus prevented us from [****]

them at once, as we could not have [****] 50. The engagement continued about [****] finding their men dropping very [****] bluffs ½ mile distant. There they stood ma[****][****] come in and give them an equal chance [****] prairie, although our force was much we[****] than theirs [****] trial at the same time we dispatched an express off to an Encampment of the Peigans for a reinforcement. We mounted our horses & proceeded to the field of battle which was a perfect level, where there was no chance to get behind a tree or anything else to keep off a ball. We commenced our firing at a distance of 200 yards but very often fired at 100. We set up a continual fire for two hours when our horses getting fatigued we concluded to await the arrival of a reinforcement. None of us as yet were wounded or killed, & nothing lost except a horse which was shot under one of our men by the name of Bourbon. Of the enemy we cannot tell how many were killed for as fast as they fell they were carried off the field. After the arrival of our reinforcement, which consisted of 150 mounted Peigans, we charged upon them again & drove them across the Maria, where they took another stand and fought for about 2 hours. Here Mr Mitchells horse was shot from under him. In this Engagement the enemy had a decided advantage over us, as they were concealed in the bushes when we were in the open prairie. We however succeeded in making them retreat from this place back on a high prairie. After we pursued them across the Maria, they rushed upon us & compelled us to retreat & recross the river. Then they had us in their [word omitted] if they had had the courage to avail themselves of their good fortune, they could have killed a great many of us when we rushed into the water which was almost deep enough to swim our horses. They were close upon us, but we succeeded in crossing the stream before they fired. This foolish retreat came near being attended with fatal consequences which we were aware of but our efforts to stop were unsuccessful. We however did not retreat far before we turned upon them again with the determination of driving them [****] in this we succeeded, by this time it was [****] we could see them no more, and we concluded [****] during the day we had 7 killed & 10 wounded [****] enemy scalped. It is impossible to tell how many [****]

were killed, but we must have killed & wounded at least 30. We however got two bodies which our Indians burned after scalping them. The Indians that were with us in this battle, deserve but little credit for their bravery, for in every close engagement the whites, who were but few, always were in advance of them. This however had a good effect, for it removed the idea that they had of our being cowards, and made them believe that we were in fact actually brave. Had it not been for the armed assistance that we gave the Paigans they would have been cut off, for I never saw Indians behave more bravely than the enemy this day, & had they been as well supplied with powder & ball as we were they would have done much more execution for necessity compelled them to spare their ammunition. As they had come a long way it was necessary to save enough to return home. Had we been aware that the Indians were enemies even after they had surrounded the Fort & surprising us in the manner they did we could have massacred a great many of them, but thinking that they were a band of Indians coming with this ceremony to trade we did not fire upon them untill the balls & arrows came whistling about our heads. Then only was the word given to Fire! But had they {possessed} courage enough at the outset to have rushed in at the gates of the Fort, we could have done nothing, but be compelled to suffer death under their tomahawks.

GRISLY BEARS

Mr Denig gave us the following Bear Story, as he had it from the parties concerned. In the year 1835 two men set out from a trading post on the head of the Cheyenne and in the neighborhood of the Black Hills, to trap beaver. Their names were Michele Carriere & B Le Brun. Carriere was a man about 70 years of age and had passed most of his life in the Indian country, in [****] occupation. One evening as they were setting their traps [****] stream of the Cheinne, which was {wooded} but [****] trees, their ears were suddenly saluted by a [****] moment a large she bear was rushing upon them [****] being a young & active man immediately grasped his gun [****] the bear through the bowels. Carriere also fired but missed the bear then pursued them but as they ran, their legs done them justice, and they escaped through the bushes, where the bear lost sight of them. They had given up the chase & were engaged again in setting their traps. Carriere who went a short distance from Le Brun, in the thicket, with his traps, came directly in front of the huge wounded bear, which with one spring, bounded upon him and tore him in an awful manner. With one strike of its paw in his face & forehead, it cut his nose in two, and one of the claws reached inward nearly to the brain, at the root of the nose. The same stroke tore out his right eye & most of the flesh from that piece of his face. His arm and side were literaly torn to pieces and the bear after handling him in that gentle manner for 2 or 3 minutes threw him upwards about 6 feet where he lodged, to all appearance dead, in the forks of a tree. Le Brun, hearing the noise, ran to his assistance & again shot & killed the bear, then helped the wounded man to the ground. Little of life, or the remains of human appearance was about him. Le Brun however made a *travaille* and carried him to the nearest trading station where he recovered slowly, but made the most mutilated looking being imaginable. I (Denig) saw the old man about 1 year after the accident and some of the wounds were yet unhealed. He recovered however to live a few years and was killed by the Blackfeet near F^t Union. Carriere in telling this story, says, that he fully believes it to have been the Holy Virgin that carried & placed him in the forks of the tree, & thus preserved his life. The bear is stated to have been as large as a common ox & weighed, according to Carriere, *1500 pounds* more or *less* (?).

When Bell was fixing his traps on his horse this morning, I was amused at Provost & Le Fleur laughing outright at him, as he first put on a Buffaloe robe *under* his saddle, a blanket over it, over that his Musquitoe bar & then his rain protector. The old hunters could not understand the reasons why a man should not travel as comfortable in this country as in any other. Then besides he took a sack of biscuit. And Provost only 1 old blanket & a few pounds of dried meat and rode off in his dirty breches. La Fleur was worse off still, but he carried a bottle of Whiskey, to mix with the salted water found in the Mauvais terre (bad grounds) among which they have to travel untill their return. Harris & I contemplated going to a quarry from which the stones for the Powder Magazine were obtained but it became too late and the horses brought for us to ride proved the wrong ones, being both runners. We went however after rabbits across the river, & returned having killed a red cheeked wood pecker & shot at & missed a rabbit. We had a sort of show by Mr Moncrevir, which was funny & well performed. The following Grisly bear story was related to me by Jean Batiste, our waiter &c. It happened about 12 years ago, when Mr McKenzie had charge of this fort and is as follows. About the season when green peas are plenty & good Batiste was sent to the garden, about ½ mile distant to gather a mess of them; he did so, but when about to the end of a row, to his astonishment he saw a furious bear gathering peas also. He dropped his large tin bucket & ran to the fort as fast as possible, told Mr McKenzie who imediately summoned several other men with guns & well mounted, they reached the garden, saw the bear, killed it, but alas Bruin had completely emptied the bucket of its peas. This and the previous stories are facts, good night, God bless you all.

AUGUST 8th, TUESDAY

[See fig. 27.]

Another sultry day. At the ringing of the bell this morning Harris went off after rabbits & shot at a young one without effect. Immediately after breakfast Mr Larpenter drove him & myself in search of Geological specimens, but we found none worth having. We killed a sperm hoodii, which although d[****]ly wounded, entered its hole and Harris had shot it, had to draw it out by the hind legs. We saw a family of rock Wrens and killed 6 of them but the first one went into a hole & we left it. I killed 2 at one shot and 4 alltogether. [****] of them after being shot went under a flat stone and then was found quite dead on Mr Larpenters turning the stone up. We saw signs of Antelopes & Hares (Townsends), and rolled a large wrock from the top of a high hill. The notes of the wrock wren are a rather sharp prolonged Cree e e e e. On our return home we heard that Boucherville and his 5 hunters had returned with nothing for me and that they had not seen Bell and his companions. We were told also that a few minutes after our departure the roarings of Buffaloe had been heard across the river, and that Owen & two men with one cart had been despatched to kill 3 fat cows, and no more. Harris was sorry that he had not been here, and so was I, as both of us would have gone and joined & seen the fun. The milk of the Buffaloe cow is truly good & finely tasted, but the bag is never large as in our common cattle, & perhaps this is a provision of nature to render the cows more able to run off and escape from their pursuers. The first day I have fully of leasure I will describe the whole process of fleecing the Buffaloe; camping in particular &c &c. Bell, La Fleur & Provost returned just before dinner. They had seen no Gros Cornes[2] & only brought the flesh of 2 deer killed by La Fleur and a young Magpie. This afternoon Provost skinned the calf that was found in one of the cows that Owen killed. It was large and we have taken its measurements. It was looked upon as phenomenon as no Buffaloe cow has ever young at this period. Their calving time is from the 1st Feby to the last May. Owen went off 6 miles from the Fort before he saw cattle. They were somewhat more than 300 in number, and that made Harris & I regret the more our having gone fruitlessly after stones. It is curious that while Harris was searching after Rabbits early this morning he heard the bellowings of the bulls & thought at first it was the growlings of a Grisly Bear, and again those of one of the Fort bulls. But he mentioned it to no one. To morrow morning La Fleur & someone will go after Big Horns again & they are not to return untill at least one or two males are killed. This evening we have gone fishing across the river and caught 10 good catfish of the Upper Missouri, the sweetest and best fish of that sort that I have eaten in any portion of the country. Our boat is going on well and looks pretty to boot. Her name is to be the *Union* in consequence of the United exertions of my companions to do all that they could for me in this costly expedition. The young Buffaloes now about the Fort have began shedding their old coats, the latter coloured hair dropping off in patches about the

2   Bighorn sheep.

size of palm and the new hair is dark brownish black. [Measurements of unborn bison calf given here.] All the hoofs had soft excresences beyond their true and horny formations which I suppose shrink in & become hard ere the animal receives birth.

AUGUST 9, WEDNESDAY

Extracts from Mr Culbertsons journal kept at F^t McKenzie Blackfoot Indian Country 1834

Friday 15 June. Blood Indians started this morning to go to war against the Crows. They had not long left when the "[****d] Bulls back's fat" son with his sister, brother, & brother in law, returned to the fort, and saw that they must go back to the camp. After I had given them some tobacco and ammunition they all started but they did not get more than 2 miles from the fort when they were all killed by the crows except one who by some means, or other, got on one of the Crow's horses & fled to the Fort. The squaw without doubt was taken prisoner. In the evening I went out and found the bodies of her brother & brother in law. She was not there. On Saturday 14th went out once & brought in the bodies and had them decently interred. The young man who escaped & was slightly wounded started to go to the camp with 3 Gros Ventres.

Tuesday 24, We were all surprised this evening at the arrival of the squaw, who had been taken prisoner, & was taken to the Crow village where she was kept tied every night except the one on which she made her escape. During the day having it in contemplation to escape, she took the precaution to conceal a knife under her petticoats, but was unfortunately out with one of the Crow women, and in stooping the knife fell out. This was reported, and as a punishment, they stripped her of all her clothing. This however did not prevent her from availing herself of a good opportunity. She started, without anything under the heaven to cover her nakedness and in this plight she traveled across the prairies for 4 days & nights, and the weather was unfortunately quite cold and wet for the season. She arrived at the Fort in a wretched condition, but a great consolation to all her friends & relations. She says that after her arrival at the village, they made her dance with the scalp of her brother tied to her hair, and clothed in the bloody shirt of husband. On Wednesday the 25th a band of about 400 Crows arrived at the Fort to day, with the intention of taking the fort by stratagem, if they could get an opportunity. But such they did not get. I would not permit one of them to enter the fort, or within firing distance, with their arms they used every artifice in their power to induce me to let a few of them enter to smoke, that their intentions were good and that they loved the white people. Finding all this of no avail, they brought their best horses to give me, for which they did not wish to receive anything but the privilege to allow a few of them to come in, but all this was in vain as I was well aware of their treacherous intentions. I divided my men in the 2 Bastions with orders to fire upon the first one that might approach during the night, and I warned the Indians of my having given such orders, meantime telling them that I did not wish to strike the first blow, but that if they commenced they would go off with sore hearts. There was an American with them who told me of their intentions, and that they were determined to take the Fort. I sent them word by him that we were ready for them, & if they thought themselves able to come and try, but when they saw the muzzles of our canons pointed at them they were not so anxious to rush—On the 26th the Crows made another attempt to get in, but after a long persuasive [t****h] they found it would not do. They crossed the river and came on the high bank opposite the Fort, and fired upon us and while some of them were yet crossing the river, I let loose a cannon ball among them, which if it done no harm made them move at a quick pace, and after a little they all went off leaving us without food of any sort but fortunately on the 30th Monday, a party of Blood Indians came in from the Crows with 15 horses, & brought us meat and will go hunting buffaloe tomorrow.

From the above the Nature of the Indians of this region, may be exemplified a thousand times better *because true* than all the trashy stuff written & published by Catlin! The young man who made his escape being wounded by the Crows, who had him several times by the hair but could not hold him,

Providence appeared to have been his protector, and his shield. He pushed one of the Crows from his horse, leaped on it, in an instant and flew with the swiftness of an arrow receiving only a slight wound on the thigh. The Crows had stolen the whole of our horses, a few days before, and promised to return them provided I would suffer them to come in. I was also informed that they had also brought pack horses to carry off all the goods after having accomplished the destruction of the Fort & massacres of its inmates.

BIGHORNS & ANTELOPES

Squires, Provost & La Fleur went off this morning after an early breakfast, across the river, after Big Horns with orders not to return without some of these wild animals which reside in the most awful portions of the Broken & highest Clay hills and stones that exist in this region of the country. They never resort to the low hills except when moving from one spot to another and cross rivers which they swim well, as do Antelopes.

ELK

Habits of the Elk as far as seen by Bell on the Yellowstone River

The males at this season are divided from and keep apart from the females. They have tremendous horns in [****t], and are often seen among bands of buffaloe. Instead of making for the woods towards the river, they run towards the hills across the prairies. They carry their heads high with their horns lying back, & handsomely balanced. The nose was [****] parallel with the back & somewhat above as if {seeking} all the while. They generally walk, when frightened trot & gallop at a steady pace like a horse. Bell saw two run over an open prairie for at least 1½ miles, then nose the hills & disappeared. One of these was slightly wounded with a ball. The Females were found lying down principally on the sand bars betwen 8 & 11 O'clock, after which they retire to the thick bushes or tall willows. When shot at in these bars they always run for the willows. 12 Females were seen in one band, & also 4 males were seen together and other groups of females of 75 &c. When shot at & wounded they hollow loud, will cross the patches of willows & take to the prairies. They are shedding their winter hair at this season. We saw upwards of 40 in this time, Deer & Buffaloes were also very abundant. These latter were seen to the number of 1000 going in a line at least 2 miles long across the prairies and we were not at any time more than 10 Minutes without seeing some of them. Saw 2 Grisly Bears and several antelopes. When we killed the first Bulls we saw 15 wolves at the 2 carcases, among them some Prairie wolves. The porcupine that resort to a ravine where about 100 trees are standing & then raise a brood or broods, for they never have more than 4 young ones, & have only that number of teats, remain then untill they have not only eat all the tender branches, but broken off the large ones and upper parts of the trunks, & killed every tree after which in their own defence they have to remove to another place. In such a ravine as I have mentioned here, no less than 13 porcupines have been killed in a season by a single hunter.

COPY OF BELLS JOURNAL ON HIS
TOUR TO THE YELLOWSTONE.

Aug 2ᵈ started at ½ past 7 this morning, saw yellow legs (Godwits and some young blue winged teals in the ponds across the first prairie. Shot 2 Curlews. Saw 2 very fine small elks. They were lying down under a bank when they got the wind of us. The sharp tailed Grous are first rate eating now as they feed entirely on Grasshoppers & berries of different sorts. When a cow is killed among many bulls they remain about her untill actualy driven away by the hunter. Calves of even one year old will rarely abandon their mother untill driven away by being frightened. Owen climbed a tree to a white headed Eagles nest and drew a young one out which fell to the ground & was caught alive, and brought to the Fort. Is it not very remarkable that Eagles of this kind should have their young in the nest at this late season when in the Florida's I have shot them of this same size in July. Shot at a wolf, which being wounded ran off about 100 yards & yelled like a dog, a very remarkable instance, as all the wolves that we have killed & heard of rarely do more than snipping at the pursuer & if running when shot at. On his trip up the Missouri after Big horns on the 7ᵗʰ Inst. Monday last, he found hundreds of cliff swallows nests with the old ones feeding their young. This is also very late and uncommon at this season. Saw a Peregrine Falcon feeding its young. La Fleur shot 2 bucks of the white tailed deer at 2 shots and they brought the meat home which proved fat & very good. Saw beaver tracks & young green winged teals. They saw hills impregnated with sulphur, & some coal on fire, and now & then giving way by hundreds of tons at a time. The above is about the amount of his journal during these 2 expeditions.

I have done scarcely anything else but write during this day. We have had an arrival of 9 Indians among whom are 5 chiefs, all Assiniboins. These 9 persons fed for 3 days on only the flesh of a single swan. They saw no buffaloe although they report them exceedingly abundant about their village about 200 miles from here. Harris is unwell complaining much of pains in his bowels, and with a [dia****]. He attributes this to his having bathed last night in the channel of the Missouri, where he says the water was cold. Bell who did the same does not complain of anything and I hope that my Friend will feel better tomorrow morning. I copy Bells a/c of the Burning hills that he saw, verbatim, thus, "In returning I saw a vein of coal on fire. We were following a path close to the fort of a high bute where on looking ahead I found the way suddenly blocked up by the Earth falling down from above and in looking up saw a vein of coal or other [a****] substance on fire. It was about 2 feet thick, about 75 ft from the bottom & 40 from the top. It was burning very slowly for about 50 yards emitting a whitish smoke something like sulphur when burning and turns the earth or rock afire quite red or of a brick colour. It would undermine the earth above which then would fall down & this was the cause of the obstruction in the path before us. In some places several hundred tons had fallen down at the same time. It must have been burning for some time as it had passed along the bute for some distance. In one place I saw banks of clay 20 feet high quite red, & some parts quite hard, while others were very scaly & soft, & were crumbling to pieces. Where the fire was burning the clay was red, varying from 1 to 3 feet in thickness. No appearance of coal presented itself where the fire was extinguished and had passed along but very distinct above, the fire, and I have no doubt there is a small quantity of sulphur, mixed with the coal & whatever it may be. In another place a short distance from the hills and in a ravine I also saw some red stones which looked very much as if they were the corners of a house, & had the remains of 3 sides yet straight. The stones varied from 5 to 20 inches in thickness, many were square, & about 18 or 20 feet high, thus it appeared to me without stopping to examine them minutely."

This Evening 12 catfish were caught, and I am off to bed, wearied of Indians, so filthy and so beggarly are all that I have had the honor to see, Chiefs & {Minnidens}. The Sperm Hoodii shot by Harris & which was an old female had her pouches distended and filled with the seeds of the wild sunflowers of this region. I will go and follow one of their holes and describe the same. God bless you all, good night

AUGUST 10, THURSDAY

Having found the Phil^a Saturday Courier dated March 15, I will copy the handsome notice published in that paper and writen in verse by Doct^r J. K. Mitchell of that city.

To Audubon

A health to thee brave wanderer, Of mountain, plain, & lake,
A blessing on thine enterprise From friendly hearts, oh take!
Speed, speed thee to the native wilds Of natures fairy brood,
To mark the living manners rise, In deepest solitude,
Not as soldiers abashed, enchained, Or stuft in glassy cage,
But bright & bold & beautiful, In love, repose, or rage.
Thou seest them as Nature sees. Where wing & limb are free
Their food, their gambols, haunts & moods,
on mountain, plain [****]
Poor children of the lonely glade, Deep wood & rocky hill,
How soon must perish every trace, Of them, but that which skill
And taste, and feeling, yet may give. That future times may see
Oh, Audubon, the living traits Preserved by art & thee!

I envy thee, I envy thee. The rapture of the wild
When screened by fragrant shrubs & flowers,
Great Natures favorite child,
Thou watchest with the Eagle Eye, And lion heart the play
Of panthers when their spotted young, Are out for sport & prey,
Or when, where rise cloud crested hills, And Eagles love to rest,
Thou sittest on that fearful steep, To watch their guarded nest,
And mark the ways of love & learn How rise the growing brood,
Nor feel one moments weariness In long, long, solitude,
Thou seest not above the things That come to common Eyes
For in this mighty loveliness To bless thy vision rise
Strange shapes & stranger incidents Unknown to soar before
Untill with wise & simple eye Thou didst the wilds explore,
Like as to him of paradise, The wondrous strangers came
To thee in grand untrodden glades To gain their primal name.
But not like him who blessed *that* scene Thou lovest still to see
In every instinct action, *thought*, The present Deity.
If passion marred *that* paradise, Let taste preserve us *this*.
And teach us still that Natures works Are {fineless} stores of bliss.
Then speed thee on, brave wanderer, By mountain, plain & lake,
A blessing on thine Enterprise, For thee & thine, oh take!

Bell and I took a walk after rabbits—saw some—Harris was, thank God, much better. The Indians having received their presents went off with apparent reluctance, for when you begin to give then, the more they pester you for more. It was discovered that the young man who has the care of the cows & other cattle has [****] truely lousy, and M^r C had him taken, marched to the bank of the river, & had his hair all cut off close—the sight of the [****] [****] his head was more than shocking. The lice had actually eaten [****] skin in various parts, and were themselves matted thick amid his dirty hair. He was however cut close, for shaving could not be done, and afterwards stripped of all his clothes, & taken to the river, where his head was covered with new soft soap which made him smart enough. He had already been bleeding at the nose,—he was scrubbed, combed &c his clothes were all thrown away into the river, and new clean ones put on, after which his head was covered with sulphur & grease, and I trust he will not forget to wash & comb himself shortly. The insects that he had about him were the largest that I have ever seen, and many jokes were passed on the poor Devil, who it appeared had not cleaned himself for upwards of a year. The horse guard brought in another specimen of the Sperm Hoodii and after dinner we are going to examine one of their burrows. We have been & have returned. The 3 burrows which we dug were as follows, straight downwards for 3 or 4 inches and gradually becoming deeper in an oblique slant to the depth of 8 or 9 inches and not more, and none of these holes extended more than 6 or 7 feet. I was disappointed at not finding nests, or rooms for stores.

SKINNING BUFFALOES BY WHITE HUNTERS

Although I have said much about Buffaloe running and butchering in general, I have not given the particular manner in which the latter is done by the Hunters in this country. I mean the white hunters, and I will now try to do so. The moment that the Bull or cow has fallen and is dead 3 or 4 hunters, covered with gun powder over their faces & hands, and with pipes lighted & smoking {first} [****] the animal erect on its body, and by drawing out each fore and hind legs apart, fix the body so that it cannot fall down again. An incision is made from immediately above the root of the tail to near the neck, and the skin is taken off in the roughest manner imaginable, downwards and on both sides at the same time. The Knives are going on all in the same time, and many wounds occur to the hands & fingers, but rarely attended to at this time. The pipe of the one has gone out & with his bloody hands he takes that of his nearest companion, who has his own hands equally [****]. Now one breaks the skull of the beast and with [bl****] [****] draws out the hot brains & swallows them with a [pecu****] [****]. Another has now reached the liver and is gobbling down his throat enormous pieces of it, whilst perhaps a third who has reached the paunch is feeding luxuriously on some, to me, disgusting looking offall. But the main business proceeds on. The flesh is taken off from the sides, the bosse & hump bones, from where these bones begin to the very neck, and the hump itself is thus destroyed for the hunters give the name of hump to the mere bones, when slightly covered with flesh, and it is cooked in this way and good when fat, young & well broiled. The pieces of flesh taken off from the sides of these bones are called ~~de pouille~~ *fillets* and are or would be the best portions of the animal if properly cooked. The fore quarters or shoulders are taken off as well as the hind ones, one after another, and the sides covered with a thin portion of flesh called the *de pouille* including in that also the fillets is taken off. Then the ribs & bosse bones are broken off at the vertabrae. The marrow bones, which are those of the fore and hind legs only, are cut out at the last. The feet are usually attached to these as well as to the quarters. The paunch is stripped of its layers of fat. The head and back bone are left to the wolves, the pipes are all emptied. The hands, faces, & clothes are all bloody, and now a good grog is much enjoyed, as the stripping of the flesh from 3 or 4 animals is truly very hard work. In some cases when I was present, and no water near, our supper was cooked without our being washed, and it was not untill we had traveled several miles the next morning before we had an opportunity of cleaning ourselves, and yet every one becomes hungry, feeds heartily and feeds the same. When the wind is high and the cattle run towards it, the hunters guns very often snap and it is during their exertions to replenish the pans, that the powder flies & sticks to the sweat every moment accumulating on their faces. But all this never stops for an instant. These daring hunters, who the moment the chase is ended, leap off their horses, let them graze and begin their butcher [****] work at once.

AUG 11th

The weather has been cold and blowy; The day has passed in con[****]tive idleness with me. Squires returned this afternoon alone, he left Provost and La Fleur behind. They had seen only 2 Big horns a female and her young.

It was concluded that if our boat was ready by Tuesday next that we would leave on Wednesday morning, but I am by no means assured of this. Harris was startled at the idea. Our boat although 40 feet long is I fear too small, "*Nous verons.*" Some preparations of packing were made, but Owen, Harris & Bell are going out early tomorrow morning to hunt buffaloes and when they return then we will see. God bless you all, good night

The activity of Buffaloes is almost beyond belief. They can climb the steep defiles of the mauvais terre in hundreds of places where man cannot follow them, and it is a fine sight to see a fine gang of them in rutting time, proceeding around the defiles at the Elevation of 3 or 400 feet from the bottom below, and from which places if an accidental slip, he goes down rolling over and breaks its neck ere he reaches the levell ground. Bell & Owen saw a bull about 3 years old that leaped a ravine filled with Mud & water at least 30 feet wide. It reached the middle at the first bound, and with the second was mounted on the opposite bank from which it kept on bounding untill it reached the top of a hill more than 30 yards high, and was lost from sight in a few minutes. M^r C tells me that these animals can sustain hunger in a most extraordinary manner, he says that a party of his men saw a large bull in a spot half way down a precipice where it had slid, from which he could not ascend or descend, or in other words could not escape. They reached this Fort and on their way back, the 25 day, they passed the Hill and there saw the Bull, still standing. The only thing that annoys them most is crossing the river on the ice. Their hoofs slide from side to side & they become frightened, & stretch their legs apart to support their body. In such situations the hunters near them to the touch, stab them to the heart, cut their ham strings when of course they become an easy prey. When in large gangs those in the center are supported by those on the sides and if the stream is not large many of them reach the land & escape.

MAUVAISE TERRE

The only idea that I can give in writing of what is here called in French, "Mauvais terre" would be to place some hundreds of loaves of sugar, from quite small & low to large & high irregularly [****]ced at top and placed somewhat apart from each other and then one who has not seen these resorts of the Rocky Mountain Ram may form an idea of their haunts and of the difficulties of approaching them, putting aside their extreme shyness & superior activity. They form paths around these broken headed cones, that sometimes are 6, 8 & even 1000 & 1500 feet high, and run round them at full speed on a tract that to the Eye of the hunter does not appear more than a few inches wide, but which in fact is from 1 foot to 18 inches wide. In some places there are hills of earth 8 to 10 feet high and covered with a platform formed of hard and to appearance shely rocky substance on which the Big Horn is seen looking down on the hunter and standing as if a statue. No one can imagine how they reach these curious places and there along with their young even when quite small. Hunters say that the females usually bare their young in such places to save them from the wolves attacks which after men seem to be their greatest destroyers. The Mauvais terres are mostly formed of greyish white clay very sparingly covered with small patches of thin grass on which the Big Horns feed, but which from appearances must be of a very scanty Nature, and yet there & there only they feed as not one was ever seen in the bottom or prairie land farther than the foot of these Extraordinary hills. In wet weather no man can climb any of them as at such times they are sliding greasy muddy grounds. Often times when a Big Horn is seen on a hill top, the Hunter has to ramble about for 3 or 4 miles before he can approach within gun shot of the game, and if once seen, to pursue them is utterly useless. The tops of some, and in fact whole hills about 30 feet high in the Mauvais terre are composed of conglomerated masses of stone, sand, clay & earth of various sorts melted together, & having a brick like appearance. In this mass pumice stone of all sizes is to be found. The whole evidently caused by volcanic action. The bases of some of these hills cover an area of 20 acres, and rise to the height of 3 or 400 feet & even to 800 & 1000. I was on some 3 or 400 feet high or, so high that the surrounding country looked far beneath me. The strata is of different coloured clays, coal and an Earth impregnated with various

salts, which appear to have been formed by fire or internal heat, the earth
or stones of which I have first spoken. Lava sulphur, salt of various kinds,
oxides & sulphates of Iron, and in the sand at the tops of the highest hills
I have found marine shells but so soft & crumbling as to fall to pieces the
instant that they were exposed to the air. I spent a short time in breaking
lumps of sand in hopes that I would find some hard enough to carry to
the F͔ but 'loves labour was lost' and I regreted this exceedingly that I had
thus to abandon it. I also found globular & oval shaped stones very heavy
& apparently composed mostly of Iron weighing from 15 to 50 pounds,
numbers of petrified stumps from 1 to 3 feet in diameter. The Mauvaise
terres abounds with them and they are found in all parts of it from the
vally to the tops of the hills, from pieces 2 inches square to stumps 3 or 4
feet high and of Equal diameter, these petrified trees appear to be most of
them cedar. On the sides of the hills at various heighths are shelves of rock
or stone projecting from the sides from 2 to 6, 8 & even 10 feet & generally
square or nearly so. These are the favorite resorts of the Big Horns during
the heat days and either here or on the tops of the highest hills they are to
be found, but of those hereafter. Between the hills where two or more are
connected together, there is generally quite a growth of cedars, but mostly
stunted, & standing quite close together, with very large stumps, & among
the stumps quite a show of grass. On the summits in some few places there
are table lands, varying from an area of 1 to 10 & sometimes 15 acres. These
are covered with a short dry wirey grass & immense quantities of flat leaved
cactus, the spines of which often warn the hunter of their proximity & the
hostility existing between them and his feet. These plains are not more easily
travelled than the hill sides As every step may lead the hunter in a bed of
this pest of the prairies. In the valleys between the hills are ravines some
of which are not more than 10 or 15 feet wide, while they make one giddy
to look in them. They are of various widths from 10 to 100 feet, the edges
lined at times with wild cherry bushes. Occasionally the buffaloes make
paths to cross them but it is rare. The only sure pass is to follow them to
their head, which is generally at the foot of some hill, & go around. These

ravines are mostly between every two hills, although like every general rule, there may occasionally be found places where 3 or more hills make only 1 ravine. These small ravines connect with one larger, the size of which bears its proportion with the tributaries, this larger one ~~connects conducts~~ runs to the river, or the water is carried off by subterranean canals. In these valleys & sometimes on the tops of hills are holes called "sink holes," these are formed by the water running in small holes, & washing away the Earth beneath the surface, leaving a crust incapable of supporting the weight of a man & if an unfortunate one ~~f**ds~~ treads on that crust, he finds himself in an unpleasant & uncomfortable predicament. This is one of the few ~~ills~~ evils that attend hunting in these lands. These holes eventually form ravines. Through these lands it is almost impossible to travel with ~~for~~ a horse ~~to travel~~ unless by good management & a correct knowledge of the country. The sides of the hills are very steep covered with loose earths & stones, such as I have spoken of before. Occassionally a bunch of wormwood scattered here and there seems to assist the daring hunter, because it is no light task to follow the Big Horn through their lands and is attended with much danger as the least Mis-step at times would send him headlong into the ravines below. On the sides of the high hills the water has washed away the earth in places leaving cones of various heights & sizes & in some places all manner of fantastic forms, are made by the same process. Occassionally in the valleys are found isolated cones, & domes destitute of vegetation, naked & barren. Throughout the mauvaise terre there are springs of water impregnated with salt, sulphur, magnesia, and all other salts of the kind. Such is the water that the hunter has to drink, & were it not that it is as cold as ice, it would be impossible to swallow it, as it operates both as a cathartic & an emectic. This is one of the most disagreeable attendants to hunting in these lands. Numerous snakes of all kinds are also found here. I have seen 1 copper head & a common garter snake.

Notwithstanding the rough nature of the country Buffaloes have paths running through it from the prairie to the river. The hunter sometimes after toiling for an hour or two up the side of one of these hills, to reach the top in hopes that when there he will have for a short distance at least, Either a level place or good path to walk on, finds to his disappointment that he has only reached a {point} scarcely large enough to stand on, & he has the trouble of descending perhaps to be disappointed again & again. Such is the deceptive character of the country. I was thus deceived time & again while in search of Big Horns. If the hill does not terminate in a point, the connection betwen it & others is so narrow that nothing but a Big Horn can walk upon it. Such is the country that the "mountain ram" inhabits and if from this imperfect description any information can be gained, I will be doubly paid for the trouble that I have had in them.

BIG HORNS

As I have before said the Big Horns are fond of resorting to the shelves or the sides of the hills during the heat of the day & when these places are shaded, here they lie, and are aroused instantly upon the least appearance of danger, & as soon as they have discovered the cause of alarm, away they go over hill & ravine, occasionally stopping to look around, & when ascending or descending there is no apparent change in their speed. They ascend & descend places when thus alarmed that it is almost impossible to conceive how & where they find a foot hold. When first seen if they do not see the hunter, or while they are looking about when first aroused is the only opportunity that the hunter has to shoot them, for as soon as they start there is no hope, as to follow & find them is a task not easily accomplished, for where or how far they go when once started heaven only knows as but few hunters have ever attempted to follow them. At any time they have to be approached with the greatest caution as the least thing will set them on the "qui vive."[3] When not found on these shelves, they are seen on the tops of the highest & most inaccessible hills looking down on the hunter apparently conscious of their security, or lying down on some sunny spot, where they can look all around them. The only times when these animals Can be approached & shot, are when on the shelves, on the cliffs within

3   To be on the *qui vive* is to be on the alert or lookout.

gun shot, & when running from one place to another. Sometimes they will remove from one spot to another only a few hundred yards and it will take the hunter several hours to approach so as to get a shot at them, after having tried a long time to approach them at the first place. They remove thus at times even without seeing the hunter, or even without any alarm, apparently following their own fancy & when the hunter approaches, he finds his game gone & he doomed to another toilsome tramp to make a second approach. I have been thus baffled two or 3 times. The less difficult hills are found cut up with paths made by these animals. They are generally about 18 inches wide; They appear to be as agile as the "European Chamois" is described, leaping over precipices from one to the other, ravines & running up & down almost perpendicular hills. The only places that I could see where they found food was betwen the Cedars. As the places where these animals are most found are barren and without the least vestige of vegetation. From the character of the lands in which these animals are found, their own shyness, watchfulness, it is easy to be seen what the hunter has to Endure, and through what scenes he has to pass to secure these "wild goats." It is one constant time of toil, anxiety, fatigue & danger. Such the country, such the animal & such the hunting

AUGUST 12, SATURDAY

Harris Bell and Owen went after buffaloes, killed 6 cows & brought them home, weather cloudy & raining at times. Provost returned with La Fleur this afternoon, had nothing, saw a Grisly bear. I caught 2 cat fish the *Union* was launched this evening and packing &c is going on. I gave a note to Moncrevier of the animals I wished him to procure for me, and am off to bed. Good night, God bless you all

AUGUST 13, SUNDAY

A most beautiful day, about dinner time I had a young badger brought to me dead. I bought it for two pounds of sugar. The body of these animals is broader than high. The neck is powerfully strong as well as its fore arms and strongly clawed fore feet. It weighed 8½ lbs. Its measurements were all taken. When the badger is interupted by its pursuer from its hole, it erects its hair and at once shows fight. A half breed hunter, I was told, by Provost who, walked to F^t Mortimer, is anxious to go with us down the river But I know the man. I talked about it to M^r C and he tells me that my only place is to pay him by the piece for what he kills or brings on board and in case he did not turn out well betwen this place & F^t Clark, to leave him on shore there. I have sent him word to that effect by Provost this afternoon. Bell is skinning the Badger, Sprague finishing the map of the river made by Squires and the latter is bringing up journals. The half breed has been here, and the following is our agreement:

It is understood that Frances Detallé will go with us & procure for me the following quadrupeds for which he will receive the following prices payable at Fort Union: Fort Clark or F^t Pierre as the case may best suit him & I.

For each Big horn male	$10
For a large Grisly bear	$20
[For each] large male elk	6
Black tailed deer male or female full grown	6
[For each] Red foxes	3
small grey fox	3
[For each] Badgers	2
Porcupines large	2

Independent of which I agree to furnish him with his passage and food as another hand on board. Whatever he kills for food will be settled when he leaves us, or if as he says, when he meets the opposition steamer coming up to F^t Mortimer.

The manner of dressing skins among the Indians is as follows. The skin is first hung over a post and all the flesh taken off with a bone having teeth like a saw, this is done by striking downward on the skin. It is then stretched on the ground & fastened with pegs. Here it remains until dry usually 1 day or two at most. After it is dried the flesh side is cut down with a knife fastened in a bone called a "gralt" which makes the skin even & takes off about ¼ of its thickness. This is done by striking down on the skin. The hair is taken off with the same instrument & in the same manner. This operation is very tedious & fatiguing. After {cutting} the skin to a proper thickness it is covered either with brains, liver, or grease, as the case may be. In this condition, the skin remains over night. The next day the skin is rubbed with a bone or scraped, rubbing the brains in, that are on it, & the rubbing is continued, in the sun or near a fire until nearly dry. Then a cord is fastened to two posts, & over this the skin is pulled & worked untill quite dry. After this with the exception of one end it is sewed together around the edges, a smoke is made of rotten wood in a hole in the ground, over this the aperture is fastened. The skin being suspended on a tripod the smoke completely fills the inside, & it is left over the fire for 20 or 30 minutes, or untill completely smoked, this renders it soft after being wet. Robes are dressed in this way with the exception that the Hair is not taken off. They are not smoked and are generally divided into two parts. A strip of about 3 inches is taken from each piece in the back the whole length of the robe. As the hump would make the robe {uneaven}, the 2 parts after dressing are sewed together and the robe is complete. The scrapings of the skin are boiled with bull berries, making a gluteneous substance, which is considered quite a delicacy among the Indians. The strips cut from the robes are sewed together & make robes for the children, caps, shoes &c. The horns are made into spoons, bowls &c. The bones are pounded up fine with a large heavy stone and boiled in water. The grease that rises to the top is skimmed off & put in bladders. That is the favorite & famous marrow grease, which is considered equal to butter. Sinews are used for strengthening these {b****s} & answers the place of thread. The intestines are eaten. The shoulder blade is used for a hoe. Nothing is lost nothing wasted, & all used. Balls are found in the stomachs of buffaloes, as in our common cattle. Robes are good only from Octr to March.

The Original Field Notebook and the Newberry Partial Copy

The field notebook is held in a private collection. Of the two hundred lined pages of this calfskin notebook, sixty-eight pages contain either journal entries, lists, notes, or sketches. The dates of the entries are July 26–28 and August 16–November 6, 1843. All the contents are in pencil and in Audubon's hand. (See figs. 28–29.)

The Newberry partial copy is written on sixteen ledger-size pages and is held in the Everett D. Graff Collection of Western Americana at the Newberry Library, Chicago. It contains journal entries for August 16–October 19, 1843. The hand appears to be the same as that of the Beinecke partial copy, that of Lewis M. Squires. Only the heading across the top of the first page is in Audubon's hand. It is a well-preserved manuscript. (See fig. 30.)

A comparison of the two manuscript journals indicates that Audubon wrote or sketched in the notebook while in the field or on a boat and later, in more comfortable or convenient surroundings, developed the field entries in greater detail and with fuller expression. I present them below with the initial version of each day's entry from the field notebook first and the later version from the Newberry copy second. I include notes about the text in square brackets to distinguish these from explanatory notes, which appear as footnotes.

FRONT MATTER THAT OCCURS ONLY
IN THE FIELD NOTEBOOK

[Inside front over, in a hand other than Audubon's]

From Gideon B. Smith

to his friend

John J. Audubon

March 13 1843

6. 10. 7. 4½

72.

Wants

1 Ax.—	Frying Pan
1 hoe—	Buffaloe Grease
Coffee—	Marrow Grease
Fish Hooks.—	Snuff!
Butter.—	Coffee Pot.
Vinegar.—	Whiskey?—
1 Bag (leather for stones &ᶜ	

1ˢᵗ DAY—[JULY 26]

Started at 6—same people, Antelopes—Grous—about 10 or 12 Bulls—A Curious Hawk—No Cows at 12.—

Blackfoot ~~Creek~~ River—Abundance of Antelopes but {unsuccessful} with them; they play and curious tricks [****] ran after a Bull—Squire in eminent danger—Another Bull killed by Owen—Owen killed a Cow, 6 Bulls standing. A Swift Fox—Harris and Bell after it—Camp at the 3 Mammelles 30 Miles from the Fort—All well—

Bulls will attack Men When the Cows are killed near them—

2ᵈ DAY [JULY 27]

Beautiful Morn—Good Night sleep, no Musquitoes. Bulls Belowing, Wolves howling, Antelopes snorting—More like a Whistling than the Deer—Snapped 2ᶜᵉ at Antelopes within 30 yards—Shot at a Hare with buck shot—Harris killed it—a fine old female—Buffalo, Cows & Bulls—4 cows killed—Lafleur made a bad shot and I could have shot a Buck had not been for courtesy—Stopped to skin a fine Cow—Bell killed one, Harris did not run the [****ken] blunder of Owen—Start for the Coupe⁴—at 2 o'clock Butchering &c—Antelopes plenty—Harris shot at a Wolf from the Waggon—Lost sight of our Cart.—The Coupe a Curious place thus—[See fig. 31.]

Camp. Scarcity of Water.—Fire of Buffalo dung &c &c Weather cool no Musquitoes—Spread all the meat on the ground—Slept in the tent—

3ᵈ DAY—[JULY 28]

Rose at ½ past 3—Cold = breakfast—Start—Start. 3 on Horseback 1 Horse following—a few Antelopes—reached the ferry at ½ past 7—found all well.— [See figs. 32–33.]

Beavers when shot swiming Sink at once if killed but their body rise again in about 20 to 30 minutes—Otters do the same but remain under for *one hour* or more—

THE PAIRED ENTRIES FOR AUGUST 16–OCTOBER 19

Started from Fort Union at 12 M. on Wednesday Aug 16, 1843 in the Mackinaw Barge Union

Shot 5 young Ducks—Camped at the foot of a high bluff—Good supper of Chickens Ducks &c

Copy of my Journal from Fort Union homeward Commenced Augt 16th at 12 o'clock, the moment of our departure.—

1843
16th August.
Wednesday.

We left at 12 in a Mackinaw Boat 41 feet long and 8 feet wide at the Bottom; and this proved sadly too small, for what with our effects and Barrels of pickled skins as well as ourselves, and Mr Alexr Culbertson, Wife and Child that accompanied us; we were packed in and under the roof as if so many

4 A "gap in the hills," according to Harris (EH 159).

herrrings in a Box. What would you have done, if my Friend Harris' opinion had been followed of having a light boat 25 feet long? but we were under way. The Channel was very narrow and the sand bars nearly extended accrosse the river. We reached the Mouth of the Yellow Stone however without any dificulty, and as we glided in, I shot 5 Young Wood Ducks.—A short time afterwards we landed to take in a curious pair of Elk Horns and at nearly Sun set we stopped and encamped at the foot of a high Bluff, which was truly dangerous if Indians had been on or near the River.—Our supper was a good one, consisting of Chickens brought by Mr Culbertson and the Young Ducks &c—The weather was pleasant and we all slept well though closely packed.—

17th Thursday

Started early. Saw 3 Bighorns—antelopes and fully 20 Deer.—22 swans.—1 wolf. Ducks. Stopped a short time on 2 Bars—Shot a female Elk (Mr C) I killed 2 Bulls.—Camp.—Buffaloe Bluff, Tamias quadrivitatus. Bear Wallow.

17TH THURSDAY.

Started early; saw 3 Big Horns, Some Antilopes and fully 20 Dear & 22 Swans, 1 Wolf and some Ducks, stopped a short time on a sand bar. Mr C. shot a female Elk, and I Killed 2 Buffalo Bulls, we encamped at this spot and called it Buffalo Bluffs.—In the evening we killed Tamias quadrivitatus,[5] and saw several Grisley Bears wallowing places.—These are large and round and about 1 foot to 18 Inches deep.—

18th Friday

Fine. Bell snapped at a superb male Elk.—4 Catfish—the 2 Bulls untouched since killed.—Stopped to make an oar—

Kayac the french Missourians name for Buffalo Bulls—Orignal french for Moose—

[Measurements of *Tamias quadrivittatus* given here.]

5   The four-lined ground squirrel of the Royal Octavo *Quadrupeds.*

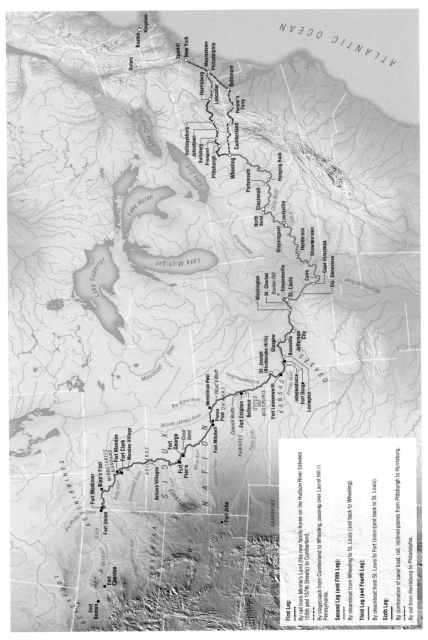

FIG. 1. The Missouri River expedition. Map by Erin
Greb, cartographer, Doylestown, Pennsylvania.

FIG. 2. John Woodhouse Audubon and Victor Gifford
Audubon, *John James Audubon*, 1841. Image #1822,
American Museum of Natural History Library.

FIG. 3. Front view of Maria Audubon's Salem,
New York, home. Courtesy of Lance Spallholz.

FIG. 4. Rear view of the Salem home (the
burn barrel would be just out of the frame to
the left). Courtesy of Lance Spallholz.

FIG. 5. The sisters Maria Rebecca (*seated*) and
Florence Audubon. Courtesy of Lance Spallholz.

FIG. 6. Edward Harris, photograph of a daguerreotype,
undated. Courtesy of the Alabama Department
of Archives and History, Montgomery.

FIG. 7. Isaac Sprague, self-portrait, undated.
Courtesy of the Wellesley Historical
Society, Wellesley, Massachusetts.

FIG. 8. *John James Audubon*, 1843. Drawn by I. Sprague, New York, Feb. 1843. Mass Audubon Collection, Museum of American Bird Art, Canton, Massachusetts; gift of Priscilla Sprague, 1974.

FIG. 9. J. C. Wild, *View of Front Street, St. Louis,*
1840. Missouri History Museum, St. Louis.

FIG. 10. John James Audubon, J. T. Bowen, lithographer, "Pseudostoma Bursarius, Canada
Pouched Rat," 1845, hand-colored lithograph, plate 44 of *The Viviparous Quadrupeds of
North America*. Courtesy of the Jule Collins Smith Museum of Fine Art, Auburn
University; Louise Hauss and David Brent Miller Audubon Collection.

PLATE LXXVIII.

CERVUS MACROTIS, SAY.

BLACK-TAILED DEER.

FIG. 11. (*opposite top*) Karl Bodmer, *Snags on the Missouri River*, 1833. Courtesy of the Joslyn Art Museum, Omaha, Nebraska.

FIG. 12. (*opposite bottom*) John Woodhouse Audubon, J. T. Bowen, lithographer, "Cervus Macrotis, Mule Deer," 1846, hand-colored lithograph, plate 78 of *The Viviparous Quadrupeds of North America*. Note that the artist John Woodhouse, has painted in his father as the hunter figure to the right. Courtesy of the Jule Collis Smith Museum of Fine Art, Auburn University; Louise Hauss and David Brent Miller Audubon Collection.

FIG. 13. (*above*) John James Audubon, J. T. Bowen, lithographer, "Spermophilus Ludovicianus, Prairie Dog," 1846, hand-colored lithograph, plate 99 of *The Viviparous Quadrupeds of North America*. Courtesy of the Jule Collins Smith Museum of Fine Art, Auburn University; Louise Hauss and David Brent Miller Audubon Collection.

PLATE LVII.

BOS AMERICANUS, GMEL.
AMERICAN BISON OR BUFFALO.

FORT UNION . U. Mo.
From the opposite bank of the Missouri .
Drawn by I. Sprague — July 1843.

FIG. 14. (*opposite top*) John James Audubon, J. T.
Bowen, lithographer, "Bos Americanus, American
Bison or Buffalo," 1845, hand-colored lithograph,
plate 57 of *The Viviparous Quadrupeds of North
America*. Courtesy of the Jule Collins Smith Museum
of Fine Art, Auburn University; Louise Hauss
and David Brent Miller Audubon Collection.

FIG. 15. (*opposite bottom*) Isaac Sprague, sketch
of Missouri River, June 1, 1843. American
Museum of Natural History Library.

FIG. 16. (*above*) Isaac Sprague, *Fort Union,
U. Mo.*, 1843. Neg. #1980.15, collection of
the New-York Historical Society.

FIG. 21. John James Audubon, *Camp at the Foot of the Three Mammelles, Thursday July [27], 1843.* American Museum of Natural History Library.

FIG. 22. Isaac Sprague, sketch of encampment on Missouri River, September 2, 1843. American Museum of Natural History Library.

FIG. 23. John James Audubon, J. T. Bowen, lithographer, "Vulpes Velox, Swift Fox," 1845, hand-colored lithograph, plate 52 of *The Viviparous Quadrupeds of North America*. Courtesy of the Jule Collins Smith Museum of Fine Art, Auburn University; Louise Hauss and David Brent Miller Audubon Collection.

FIG. 24. Isaac Sprague, sketch of encampment on Missouri River, May 26, 27, and September 16, 1843. American Museum of Natural History Library.

FIG. 25. John Woodhouse Audubon, *John James Audubon*, 1843. Image #1498, American Museum of Natural History Library.

FIG. 26. (*opposite*) John James Audubon, first page of the Beinecke partial copy of Audubon's original 1843 journals, August 5, 1843. General Collection, Beinecke Rare Book and Manuscript Library, Yale University.

Aug 5
Saturday

The weather was cool, the wind at North East & the morning menacing rain. We moved towards the soon as now come [...] turned on our way. Harris shot a pigeon of the sort, & [...] could have camped when had we not been afraid of his running into some hole. We saw only a few holes, antelope & wolves, the White horse that had gone out as a hunter, returned as a pack horse, loaded with the entire flesh of a buffaloe cow and our 2 mules drew the heads of 4. This morning, at day light when we were called to drink our coffee, there was a buffalo bull feeding within 20 steps of our tent, and it removed slowly towards the [...] as we busied ourselves to make preparations for our departure. We reached the Fort at 12 & had a good dinner. Squires, Provost & La Fleur had returned, they had wounded a Big horn & [...] Came Bell returned this afternoon, they had seen 20 Cocks of the plains, but brought an Elk skin female, and a female porcupine as well as a White headed Eagle. Buffaloes become so poor during hard winters when the snows cover the ground to the depth of 2 or 3 feet, that they lose their hair, become covered with [...] & cures, on [...] the magpies feed, and the poor beasts die by thundreds. One can hardly conceive how it happens, that notwithstanding their general deaths and the immense numbers that are murdered daily on the immense wastes, called prairies, that so many are yet every day to be found. Sprague never spoke to me or was retained, but this evening Bell related all his adventures to every one [...] We all [...] off to bed pretty well fatigued. Bell [...] & Squires [...] I had a couple more Big Horns skulls [...]

Swift Fox Measurements of the young swift, male [...]
From nose to root of tail 12½, tail [...]
at shoulder 7¾, width by [...] tread [...]
Ear from opening 1¾, Nose to ear [...]
between ears [...] 1¾, Letter [...]
[...] the skin of this animal [...]
[...] did not run best of [...]
[...]

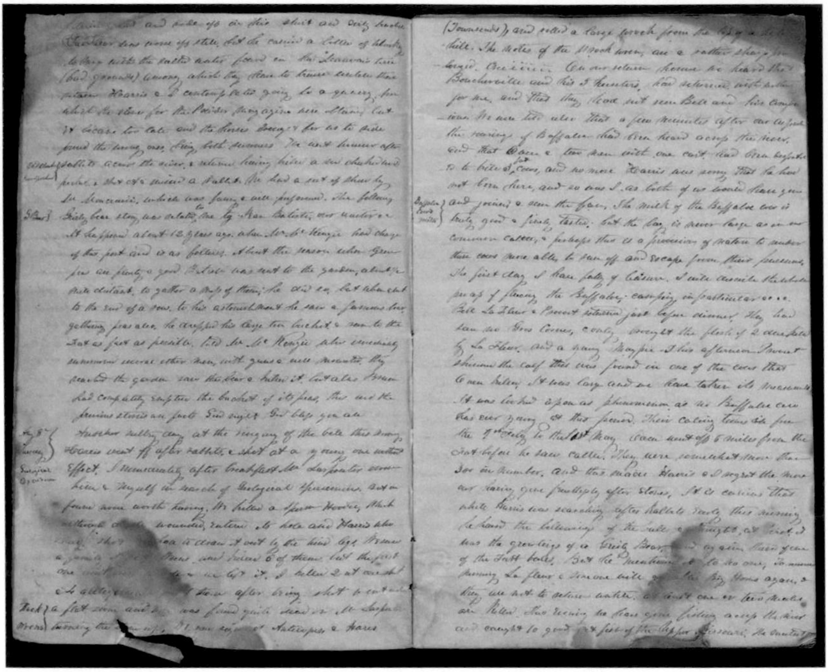

FIG. 27. John James Audubon, pages 8–9 of the
Beinecke partial copy of Audubon's original 1843
journals, August 8, 1843. General Collection, Beinecke
Rare Book and Manuscript Library, Yale University.

FIGS. 28–29. (*opposite*) John James Audubon,
original field notebook, 1843. Private collection;
photos courtesy of Eugene Beckham.

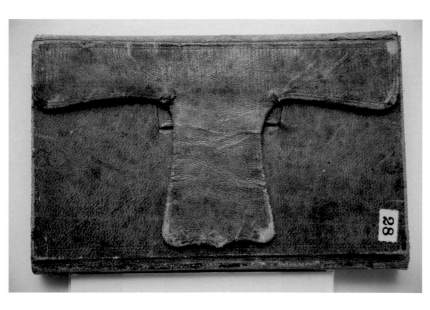

28

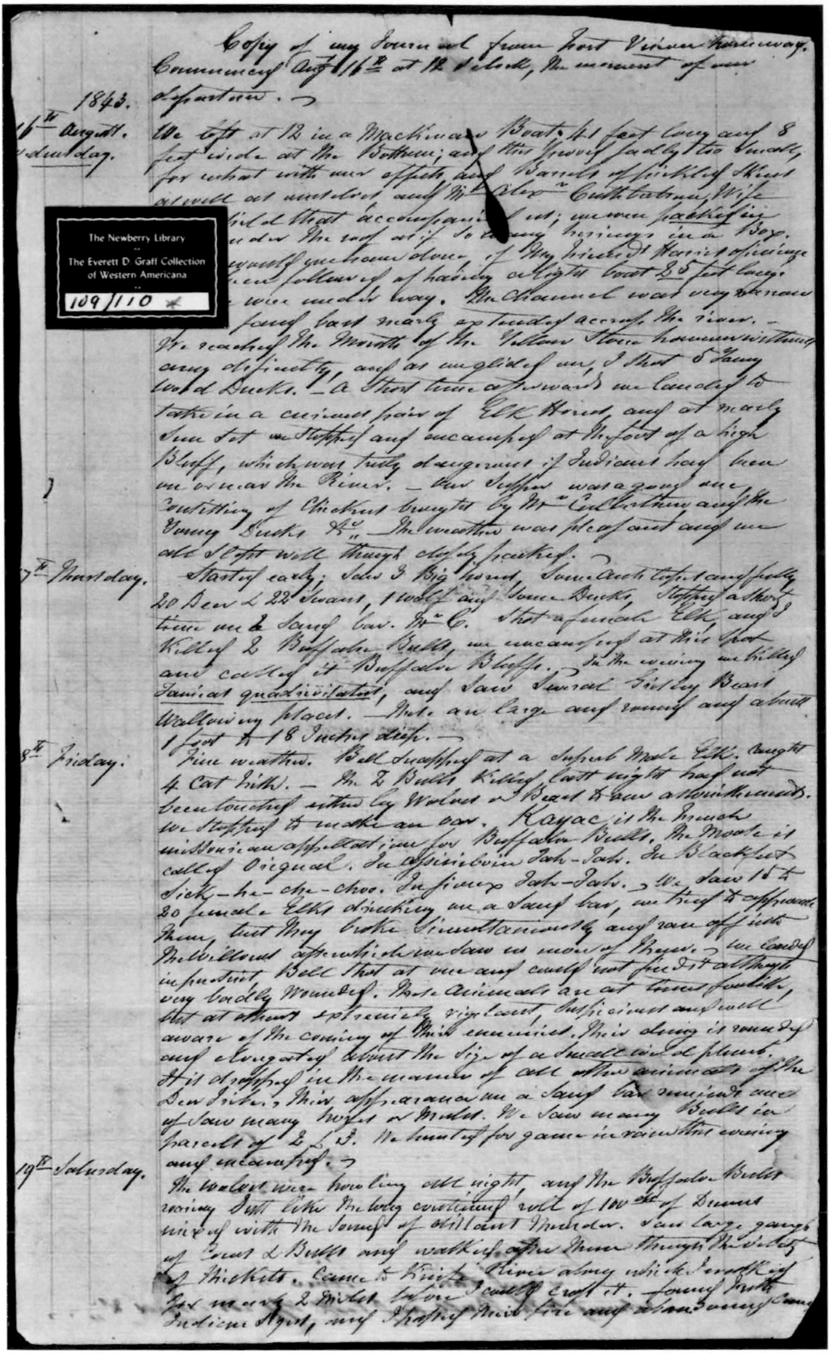

FIG. 30. John James Audubon, "Copy of my Journal from Fort Union homeward." MS, the
Everett D. Graff Collection of Western Americana at the Newberry Library, Chicago.

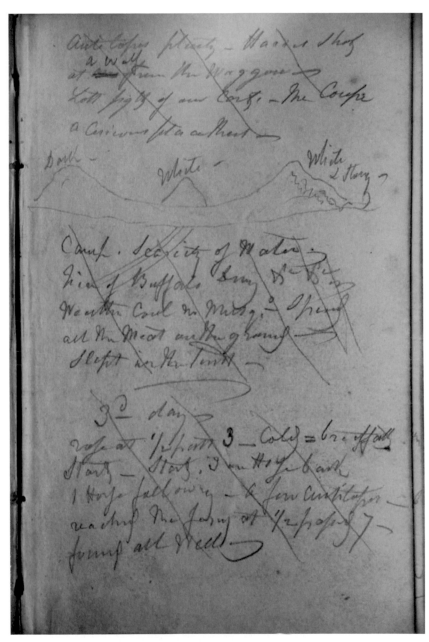

FIG. 31. Audubon's field sketch of "The Coupe," July
2, 1843, original field notebook. Private collection.

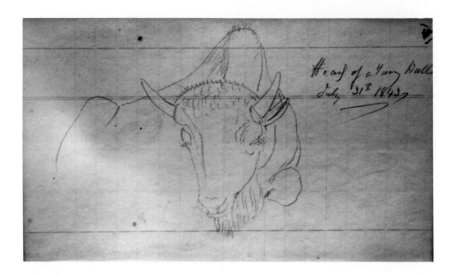

FIGS. 32–33. Audubon's two field sketches
of dead bison, July 31, 1843, original field
notebook. Private collection.

FIGS. 34–38. (*above and following spread*) John James Audubon's powder horn from the Missouri River expedition, 1843. Private collection; photos courtesy of Eugene Beckham.

FIG. 36

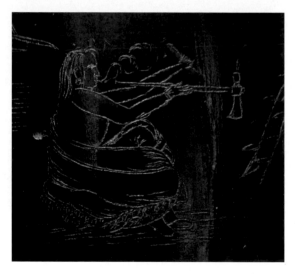

FIG. 37

FIG. 38

FIGS. 39–42. (*above, opposite, and following spread*) Edward Harris's powder horn from the Missouri River expedition, 1843. Photos courtesy of the State of Alabama Department of Archives and History, Montgomery.

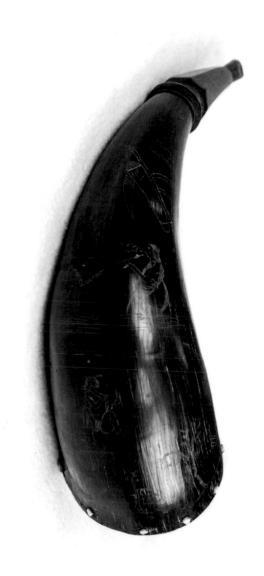

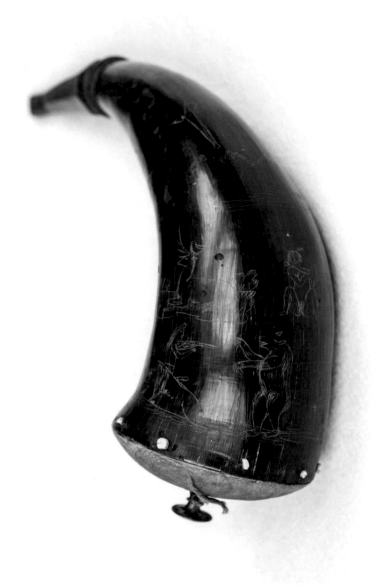

FIG. 41

FIG. 42

FIGS. 43–44. (*opposite and above*) The Catlin powder
horn and sketch of the engravings. Reproduced with
permission from the Luzerne County Historical
Society and Museum, Wilkes-Barre, Pennsylvania.

FIG. 45. George Catlin, *Buffalo Chase*. Reproduced
with permission from the Luzerne County Historical
Society and Museum, Wilkes-Barre, Pennsylvania.

FIG. 46. The fourth St. Louis–style powder horn. The powder horn came into the
possession of the current owner (Paul B. in eastern Texas) through his mother,
who traded an antique bed for it. The trade occurred in the St. Louis area in the
1960s. No other information is available on the horn's origins. Photo courtesy
of Rick Sheets, journeyman horner, Honourable Company of Horners.

*In Assiniboins Tah-Tah—In Blackfeet Sick-e-chi-choo. In Sioux Tah-
Tah—15 to 20 female Elks drinking tried to approach them, but they broke
and ran off to the Willows and disapeared—Landed in Pursuit. Bell shot
at one but did not find it although very badly wounded—These animals
are at times foolish, but at other Vigilant, suspicious and well aware of the
coming of their enemies—Dung = many Bulls in parcls of 2 or 3—Vain
search for game—Camped*

18TH FRIDAY.

Fine weather. Bell snapped at a superb Male Elk. Caught 4 Cat Fish.—The
2 Bulls Killed last night had not been touched either by Wolves or Bears
to our astonishment. We stopped to make an oar. Kayac is the French
missourian appellattion for Buffalo Bulls. The Moose is called Origual.
In Assiniboin Tah-Tah. In Blackfeet Sick-he-che-choo.—We saw 15 to 20
female Elks drinking on a sand bar, we tried to approach them, but they
broke simultaniously and ran off into the willows after which we saw no
more of them.—We landed in pursuit, Bell shot at one and could not find
it although very badly wounded. These animals are at times foolish, but at
others, extremely vigilant, suspicious and well aware of the coming of their
ennemies. Their dung is rounded and elongated about the size of a small
wild plumb. It is dropped in the manner of all other animals of the Deer
Tribe.—Their appearance on a sand bar reminds one of so many horses or
mules. We saw many Bulls in parcels of 2 & 3. We hunted for game in vain
this evening and encamped.—

19th Saturday

*Wolves howling, and Bulls roaring. (just like the long continued roll of 100$\underline{ds}$
of Drums—Saw gangs of Buffaloes—Walk after Cows & Bulls along the
River—Headed Knife River 1½ mile, fresh signs of Indian, Burning Ashes.
1 Cow killed by M$\underline{r}$ C, and I knocked her down with 2 Balls—abundance
of Bear Tracks—*

[Measurements of a bison cow given here.]

Saw a great number of bushes bearing the berries of which Mrs C. has given me a necklace.—abundance of Buffaloes on the Prairies.—Mr C. Killed another cow—I went to see it, and had a severe Fall on a log.—Bell killed a Doe and wounded 1 Fawn—

19TH SATURDAY.

The wolves were howling all night, and the Buffalo Bulls roaring just like the long continued roll of 100^{ds} Drums mixed with the sound of distant Thunder. Saw large gangs of Cows & Bulls and walked after them thrugh the sides of Thickets. Came to Knife River along which I walked for nearly 2 miles before I could cross it. Found fresh Indian signs, and I passed their fire and abandoned camp. The ashes were still smoking, and I found a bag containing some tabako that had been forgotten.—Saw Mr C. going on before me, and presently heard the report of his rifle.—He had wounded a Cow and I shot 2 balls in her to dispatch the poor beast. It was tolerably fat, and made us good food. We saw abundance of Bear tracks, saw great many bushes bearing the berries of which M^{rs} C: has given me a necklace, Abundance of Buffaloes on the Prairies. Mr C. Killed another Cow. I went to see it; and had a severe fall on a log. Bell Killed a fine Doe and wounded her Fawn.—[Measurements of *Tamias quadrivittatus* given here.] Lives in burrows in high clay cliffs and ravines, ascends trees where they probably nestle.—very active and gaily in movements, abundant about here.—

Sunday, August 20

Tamias quadritatus runs up trees—abundance of them in the Ravine. H. killed one—B. wounded an Antelope. Saw many thousands of Buffaloes. The roaring of these Animals resembles the grunting of Hogs, with a rolling Sound from the throat, that when loud is heard for Miles probably. They bellow with the head down, tearing the Earth up the while. Mr C. killed 2 Cows—Sprague killed 1 Bull & made 2 sketches after death. The Men killed 1 Cow, but her Bull would not leave her although shot 4 times through. Stopped by the Wind all this day—Suffered much from my fall.—

20TH AUGUST, SUNDAY.

Bell wounded an Antilope, saw many of those truly beautiful animals, saw thousands of Buffalo. The roaring of them resembles the gruntings of Hogs with a heavy rolling sound from the throat. M̲r̲ C Killed 2 cows. Sprague killed 2 Bulls and made 2 Sketches of them as they laid dead on the ground. The Men killed 1 cow, but her Bull would not leave her although shot through the body 4 times.—Stopped for wind all day; I suffered greatly from my fall.—

Monday, August 21

Abundance of Buffaloes over all the Bars and Prairies—Many Swiming. Stopped by the Wind at 8 o'clock.—Breakfasted where tracks of Bears. hundred of Buffaloes, We left our safe anchorage & good hunting ground too soon—The mom̲t̲ we had passed the point below us the wind blew high, and we were obliged to land again on the opposite shore where the wind has annoyed us—Bear tracks led us to search for them but in Vain—Collected seeds—Shot at a Rabbit, but have done nothing. Saw many young & old Ducks. Malards—Black & Gadwalls—I shot a Bull and broke his thigh. Shot at him 13 times afterward.—Camped at the same place—

21ˢᵗ MONDAY.

Abundance of Buffaloes on the Sand Bars as well as on the Prairies and many swiming from side to side.—Stopped by the wind at 8, many Bear tracks around us. Searched for these animals in vain, collected fuel and shot a small Rabbit, but nothing done.—M̲r̲ C and I walked up a small creek with many Ducks, Mallards, Gadwalls &̲c̲ old and young. I shot at a Bull in the thigh and broke it but not withstanding, it was shot at afterwards no less than 13 times, Bell assisting us.—remained here tonight.

Tuesday, August 22

Left early and travelled about 12 miles.—Went hunting Elks.—M̲r̲ C. Killed a Deer and brought the meat with Squire on their backs—I saw nothing, but heard 4 shots running which I thought were from Harris—I ran for upward of a mile to look for him, hallowing the whole distance but heard nothing of him. Sent 3 Men who halowed also, but came back without further intelligence.

The fuss was 4 shot at an Elk and missed—Bell shot a female and brought a portion of the meat—We Walk^d to the little Missouri and I shot the 4th Bull this trip. We crossed this stream and brought the Nerf.—Saw Ducks—

Started this Afternoon and went below the little Missouri, *returned to the Bull and took his Horns &^c—On going to the Boat Sprague saw a G. Bear. We went toward the spot. The fellow had taken under the high bank, and it was killed in a few seconds. Mr C. shot it first through the neck. Bell elsewhere and I in the belly. Sprague said that he never promised us* anything. *Said that if he saw another Bear he would not Mention it to us &^c &^c and was quite abusive in other words.—I told him that he might go any direction he chose and he said he would depart tomorow morning, but of this I am doubtfull.—*

22^d TUESDAY.

Left early and travelled about 12 miles when we stopped for wind. Went hunting Elks. M^r C killed a Deer and he and Squire brought the meat on their backs. I saw nothing but heard four shots running, which I thought were from Harris. I ran for upward of a mile hollowing the whole distance but nothing of him. Sent 3 men who hunted for him but brought no intelligence. The firing was at an Elk which was missed. Bell shot a female and brought a portion of the meat. Mr C., Squire and I walked to the little Missouri, and I shot the the 4th Bull I have killed this trip. We crossed the stream by wading and brought the "*Nerf.*"[1] Saw some Ducks.—We started in the afternoon and landed below the little Missouri—returned to the Bull and took off its horns &^c. On returning to the boat, Sprague informed us that he thought he had seen a Grisley Bear. We went toward the spot. The fellow had taken its path below the high Bank, and we all fired at it

---

1   The *Oxford English Dictionary* gives "nerf" as an obsolete term for a nerve or tendon. The context here suggests that it refers to a choice piece of meat, possibly from along the spine, like the loin or backstrap of a deer.

in so quick a manner that it was killed in a few seconds. Mᴿ C. shot it first through the neck, Bell 2 Balls elsewhere and I 1 in the belly. Sprague abused me greatly, but let it pass! We had sat about a dozen steal traps to catch Prairie Marmots as there was a large village of them but caught none.— These animals are extremely cunning and in all probability did not go out of their burrows that [****] The wind was high and cold. Provost skinned the Bear which proved poor and about 3 or 4 years old, weighd probably 400ᴸᴮ. Bell shot 2 Prairie Marmots. Their notes resemble precisely those of the Arkansas Flycatcher. Left this afternoon and travelled about 10 miles. Saw another bear on the bank which was bent on swiming to a dead Bull fastened on a log.—He saw us, reared on his hind feet, gave a grunt, leaped on the upper bank and disappeared in a moment. Several wolves were also desirouss to go to the drowned Buffaloes but we frightened them all off. We saw wild pigeons and encamped in a bad place with appearance of rain.—[Measurements of male grizzly bear given here.] When Buffalo bulls are in large herds their belowing, may be heard probably for miles. They below with the head inclined downward, pawing and tearing up the Earth all the while.—

23d Wednesday

Provost skinned the Bear. No Prairie Dogs caught.—The Wind high and Cold.—2 Prairie Dogs shot.—Their notes resemble precisely those of the Arkansa fly catcher!—Left this afternoon and travelled about 10 miles— Saw Another Bear, its movements—Several Buffaloe Drowned. Wolves. Wild Pigeons.—Camp in a bad place, appearance of rain.—

23 AUGUST WEDNESDAY.

Provost and Bell finished skinning the bear and its fine skin was put in a Barrell with strong pickle made of salt and water, an excellent method to preserve all the skins of larger animals.—

The wind continued high and cold. We left and only travelled about 10 miles.

24th Thursday

A Bad night of wind and cloudy—left early as it became calm.—passed L'Ours qui dance—travelled about 20 miles and stopped by the wind—hunted but found nothing. Harris Caught 1 fish. {Novoborcis} 7 Bottles of oil from our Bear! Heard what was thought Guns but I thought that it was the falling in of the Banks.—Then the wolves of both Kinds (large & small) howled so curiously that it was supposed they were Indian Dogs. We went to Bed all prepared for Action in Case of an attack. Pistols Knifes &c but I slept well though very cold.—procure curious stones with impressions of shells.

24th THURSDAY.

A bad night of wind and cloudy. Left early as it became calm. Passed the Hills called emphatically "Lours qui dance," The Dancing Bear Hills, thus called because it is reported by the Indians that a bear was seen to dance as if music had been played to it on the occasion? All of which is a mere superstitious legend of the poor deluded Indians.—We travelled about 20 miles and were again stopped by the wind. Hunted but found nothing. Harris caught a specimen of {Vespestatio muoboracencis}.—The fat of the bear was rendered and produced only 7 Bottles. We heard what was thought reports of guns, but I thought that it only the noise produced by the falling in of the Bank of the river, which indeed at times any one not accustomed to such sounds would easily take them for gun shots. Then the wolves of both kinds, large and small (that is the common and the Prairie Wolves) howled so curiously and were in such large numbers that it was supposed the hubbub was the noise of a large party of Indian dogs along with their red masters. We all went to sleep however well prepared for action in case of need, our Guns and pistols loaded with Balls and our knifes all close to our bodies, but I slept well though cold. We procured some stones with impressions of curious shells.—

25th Friday

Fair, but foggy did not start early.—quite calm.—passed the 2 Rickaris Winter Villages.—Many Eagles & Peregrine Falcons.—Shot another

Bull.—Bell gives little credit to any other than himself. Sprague has not open his lips to me, and I hope he will remain with the Mandans! Passed the Gros Ventre Village at 12.—No Game about this place.—La Main Gauche an Assiniboin Chief of great renown Lost 70 Warriors killed and 30 Wounded the Year following the Small Pox on the Prairie opposite, and lost his renown—The Gros Ventre are a courageous Tribe.—Reached the Mandan Village, hundreds of Indians swam to us with Turbans around their Heads—our old Friend the 4 Bears met us on the shore—I gave them 8 pounds of Tabacco—he came on board and goes down with us to the fort—reached Fort Clark at 4. received 3 presents froom M͟r C.—Sprague attempted to stay there but was defeated by M͟r C's contrivance.—M͟r C. & Squire rode out in the dark with 4 Bears to the Gros Ventre Village and return͟d at 11. Curious Incidents—gave 10ᴵᵇ tabacco to a chief called The Iron Bear.

2 5 t͟h͟ FRIDAY.

Fair and calm but foggy, and therefore did not start early. Passed the 2 Ricaree winter villages but not an Indian to be seen.—Many Eagles and Perigrine Falcons.—Shot another Bull. Bell gives little credit to any person but himself.—Sprague has not spoken to me since the death of the Bear! We passed the Gros Ventre village at 12. No game about this place—"*La Main Gauche*" an Assiniboin chief of great renown has had a Battle with the Sioux and lost 70 warriors killed and 30 wounded, the year following the Small Pox on the Prairie opposite and lost his renown.—The Gros Ventre are a courageous nation. Reached the Mandan Village. Hundreds of Indians swam out to us with pieces of cloth tied around their heads in the manner of Turbans. Our old friend the *Four* Bears met us on the shore and remembered us all at once. I gave him 8ᴵ͟ᵇ of tabaco and he came on board of us to be taken down to Fort Clark, and we reached that place at 4 of the afternoon. I received 3 presents from M͟r C. Sprague attempted to remain there but was defeated by M͟r. C. contrivance. M͟r C., Squire and "Four Bears" rode out in the dark to the Gros Ventre's village—and returned to the boat at 11.—We had some curious incidents that evening and during the night.—We gave 10$\underline{\text{lb}}$ pound of Tabaco to a chief called the *Iron Bear*. I bought a Basket, visited several lodges and eat of several Indian Messes, some of which were pretty good. We had presents of Corns, dried green pumpkins &$\underline{\text{c}}$ and my Swift Fox was brought on board. This animal was presented to me when on my way up the river, by M$\underline{\text{r}}$ Chardon one of the Company's partners. It was kept in a garret where it fed itself by procuring rats only, but those animals are now very attendant at all the forts of the Company having been brought up there among the goods of their anual cargoes.

26th Saturday

fine but so much as windy started early. Landed to Breakfast—Sprague's surliness. The wind rising—a Canoe passed us with 2 Men of the Opposition.—Stopped by the Wind for 4 hours—Diner—Started at 3. passed the Bute Quarré at ¼ past five.—followed all the while by the Canoe.—Those 2 fellows are afraid of Indians and want to come on board of our boat. Landed for the night, and went on top of one of the Butes from which a fine extensive view is found. Saw a Band of Buffaloes—Approached them but by accident did not kill a Cow. Harris whom we thought far off, shot at them first, and Moncrevié and us lost our chance.—Elks whisling.—The Canoe Men landed close to us. Saw many swans.—

26$\underline{\text{th}}$ AUGUST, SATURDAY.

Fine but somewhat cloudy, and I gave orders to start very early, and M$\underline{\text{r}}$ C was sorry for it, as he thought that had we remained a few hours longer, I might have had some fine presents.—Landed to Breakfast.—Sprague was very surly. The wind rose and a canoe passed us with 2 men belonging to the opposition Company. Stopped by the wind for 4 hours, dined and started again at 3, passed the "Bute Caree" followed all the while by the canoe men; those fellows undoutedly afraid of the Indians were glad to place themselves under our protection as it were. They were anxious to come on board of us but it was no go. Landed for the night and went to the top of a "Bute" or

Hill from which we had a fine extensive view of the country around. We saw a band of Buffaloes which were approached but by some accident did not kill a cow. Harris whom we thought far off, shot at them first, and Mr Moncrevié a Clerk of the Company taken in at Fort Clark and our selves lost our chances. We heard Elks whistling. The canoe men landed close to us. We saw many swans.—

27. Sunday

Started early in Company with the canoe.—Wolves 4. Bulls 6. 4 together to our sorrow, as those are not fit to eat at this season, being poor & stinking. The wind rather high.—Saw a Gang of Cows. Landed badly in the wind, assisted by the Canoe Men.—Vain search after the Cows, although we saw their wet tracks *and followed those for 1 mile at least. Gave up the chase—On returning to the boat missed her because she had removed below.—Walked up and then downward and at last reached her.—Mr C. worked at the Parflesh with Golden Eagle feathers killed by himself.—*

Stopped by the wind at 12.—Walked off and saw Buffaloes—but the wind was adverse.—Bell & Harris killed a Cow, a single one that had been wounded whether by shot or by an arrow no one can tell.—The Cow sound however.—We saw a bull on a Sand Bar.—The fool took to the Water and swam so as to meet us. We shot at it about 12 times about the head.—I shot it through the Nose. *He was shot by Sprague thrugh one eye.—Bell and Harris thrugh the head &^{c} and yet this Bull made for our boat and came until Mr C. touched it with a poll.—he turned round and went off across the River, but acted as if Wild and Crazy. He ran on the bars and at last swam again to the opposite shore, in my opinion to die.—But M^r C says he may live for a month!—We landed in a good harbour on the east side about 1 hour before sun down.—I was amused at Sprague bringing some plants on board and hiding them sudenly although close to me at the time.—he is in fact as strange a being as the Bull we have just now shot at!*

27 AUG, SUNDAY.

Started early in company with the canoe, saw 4 wolves, 6 Buffalo Bulls, but these animals are unfit to eat at this season being very poor and stinking. The wind rather high. Saw a gang of Buffalo cows; landed in the mud and followed their tracks for more than one mile, and gave them up. Meantime our boat had been removed and we had some trouble in finding her.—M͟r C worked at the par flesh with Golden Eagle tail feathers killed by himself. Stopped by the wind at 12. Walked off and saw buffalo but the wind was adverse. Bell and Harris killed a cow that had been wounded, but sound. We saw a bull on the sand bar, which took to the water, and swam directly toward us. We shot at him about 12 times. I shot it through the nose. Sprague shot it through one eye. Bell and Harris shot it through the head, and yet the Bull made for our boat, until M͟r C touched it with a pole [****t] it swam accrose the river, but acted as if crazy. It ran on the bar for a while, then recrossed the stream and laid down to die, although M͟r C said it may live for a month. We landed on the East bank to spend the night. I was amused at seeing Sprague bringing some plants on board and hidding them. He is in fact as strange a youth as the Bull we shot at.—

28. Monday.

A Gale all night and this morning also.—Bulls have been shot through both eyes.—3 Balls through the heart and 7 shots besides, when to kill them the knife was obliged to be employed. On the Ice after having cut both ham strings, they have crawled on their forefeet for hundreds of yards.—

We are in a good place for hunting and I hope to have more to say anon.—The Men return͟d and told us of many Bear tracks &c and 4 of us started—Such a Walk I do not remember. It was awfull—Mire, Willows—Vines, &c &c &c—We returned much fatigued and having seen nothing.—The Wind blowing a Gale.—a Curious story about the Buck M͟r C shot on Sunday last, having found out that it was a Black tailed Deer Male.—

28th AUGUST, MONDAY.

It blew a gale all night and continues now.—Buffalo Bulls have been shot through both eyes, 3 Balls through the Heart, and 7 shots besides, when to kill them the Knife has been obliged to be used.—

On the Ice after having cut both Hamstrings, they have been seen to crawl with their fore feet for hundreds of Yards.

29th Tuesday.

Heavy Wind all night. Bad dream about my own Lucy—Caught Catfish and walked along the shores—2 Deer on the other shore.

Cut a Cotton tree to breake the force of the Waves—Sprague apologizing to me, we shook hands and now all will be forgotten.—The Weather sultry. Beavers during Winter often times come down amid the Ice, but enter all the small streams they meet with at once.—Apple River or Creek was formerly a good place for them as well as Cannon Ball River.—Saw a Musk—rat this morning swiming by our barge.—We run {and} about 1 Mile below and on the opposite side—slept on a Muddy Bar with abundance of Musquitoes.—

29 TUESDAY.

The gale still continues. Had bad dreams about my Lucy.—Caught *cat fish*, and walked along the shores. Saw 2 deer on the opposite side. It was discovered to day that the deer killed by Mr C last Sunday was a buck of the black-tailed deer.

Heavy wind all night. Bad Dreams again about my Lucy. Cut a cotton tree to break the force of the waves. Sprague apologising to me, we shook hands and now all will be forgotten, on my part at least. The weather sultry

Beavers during winter come down on the ice, but enter all the open streams they meet with at once.—Apple River or Creek was formerly a good place for these animals as well as Cannon Ball River.—Saw a musk rat this morning swiming along side of our Boat.—We ran on about 1 mile below and slept on a muddy Bar with abundance of musquitoes.—

30ͭͪ Wednesday—

Started at day light and fell down about ½ Mile.—Mͬ C. and I went off to the Prairies through the most dammable ground I ever saw but we reached the high prairies by dint of Industry through swamps and mires—We at last reached the prairies, and saw 2 Bulls, 2 Calves, and 1 Cow.—We killed the Cow and the larger Calf, a beautiful Bull.—returned to the boat through the most abominable swamp I ever travelled in.—reached the Boat at 1 o'clock, thirsty and hungry enough.—

Dispatched Bell and the whole of the men after the meat and the skin of the young Bull.—I shot the Cow in the {head} but missed the second Calf by shooting above it the second shot, although my first will oblige it to die before this day's sun set.—We sent the Men and Bell after the meat of the Cow and the skin of the Calf which was brought on board.—We started and travelled about 10 Miles.—

30ͭͪ AUGUST, WEDNESDAY.

Started at day light and went down the stream about ½ a mile. Mͬ C & I went off to the Prairies through the worst grounds we had yet visited; but we did reach the high Prairies by dint of great Industry and no small labour through swamps and mire.—When there, we saw 2 Bulls, 1 cow and her 2 Calves. We killed the cow and the larger Calf, a beautiful bull.—Returned to the boat by a far worse way than the one we had passed when going out. Reached the boat at 1 o'clock, thirsty and hungry enough.

Dispatched Bell and all the men after the meat, and the skin of the Calf, all of which was brought on board. We started and travelled about 10 miles.

31. August. Thursday.

Started early, fine & Calm.—large flocks of Ducks—Swans, 4 Wolves.— passed Mr Primaux' Winter trading house, reached Cannon Ball River at ½ past 12. No Game, Water good tasted but warm.—Dinner on shore.— Saw a Rock Wren on the Bluffs here.—Saw the Prairies on fire. Signs of Indians on both sides. Weather Cloudy & hot, reached Beaver Creek.

Provost went after Beavers. No signs.—Caught 14 Cat fish.—Power of
Buffaloes through the Mire.—

31ˢᵗ AUGUST, THURSDAY.

Started early being fine and calm. Large flocks of Ducks (probably some
millions together, and all of the Pintail species). Saw swans and 4 wolves.—
passed Mr Primaux winter trading house.—reached Cannon Ball River at
½ past 12. No Game. Water good but warm. Dinned on the shore, saw a
Rock Wren here, and left our best axe. The prairies in the distance were all
on fire, splendid sight, though so often seen previously by me.—Sight of
Indians on both sides the River. Weather cloudy and sultry. Reached Beaver
Creek. Provost went after Beavers but found no signs. Caught 14 cat fish
all most excellent. Then we witnessed the power of Buffaloes upon passing throught miry places which they do to an unacounttable and almost
miraculous extent.

1ˢᵗ *September Friday.*

Hard rain most of the night.—uncomfortably hot—left our encampment at
8 o'clock, saw cattle and landed, went off & found only 4 Bulls.—returned
to Boat & started.—After being landed for the night on a large sand bar
connected with the main, saw a gang of Buffaloes. Mr C. and Another
went off, Shot at 2 Cows, Killed one but lost her as she floated down the
stream and it was near dusk.—

A heavy cloud arose to the West. Thunder was heard, but the Moon
and Stars shone brightly. However after midnight it came on rain, and
the weather afterwards grew fine again at Intervals.—The Musquitoes
too abundant for comfort.—

SEPᵗ 1ˢᵗ FRIDAY.

Hard rain most of the night, and uncomfortable Camp, which we left at 8
o'clock, Saw Cattle & landed, went off and found only 4 Bulls, returned to
Boat and started. After being landed for the night in a large sand Bar connected with the main land, saw a gang of Buffaloes. Mr C and another man

went off. They shot at 2 cows and killed one, but she was lost as she floated away down stream, and it was now too dark to follow after her.—A heavy cloud arose from the west. Thunder was heard; but the moon and stars shon brightly. After a while however it became rainy, and grew fine at Intervals. The musquitoes bad as usual along these desolate shores.

2ᵈ *Saturday.*

Fine but Windy, went down about 10 miles and stopped for the wind.—No fresh meat on board. Saw 8 Wolves, 4 White ones.—Went to the Prairies, walked 6 miles but saw only 3 Bulls—The Wind a Gale. Saw abundance of black breasted Prairie larks and a pond with Black Ducks.—Returned to the pond after Dinner, and killed 4 Ducks, lost 2. The Wind blowing a Gale.—

SEPᵗ 2ᵈ 1843. SATURDAY.

Fine but windy, went down stream about 10 miles when we stopped again for the High Winds. We had no fresh meat on board, and uneasily perceived the effect of this on the faces of the men of our Crew.—We saw 8 wolves, 4 pure white and remarkably large. Went to the Prairies, walked fully 6 miles and saw only 3 Bulls. The wind was now a Gale. Saw abundance of black breasted Prairie Larks and a pond containing some Black Ducks (Anas fulca) returned to the pond after dinner, killed four Ducks and lost 2 others, the wind continuing a gale.—

3ᵈ *Sunday.*

Beautiful, Calm, but Cold.—Left early.—at 12 put a shore to Kill a Bull, having no fresh meat! He took the wind and ran off, touched on a Bar, I went overboard to assist. Water pleasant, weather very hot.—

Harris shot at a Prairie Wolf. At ½ past 4 saw 10 or 12 Cattle, Mᵣ C., B. & a Canoe Man went off, but the latter shot sadly too far.—The Cattle took the River and we went after them—The other Canoe man landed and ran along the shore but could not head them—he shot, and as the Cattle got to the Bank We sent them a Volley but inefˡʸ, and we are again

under way—Bell and M͟r C got well mired.—No fresh meat for another
day—Stopped for the night at the mouth of the Morau River. Wild Pigeons.
No fish. Sandpipers.

SEP͟ᵗ 3ᵈ SUNDAY.

Beautifully calm but somewhat cold, left early at 12 put a shore to kill a Bull
having no fresh meat. He took the wind, and ran off. Our Boat grounded
on a Bar. I leapᵈ overboard to assist, the water felt pleasant and the weather
had become much warmer.

Friend Harris shot at a Prairie Wolf but missed it, at ½ past 4 saw 10 or 12
cows, M͟r C & B and a canoe man went in pursuit, but the latter shot sadly
too far. The cattle came towards us. When seeing us they took to the water
and swam accross the stream. We rowed after them as fast as possible.—
The 2ᵈ man in the canoe landed and ran after them, but all ineffectually, for
although we sent them a volley of Balls, not one comitted suficient damage
to stop any one of them. I saw the gang rising on the hills and the whole
group was soon out of our sight. We are again under way. Took on M͟r C
& B and we are one more day without fresh meat on board. Bell and M͟r C
got well mired.—We stopped and spent the night at the mouth of Morau
River. Wild pigeons, no fish, some sandpipers.

4ᵗʰ Monday.

Cool night, fine this morning. Wind rose early.—Stopped by the wind at
11. M͟r C. Moncrev, and Bell gone shooting. Many signs of Elk &͟ᶜ, Wild
Pigeons.—A Bad place for hunting, but good for safety. Found Beaver
tracks and small trees cut down by them. Provost followed the Bank and
found their Lodge which he says is an old one. It is at present a mass of
sticks of different sizes matted together and fresh tracks are all around it.
To dig them out would prove impossible and we hope to catch them in
traps to night.—Beavers often feed on Berries when they can reach them,
especially Buffaloe Berries—(Shepperdia)

M͟r C. Killed a Buck and we have sent Men to bring it entire.—The
Beavers in this lodge are not residents but "vagrant Beavers." The Buck

was brought in and proves in the Grey. It is of the same kind as at Fort Union, having a longer tail We think than the kind found East. Its horns were very small, but It is skinned and in brine.—We removed about 100 yards below—in about as bad a spot for Wood to Cook &c.—

Provost and I went to set 2 traps for Beavers.—he first cut 2 sticks dry about 8 or 9 feet long. We reached the river by passing through the tangled woods. He pulled off his breeches and waded about with a pole to find the depth of the water, and having found a fit spot, he dug away the mud in the shape of a half circle, placed a bed of Willow branches at the bottom and placed the trap set over them.—he had 2 small Willow sticks in his Mouth. He split one end of one Dipt it in his horn of Castorum or Medicine as he calls this stuff and left on the end of it a good mess, after which it was placed in front of the Jaws of the Trap next the shore, he then fastened the Chain of the trap solidly planted a few sticks on each side the trap, and then the business ended. The second one was worked the same, except that he put no bed of Willows under it.—Beavers when caught in shallow water when they cannot get drowned are frequently attacked by the others, and in doing this are themselves the losers of their lives by it, as they frequently are caught in other traps placed close by.—

Mr C. & Bell returned without having shot, although we heard one report while setting the traps. Elks are abundant here, but the bushes crack and make so much noise that they hear the hunter and fly before them.—Bell shot 5 Pigeons at once. Harris and Squire both poorly, having eat indi[****] of Buffaloe Berries.—We are going to remove about 6 or 700 yards below where we are now and then spend the night. I hope I may have a large Beaver tomorrow.—

SEP^r 4th MONDAY.

We had a cool night, fine this morning. The wind rose early. Stop^d by it at 11. M^r C, Moncrevie, and Bell went shooting. Many signs of Elks &c Wild Pigeons. A bad place for hunting, but good as regarded safety. Provost found Beaver tracks and small trees cut down by them and a Lodge which he says is an old one. It is at present a mass of sticks of various sizes matted together; and fresh tracks are all around it. To Dig them out would prove impossible, and we hope to catch them in traps tonight. Beavers often feed on berries when they can reach them especially Buffalo Berries (Shepperdia). M⍦ C killed a Buck, and we have sent men to bring it entire. The Beavers of this Lodge are not Residents, but Vagrant Beavers. The Buck was brought in and proved to be in the grey. It is the same kind as that found at Fort Union having a longer tail we think than those found East. Its horns were very small. It is skinned and in brine. We removed about 100 yards below in a bad place for Wood to Cook &⍦. Provost and I went to set 2 Trapps for Beavers. He first cut 2 sticks dry about 8 or 9 feet long. We reached the River by passing through the tangled woods. He pulled off his breeches, and waded about with a pole to find the depth of the water, and having found a fit spot, he dug away the mud in the shape of a half circle, placed a bed of willow branches at the bottom, and placed the Trap *set* over them. He had 2 small willow sticks in his mouth, he split one end of one, dipt it in his horn of Castorum, or medicine as he calls this stuff and left on the end of it a good mess of the same, after which it was placed in front of the Jaws of the trap next the shore. He then fastened the chain of the trap solidly planted a few sticks on each side the trap, and then the business ended. The second one was worked in the same manner, except that he put no bed of willow under it.—Beavers when caught in traps in shallow water where they cannot get drowned are frequently attacked by others and in doing so, are themselves the losers of their lives by it, as very frequently they are caught in other traps, placed close by.—

M⍦ C & Bell returned without having shot, although we heard one report while setting the Trapps. Elks are abundant here, but the bushes are thick and crack so easily that the noise this produces make these wily animals fly before the hunter. Bell killed 5 Wild Pigeons at one shot. Harris & Squire both poorly having eat too many Buffalo berries.

We are going to remove about 6 or 700 yards and there spend the night. I hope I may have a fine Beaver tomorrow.—

5ᵗʰ *Tuesday.*

At day light after some discussions about Beaver lodges—H. B. & I with Provost and 2 Men went to the Traps—Nothing caught.—We now had the Lodge demolished outwardly, i.e. all the sticks removed, under which was found a hole about 2½ feet in Diameter, in which H. & B. & Moncrévie entered. There was nothing in.—The Beaver had gone to the river through its hole and H. saw it. It was also seen by the people at the Boat.—We had some of the Cuttings Cut off. I brought 2 large specimens, and a packet full of chips.—Before Beavers cut the tree they long for, they cut all the small twigs and sapplings around.—The chips are cut above and below and then split off by the animal—The trees lay cut down in every direction—

We left our camp at ½ past 5. Squires not well.—Phisicked him

The lodge we examined was not finished though about 6 feet in Diameter.—See Provost account—Moncrevié says that they have chambers all around. Provost says he never saw one.—When these Animals in dragging a stick find it fastened by brushwood, they cut it again so as to enable them to drag it on—H. found one of those sticks.—They were cut of different lengths, but it seems that they give the preference to those of a moderate size & cut towards the top of the Tree or its branches.—

We saw a Pigeon Hawk giving chace to a Spotted Sandpiper on the wing. When the Hawk was about to seize the little fellow it dove under and escaped.—This was repeated 5 or 6 times to my great surprise and pleasure.—The Hawk was obliged to relinquish the prey.—

As the wind blew high we landed to breakfast on a fine beach portions of which appeared as if paved by hand of Man.—The Canoe Men killed a very poor Cow which being alone had been wounded.—The wind fell sudenly and we proceeded on—Stopped by the wind at 12. Mʳ C. off & returns having killed a Young Bull Elk. Dined and walked after the Meat and skin, took measurements. On returning saw 2 Elks driven to the Hills by Mʳ C. & Bell.—Met Harris.—Started a Monstrous Buck Elk from its couch in a bunch of Willows.—Shot at it running 80 yards off, not touched. Meantime Provost had heard one from our Dinner spot. Loaded

his rifle & came within 10 paces, when unluckily the Gun snapped.—We
hope however that M͟r C. or Bell will yet kill a famous one this afternoon.

Nothing more done. Harris found a Dove's nest with a young just
hatched and the other egg cracked by the young inside.—The nest was on
the ground! Curious all this at this late season and in a Woody part of
the Country, saw a Bat.—

SEP͟t 5͟th, TUESDAY.

At Daylight after some discussions about beaver Lodges, Harris, Bell & I with
Provost and 2 men went to the Trapps. Nothing was caught. We now had the
Lodge demolished outwardly, i.e., all the sticks were removed, under which
was found a hole 2½ feet in diameter, into which Harris, Bell and Moncrevié
entered. There was nothing in it of course. The Beavers had gone through
its hole on the water side, and Harris saw it. It was also seen by the people at
the boat. We had some of their cutting procured and I brought 2 fine large
specimens, and a pocket of long chips. Before Beavers cut the tree they long
for, they cut all the twigs and sapling around. The Chips are cut from above
and from below, and are then split off by the animal. The trees lay cut down
in every direction. The Hole or lodge measured about 6 feet in diameter, but
there was [not] the least appearance of rooms or compartments as usually
described in books. We left at ½ past 5 A.M. Squire not well, Phisick him.—
This lodge was not finished. See Provost a/c. Moncrevie told us that when
completed they have chambers all around; but I know better. When beavers
in dragging a stick or part of a tree, they meet with obstructing branches of
others, they cut it again and are thus enabled to drag them along which they
do assisted both by their fore feet and their strong teeth.—

We saw a Pigeon Hawk giving chase to a spotted sand piper on the wing,
and whenever the Hawk was on the point of catching the bird, the latter
dove and thus escaped. We all saw this repeated 5 or 6 times.—

As the wind blew high, we landed on a beach when it had the appearance of having been paved by the hand of man.—The men killed a very
poor cow, which being alone had been wounded previously. The wind rose

again, and again we landed at 12.—M.ͬ C. went off and soon returned having killed a Young Buck Elk. Dinner over, went to the animal, measured it and it was skinned. On returning we started a Monstrous Buck Elk, ~~which I shot~~ from its bed of willows.—Shot at it running about 80 yards off but did not touch it. Meantime Provost had heard another from our Camp, loaded his Riffle and approached it to within 10 paces when unluckily his gun snapped. Harris found a Dove's nest on the ground with a young and the other egg about to hatch. Saw a bat this evening.

[Measurements of young male elk given here.]

6ͭͪ Wednesday.

Wind blowing harder, ransacked the point both below & above but saw only 2 Wolves, 1 a Dark grey one, the largest I have yet seen.—Harris shot a young of the sharp-tailed Grous.—B. 3 Pigeons. Provost went off about 4 Miles after Elks in the 2ͩ point below—hopes of his being fortunate.— Nothing done. Sprague found another nest of Dove on the ground with very small young.—The Common blue Bird was seen.—Also a whippoorwill & 1 Night Hawk. Wind high at South.

SEP.ͭ 6ᵗʰ, WEDNESDAY.

Wind blowing harder, ransacked the point above and below but saw only 2 wolves, 1 a dark gray one, The largest I ever have seen. Harris shot a young of the sharp-tailed grous. Bell shot 3 pigeons. Provost went after Elks on the 2ͩ point below, but returned without having seen one.—

7ͭͪ Thursday:

About 11 o'clock last night the wind shifted sudenly to northwest and blew so violently for a few minutes that we all got out in a hurry. Mͬˢ C. with her Child in her arms made for the willows and had a cover for her babe in a few minutes!—Our Guns and aminition &ͨ were brought on shore, as we were afraid of our boat sinking. We returned on board after a while but I could not sleep, becoming Sea Sick, returned to shore and laid down after

mending our fire. It rained hard for 1 or 2 hours.—Then the sky became
perfectly clear and calm. I then went on board and we slept until 8 o'clock.
The wind is now high, the sky cloudy, and we have removed our boat a few
[Measurements of a young male elk are given here.] Yards, but I fear that
we are here for another day.—Bell shot a Caprimulgus,² so small that I*
have no doubt is the one found on the Rocky Mountains by Nuttall, after
whom I have named it—These birds are now travelling South. M͟r C. &
I walked up the highest Hills of the Prairie but saw Nothing.—The River
has suddenly risen about 2 feet.—The Water rises at the rate of 8 Inches in
2½ hours. The wind has somewhat moderated. The Little Whippoorwill
proves an Old Male, but it is in Moult.—

Left our Camp at 5 and went down rapidly to the foot of an Island 4
Miles below.—M͟r C., B., Provost & Harris went off to look for Elks, but
I fear for no good as we see no Tracks nor none of their Beds.—About 10
o'clock Harris called me to hear the Notes of the new Whippoorwill.—We
heard two at once and the sound was thus Oh-will! *repeated often and*
quickly as in our common species.—

The night was beautiful but cold.—

7͟t͟h SEP͟t, THURSDAY.

About a 11 o'clock this night the wind shifted sudenly to north west and
blew so violently that we all went on shore. M͟rs C. with her child in her
arms made for the willows and had a cover for her babe in a thrice. So much
for Indian Industry, and affection to his progenitors but we were much
afraid of our Boat sinking, and therefore all was brought from her that
could easily be removed, Consisting of our Amunition, Guns, &͟c &͟c but I
could not sleep because I was very sea sick, and at last went to the shore.
However the weather cleared up again and became quite calm. I returned
on board and to the astonishment of all, I slept until 8 o'clock. The wind
rose again, and the sky became clouded, and we removed our Boat a few
yards below, but {only} into a worse situation.—The men in the canoe had

2   The nightjar family of birds, including Nuttall's or common poorwill.

pulled her high and dry.—Bell shot a Whipporwill so small, and so like the one described to me by Thos Nuttall that I take it for granted it is the same and have published it accordingly. It was a Male and I have never seen the female. Mr C & I walked up to the top of the highest Hills of the Prairies bank of the River but saw nothing.

Left our Camp this afternoon at 5, the River having sudenly raised fully 2 feet, and went down rapidly to the foot of an island about 4 miles below. Mr C, Bell and Harris went off with Guns but saw nothing but old beds of Elks and of Does. Harris called me to hear the notes of the Whippoo[rwill]. We heard 2 at once, and their notes sounded thus oh—will.—repeated often and quickly as is the mode of our more common birds.—The night was beautiful but cold.—

8th Friday.—

Cloudy and remarkably cold.—The Water has risen 6 feet ¹⁄12 since 12 o'clock Yesterday, and the Water is muddy and thick.—Started early.— The effect of sudden rises in this river is wonderful upon the Sand Bars, which no sooner covered by a foot or so of water, at once breake up, causing very high waves to run, through which no boat could run without eminent danger.—The Swells are felt for upwards of 50 Yards and are felt as if small waves at Sea!—Appearances of Rain—Strong current, reached Fort Pierre at ½ past 5. All well.

8th SEPr, FRIDAY.

Cloudy and remarkably cold. The water has risen 6 feet ³⁄12 since 12 o'clock yesterday, and the water is muddy and thick. Started early. The effect of the rising sudenly in the River is wonderful on the Sand bars, which no sooner covered by a foot of water, at once breake up causing very high waves to run through which no boat can pass without being in peril. The swells are felt for upwards of 50 yards, and are felt as if small waves at sea.—Appearances of rain. Strong current, ran 82 miles, and reached Fort Pierre at ½ past 5. All we[ll.]

9ᵗʰ *Saturday.*

Rain all night, Breakfasted at the Fort.—a few presents from Moncrevier and Mʳ C. Exchanged our Boat for a larger one.—Mʳ C. obliged to leave us and to go to the Platte establishment.—Dog Feast, &ᶜ—

9ᵗʰ, SATURDAY.

Rained hard all night, breakfasted at the Fort. Received a few presents from M. Moncrevié. Mʳ C. exchanged our Boat for a larger one. Mʳ C. is obliged to leave us to go and take charge of the Platte River establishment. We had a Dog feast with the Mandans and other Tribes.—

10ᵗʰ *Sunday.—*

Cloudy—recᵈ 2 pʳˢ of Mocassins from Bowis.—exchanged Camera lucida with Moncrevie—Weather Cloudy—Provost working at the Brine.—

The Parflesh given to me by Mʳ C. was the Couteau plain the sang—it was presented to Mʳ C. by L'Ours de fer!—

[The following two paragraphs are not in Audubon's hand]

Explanation of the Painting on Buffaloe Robe sent from a Sioux at Ft. Pierre. 'The small Blk marks denote horse tracks when at war. The round green marks are Indian Shields. The Horses are mounted by Indians— when riged for war

The first row of Horses is the rush of the Sioux on a Gros Ventre Village— The Bal of Horses are the Sioux rushing on their Enemis Horses. The Red Spots on the Horses denote wounds.[']

10ᵗʰ, SUNDAY.

Cloudy, received 4 pairs of moccasins from Mʳ Bowis. Exchanged my Camera Lucida with Moncrevié for Buffalo Robes.—Provost reboiling the Brine and adding salt to it.—

The Par flesh given to me by Mʳ C. was that of *The Couteau plain de sang.* It was presented to Mʳ C by *L'ours de fer.*[3]

3   "L'ours de fer" is the French rendering of Iron Bear, the Mandan chief referred to in the

11<u>th</u> *Monday.—*

Cloudy—Men at work on the new Boat.—Rained all day—The Wind shifted to every point of the compass.—nothing done.

1 1<u>th</u>, MONDAY.

Cloudy. The men at work with new Boat. The wind shifted to every point of the compass, nothing done.

12<u>th</u> *Tuesday.*

Rather Clear this morning, but rainded by 10 o'clock. Nothing done.—

[The following name and address are in a hand other than Audubon's]

Address J. Longhborough
Liberty Clay County M<u>o</u>

1 2<u>th</u>, TUESDAY.

Rather clear this morning, but raining by 10, nothing done.

13<u>th</u> *Wednesday—*

Raining. Nothing done.—Birds moving S.W.

1 3<u>th</u>, SEPR 1843, WEDNESDAY.

Rained all day, nothing done, Birds moving S.W.

14<u>th</u> *Thursday*

Cloudy & looking for Rain—M<u>r</u> L. making ready to start for Fort Union & ourselves for down the River.—

M<u>r</u> Laidlaw left for Fort Union at ¼ past 11. We left at 2 this afternoon, landed at the Farm and procured a few Potatoes, some Corn and 1 Pig.—We had 2 Pigs before.—Crossed the River and land<u>d</u>—

August 25 entry. Presumably "Couteau plain de sang" is a Native American name; I have found no other reference to this name.

14th, THURSDAY.

Cloudy and appearance of rain. Mʳ L making preparations to start for Fort Union, and ourselves for proceeding down the River.—Mʳ Laidlaw left Fort Pierre for Fort Union at ¼ past 11. We left at 2 same day. Landed at the Farm, procured some corn and a few Potatoes and 1 Pig. We had 4 others on board.—We crossed the River and landed for the night.—

15th Friday.—

Foggy morning—reached Fort George.—Trade of Harris—Grey Wolf. Mʳ Illingsworth—left at ¼ past 10.—Wind ahead. Obliged to stop for wind at 2. Indian & Buffaloe signs pretty fresh.—Nothing killed.—

15th, FRIDAY.

Foggy morning, reached Fort George of the opposition as it is called. Harris traded for some Indian curiosities. I saw a beautiful grey wolf belonging to the Fort, but the owner was absent. We were kindly received by Mʳ Illingworth, left again at ¼ past 10. Wind ahead, and obliged to land at 2. Indian and Buffalo signs abundant and fresh, nothing killed.—

16th Saturday.

Windy until near day light. Started early. passed Ebbets New Island—Bell heard Parrokeets—perfectly Calm—Mistake of the Great Bend, found Arvicola Pennsylvanica—Landed at the Great Bend for Black-tailed Deer and Wood for oars.—have seen nothing worthy of our attention—Squire put up a board on the center of our old Camp of the 6 Trees, which I hope to see again.—The Deer are lying down and we will not go out again before nearly Sun Set.—The Notes of the Meadow Lark here is now unheard—I saw fully 200 flying a hundred yards {or more} due South.—Collected a great deal of the Yucca plant—

16th, SATURDAY.

Windy in the night. Started early. Passed Ebert new Island. Bell heard Parrokeets, perfectly calm, and we suppose ourselves nearer the Great Bend than we were. Found arvicola pennsylvanica[4] on a sand bar, and caught it alive.—Landed at the Great Bend for Black tailed Deer and wood for oars. Saw nothing worthy our attention. Squires put up a board on the center of our old camp of the 6 trees. The Deer was lying down and we started not one. The Meadow Larks notes are heard no longer. I saw about 200 flying high and going due South. Collected a great deal of seed of the Vacca plants.—

17th Sunday.

We had a hard Gale last night with rain for about one hour.—This morning was beautiful. We started early but only ran for 2 hours, when we were forced to stop for the wind, which blows a gale.—Provost saw fresh signs of Indians; and we are told that there is a few lodges in the bottom of the bend about 2 miles below us.—The wind is north and quite cold, and the contrast betwen to day & Yesterday is great. Went shooting & killed 3 Sharptailed Grous—I shot 2.—left our Camp at 3¼ o'clock. The Wind much luld

Saw 10 or 12 Antelopes on the Prairie where the Grous were—Camped about 1 Mile from the spot where we were landed in May last at the end of the Great Bend. Evening calm & beautiful.—

17th, SUNDAY.

We had a hard gale last night and rain for about one hour. This morning was beautiful, and we started early, but only travelled for 2 hours when we were forced to stop as it blew a gale. Provost saw fresh signs of Indians, and we were told that a few loges are in the bend below us. Went shooting and killed 3 Sharp-tailed Grous. I shot 2. Left our camp at 3¼, the wind having lulled away. Saw 10 or 12 Antelopes on the Prairie, camped about 1 mile from the spot where we were landed in May last at the end of the Great Bend. Evening Calm and beautiful.

18th Monday.—

The weather cloudy and somewhat windy—Started early. Saw 1 Fish Hawk. 2 Gulls.—White headed Eagles.—yesterday saw abundance of Golden

4   Wilson's meadow mouse of the *Quadrupeds*.

Plovers. Make slow head way at our late breakfast, having landed to warm our coffee.—The Sharp taild Grous feed on rose Berries and the Seeds of the wild Sunflower & Grass Hoppers. Stop^d at 20 Minutes past 9—Wind very high.—Many dead Buffaloes in the ravines—and large roads over the Prairies—Harris & Bell & Sprague gone with Guns—Started but was pushed ashore again by the wind. Curious spot, outlines by Sprague.—The wind shifted for a few minutes but has now lulled down.—Saw Say's Fly Catcher and a Grosbeak.—Saw 2 of the Common Titlark. Left again at 2 with a better prospect.—Landed at Sun Set on the W. Side—Signs of Indians.—Wolves howling—found one dead on the shore, a beautiful grey one but too far gone.—These Animals feed on Wild plumbs in great quantities and pass them without scarcely defacing their appearance or discoloring the skin—tried to shoot Doves for our Fox and Badger without success—Pea vines very scarce.—

18th, MONDAY.

Weather cloudy and somewhat windy. Started early, Saw 1 Fish Hawk, 2 Gulls with White-headed Eagles. Abundance of Golden Plovers all moving South. The Sharp-tailed Grous feed at this season on Rose berries, seeds of the Wild Sun flowers and Grass-hoppers. Stopped at 20 minutes past 9, wind very high. Many dead Buffaloes in the Ravines, and large road over the Prairies. Harris, Bell and Sprague gone off with Guns. Started but were forced ashore again by the wind, Saw Says Flycater and a Grosbeak. Left again and went on until Sun Set when we landed for the night, on the West side. Signs of Indians, and wolves howling, found one dead on the shore, a Superb grey one but too far gone by putridity. Male animals feed on wild plumbs in great quantities, and pass them without scarcely defacing their appearance or discoloring the skin. Tried to shoot Doves for my Fox and Badger without success. Pea-vines very scarce.—

19th Tuesday—

Dark & drizley. did not leave until 6. reached Cedar Island and landed for wood to make oars—Bell gone with Gun.—Wind North. found no fit

Trees and left. passed the burned Cliffs, and got on a Bar.—The weather finer and wind behind us.—Wolves will even eat the Frogs found along the shores of this river!—Saw 5 all grey shot at them but killed none on account of the rowers, whom we dared not call to to stop.—obliged to stop on account of wind at 3 o'clock under a poor point, no Game.—

19ᵗʰ, TUESDAY.

Dark and drisly. Left at 6, reached Cedar Island and landed to procure wood for new oars. Bell gone with Gun. Found no fit Trees and left, passed the burned cliffs. Wolves will eat all the Frogs they can catch along the Banks of the River, and these are abundant. We Saw 5 Wolves, all grey. Shot at them but missed on account of the Rowers whom we do not dare to stop. Forced to land at 3 this afternoon under a poor point, no game.

20ᵗʰ Wednesday—

Wind very high.—Tracks of Wild Cats along the shore—The motion of the Boat makes me Sea Sick.—Sprague saw 1 Sharp tailed Grous.—We left at ½ past 12—The Wind having sudenly shifted. Saw an Immense number of Pintail Ducks but could not get near them—Stopped on an Island to procure Pea vines for my young Deer, and did so—The wind behind us.—our Camp last night was only 2½ Miles below White River. Stopped on a Bar for nearly ½ hour.—Shot 2 Blue winged Teals but did not get them—Camped opposite Bijou's Hill—

20ᵗʰ, WEDNESDAY.

Wind very high. Fresh tracks of wild cats along the shore. The motion of the boat rendering me almost Sea Sick.—Sprague Saw 1 Sharp-tailed Grous, left at ½ past 12. The wind having sudenly shifted, Saw an immence number of Pintail Ducks but could not approach them. Stopped on an Island to procure Pea vines for my Deer and did so. The wind behind us. Our camp of last night only 2½ miles distant below *White River*. Stopped on a bar for about ½ an hour. Shot 2 blue-winged Teals, but did not get them. Camped opposite the "Bijou Hills."—

21ˢᵗ *Thursday.*

Windy & raining most of the night—Startd early, weather cloudy & cold—
Landed to examine Burnt Hills.—Again on an Island for Pea vines—Fresh
signs of Indian Horses, Mules &ᶜ—At 12 saw 1 Bull, on our side the River
and in a few moments after 8 or 10 Cattle on the other side—Landed and
Provost, H. & B. have gone to try to procure fresh meat.—These are the
first Buffaloes seen since Fort Pierre.—

The Hunters only Killed 1 Bull. No Cows amongst 11 of them—and
this is strange at this season.—having detained 2 hours or more.—Saw
3 more Bulls in a ravine.—Stopped to Camp at the lower end of Great
Cedar Island at 5 o'clock—fresh signs of Buffaloe & Deer.—good timber
for oars.—raining hard. rained all night.—

2 1ˢᵗ, THURSDAY.

Windy and rainy most of the night. Started early, weather cloudy and cold.
Landed to examine some burned Hills. Again on an Island for Pea-vines.
Fresh signs of Indians Horses & Mules, at 12 o'clock saw 1 Bull on our side,
and a few moments after 8 or 10 on the other. These latter were mixed cows
and bulls together. Landed, and Provost, Harris and Bell went off to try to
procure fresh meat. These are the first Buffaloes we have seen alive since
we left Fort Pierre.—

The Hunters only killed 1 Bull, out of 11. [No] Cows, and this Strange
at this season, having lost 2 hours or more, went off, and saw 3 more Bulls
in a ravine. Stopped to camp at the lower end of Great Cedar Island, at 5
o'clock Fresh signs of Buffaloes and Deer, good timber for oars. Raining
hard and rained all night.

22ᵈ *Friday.—*

raining. Left at ¼ past 8, wind a head.—Distant Thunder, very uncomfort-
able night, wet & dirty.—Left and went down on the opposite side about 1
mile, when we were again obliged to come to for the Rain.—Played at Cards
for a couple of hours.—No chance to cook on account of the weather—

We dropped a few miles and Camped for the night in the mud but made a roaring fire to Cook by.—Wolves howling in numbers, owls hooting also.—Raining, We played Cards to 9 o'clock to kill the time.—Our Boat a quagmire.—

22, FRIDAY.

Raining, left at ¼ past 8 wind ahead. Distant thunder. We had a very uncomfortable night indeed, being wet and dirty. We removed only 1 mile on the other side and landed on account of the rain. Played at Cards, no chance to work on the shore on account of the weather. We dropped a few miles and camped for the night, in the mud, but made a roaring fire to cook by.—Wolves Howling all around us, and Owls hooting. Raining hard, played Cards until 9 o'clock to kill the time; our boat a very quag-mire.

23ᵈ *Saturday.*

Cloudy morn—left at 6, 5 Wolves on a Sand Bar—Red Shafted Woodpeckers—Yesterday saw 2 house Swallows, have made a good run of about 60 miles.—At 4 this afternoon we took in 3 Men of the steamer New Haven belonging to the Opposition, which was fast on a Barr about 8 miles below—We reached Poncas Island & landed for the night—at Dusk, the steamer came and stopped above us, and we found Messʳˢ Cutting, Taylor [****] and I had the great pleasure of receiving Victor & Johny's letter of the 22 July. I am sorry that my Lucy had not written a line.—

23ᵈ, SATURDAY.

Cloudy morn. Left at 6. Saw 5 Wolves on a sand bar, Red Shafted wood peckers, 2 House Swallows.—made a good run of about 60 miles. At 4 this afternoon we took in 3 men belonging to the steamer New Heaven belonging to the opposition, which was fast on a Bar 8 miles below. These men were hunters but they had not shot any thing. We landed on Poncas Island & landed for the Night. The steamer came up and landed above us, and we found Messʳˢ Cutting & Taylor, and I had the great pleasure of receiving Victor & John's letter of the 22 July. I am sorry that my Lucy had not written also.

24th Sunday.

Cloudy, Windy & Cold.—The steamer and we left as soon as we could see.—Saw a Wolf on a Bar and a large flock of White Pelicans which we took at first for a Keel Boat. Killed none.—Passed the Poncas, L'Eau qui cours, Manuel & Basil River by 10 o'clock.—Landed just below Basil River. Stopped by Wind—Hunted. I shot 1 Raven, 1 T. Buzzard, 4 Wood Ducks—Harris shot 1. We lost 4 more.—Ripe plums abundant—Gar fish in this Creek. found feathers of Wild Turkeys.—Signs of Indians, Elks & Deer.—Provost and men made 4 new oars.— Went to bed early—

24, SUNDAY.

Cloudy, windy and cold. The steamer and us started as soon as we could see, saw a Wolf on a bar, and a large flock of White Pelican which we took at first for a Keel-boat. Killed none. Passed the Poncas, L'Eau qui coure, Manuel and Basil Rivers by 10 o'clock. Landed just below Basil River, stopped by the wind. Hunted, I shot 1 Raven, 1 Turkey Buzzard, 4 Wood Ducks. Harris shot 1, and yet we lost 4. Ripe Plumbs abundant and good. Garfish in this Creek. Provost and the men made 4 new oars.—

25. Monday.

Blowed hard all night, began raining before day—All Wet and Nasty.— Started ¼ past 10. Passed Bon home Island at 4 and landed for the night at 5, 15 Miles below the [****]

25th, SEPr MONDAY.

Rained and blowed Hard all night. Wet and dirty. Started ¼ past 10. Passed Bonhomme Island and landed at 5 for the night.—

26th Tuesday.

Cold & Cloudy—Started early.—Shot at a Pelican—Landed to see the fortification Hasty Jack's River at 11—Abundance of Wild Geese—Bell killed a Young White Pelican—Weather fairer but coldish.—Sprague killed a Goose but it was lost.—Camped a few miles above the Vermillion River.—Harris saw Racoon tracks on Basil River.—

26th, TUESDAY.

Cold and cloudy. Started early. Shot at a Pelican, passed Hasty Jack's River at 11. Abundance of Wild Geese. Bell killed a young Pelican, weather fairer but cold. Sprague killed a goose but it was lost. Camped a few miles above Vermillion River. Harris saw Raccoon tracks, on Basil River.—

27th Wednesday.

Cloudy but Calm, many Wood Ducks, saw Racoon tracks again this morning.—passed the Vermillion River at ½ past 7.—The Badger got out of his Cage last night, and we had to light a candle to secure it.—We reached the Fort of Vermillion at 12. Kind reception of Mr Pascal—previous to this we met a Barge going up owned & commanded by Mr Sybill and found our good Hunter Michaux. He asked me to take him down, and I promised him $20 per month to St Louis.—We got 2 barrels of superb Potatoes, 2 of Corn, and a good fatt Cow.—

27th WEDNESDAY.

Cloudy but calm, many wood ducks. Saw more Raccoon tracks. Passed the Vermillion River at ½ past 7. My Badger got out of his cage in the night, and we had to light a candle to secure it, which was effected by one of the men throwing a Buffalo Robe over it. We reached Fort Vermillion at 12, and were kindly received by Mr Pascal. Previous to this we met a barge going up owned and commanded by a man named Sybill, and found our good old Hunter Michaux on board. He asked me to take him down and I promised him 20$ per month to St Louis. We got 2 barrels of most superb potatoes, 2 of Corn, and a fine young fat cow.

28th Thursday.

A beautiful morn. Gave our order on the Company for the 2 Barrels of Corn and 2 of Potatoes, for which I only charged him no less than $16!! Left

at 8. Young married Squaw, the Man who brought me the Young Calf at
Fort George. Antilopes are found rarely about 25 Miles from this Fort.—
Left at 8 and landed 15 Miles below on Elk point. Cut up and salted
the Cow.—Hunted. Provost saw 3 females but the order was not to shoot
except at Bucks.—I saw a prodigious large one who stood before me within
5 or 6 steps, but my Gun snapped and away he ran and Jumped in the
River and swam accross carrying his horns flat down and spread on each
side of his back.—The neck looked to me as about the size of a Flower
Barrel—Harris killed a female Turkey. Bell and the others saw plenty
of them but did not shoot any. Elks were the order of the day—I cannot
eat Beef after having been fed on Buffaloes.—On getting in the boat this
evening I had a severe fall which has b[****] my Knee and my Elbow.—

28th THURSDAY.

Beautiful morn. Gave our order on the Company for the amount of our
purchases, say 16$ for the whole. I met here the young man who brought
me the Young Buffalo calf at Fort George when our passage upward.
Antilopes are found very scarce within 25 miles, after which no more are
seen. Left at 8 and landed 15 miles below on Elk Point. Cut up and salted
the Cow.—Hunted, all hands; Provost saw 5 large females but the order
being positive not to shoot except at males. I saw a Prodigious Large Buck
who stood before me within 5 or 6 steps, but my Gun snapped and away
he went, runing towards the River, and reached the Bank, and saw it swiming accross with its great antlers flat and on each side of his back. Its neck
when I first saw the animal, appeared as if about the size of a flour Barrel.
Harris shot a female turkey. Bell and the others saw plenty of these but did
not shoot at any as Buck Elks were the order of the day. The fall I had a few
days previously on my Knee, rendered me almost lame.—

29th Friday.

Rained most of the night, and rains & blows at present. Crossed the river
and have encamped at the Mouth of the Ioway River, the boundary line

of the Sioux & Omahas. Harris shot a Wolf but missed it—Rained &
blowed all day. My Knee too sore to allow me to walk.—

29th, FRIDAY.

Rained almost all night. We crossed the River and encamped at the Mouth
of Ioway. The boundary line of the Sioux and Homahas.

30th *Saturday.*

Hard rain all night. Water rose 4 Inches.—found a new species of large
Beans in the Wild Turkey.—Musquitoes rather troublesome. Sun shin-
ing by 8 o'clock and we hope for a good dry day!—Whip-poor will heard
Night before, and Night Hawks seen flying.—No Squirrels are found
above the Bluffs or Belle Vue.—Saw a longtailed Squirrel that ran on the
shore at the cry of our Badger. Michaux, H. & B. landed to bring a set of
Elk Horns killed by Michaux last week.—Abundance of Geese & Ducks.
Weather clouding over again.—at 2 we were struck by a Heavy gust of
Wind, and were obliged to land on a weather shore. The Wind continued
heavy and the motion of the Boat was too much for me.—I supped on
shore and my man Michaux made a good camp of blankets and robes
and we sleep together.—

30th, SATURDAY.

Rained all night, the water rose 4 Inches. Found a new species of large bean
in the wild turkey. Musquitoes very troublesome. Sun shining by 8 o'clock
and we hoped for a dry day. Whip-poor-will heard night before last. Night
Hawks seen flying in great numbers. Saw a long-tailed squirrel that ran
along the shore at the cry of our Badger. Michaux, Harris and Bell landed
to bring a pair of Elk Horns Killed by Michaux last week. Abundance of
Geese and Ducks. Weather cloudy again. At 2 were struck by a Gale of
wind and we were obliged to Land on a weather shore. The motion of the
boat was too much for me, and my man Michaux made a camp of Robes
&c and we slept together.

Oct.ʳ 1.ˢᵗ Sunday.

The wind changed & lulled before morning and we left at ¼ past 6.—The weather had some better look but it rained several showers. passed the Sioux River at 20 m. past 11.—heard a pileated Woodpecker & saw fish Crows.—Geese very abundant—landed below the Sioux river to hunt Turkeys having seen a large cock on the Bluffs.—Bell Killed 1 female & Harris 2 Young Birds.—These will keep us agoing for some days—Stopped again opposite Floyd Bluff and Grave, by the wind.—~~land~~ Started again, and ran about 4 miles and obliged to land in a rascally place at 12 o'clock. had hail, & rain at intervals.—Stopped in the Mouth of the Omaha River 6 miles from the Village. Abundance of Geese.—The wind lulled and the stars are shining.

OCT.ᴸ 1ˢᵗ 1843, SUNDAY.

Hard rain all night. The wind lulled before morn and we left at ¼ past 6. The weather promised fair, but it rained several showers. We passed the Sioux River at 20 m. past 11. Heard a piliated Woodpecker & saw fish crows. Geese very abundant; landed below the Sioux River to Hunt Turkeys, having seen a large cock on the Bluff. Bell killed one female and Harris 2 young birds. These will keep me in food for some days, as I cannot eat Beef at all.—We stopped again opposite Floyds Bluff and Grave, by the wi[nd.] Started again and ran about 4 miles and were forced to land in a very ugly place at 12 o'clock, had Rain and Hail at Intervals. Stopped in the mouth of the Homaha River 6 miles from the village. Abundance of Geese. The wind lulled and the stars are shining.

2.ᵈ Monday.

beautiful but Cold. The Water rose 9 Inches and we travel well. Started early.—Stopped at 8 by Wind. Vile place but plenty of Jerusalem artichox. fried a mess and found them good.—Started again at 3 and made a good run to Sun Down—good camp. Abundance of Geese.

2ᵈ, OCTʳ, MONDAY.

Beautiful but cold. The water had risen 9 inches and we travelled well having started early. Stopped at 8. Vile place but abundance of Jerusalem Artichox, fried a mess of them and found them excellent. Started again at 3 and made a good run to Sun down.—a good camp and abundance of Geese.

3ᵈ Tuesday.

beautiful & Calm—Started early.—Saw 3 Deer on the shore.—A Prairie Wolf that was a long time under a Bank before he found a place to get up. Shot at by Michaux and missed.—Geese, Sand hill Cranes and another Wolf, passed the little Sioux River.—Saw 3 more Deer and another Prairie Wolf, 2 Swans—Pelicans and abundance of Geese & Ducks—passed Soldier river at 2 o'clock. We were caught by a Snag that scraped our roof and tore it a little. Had we passed 2 feet nearer it would have ruined our Barge.—We passed through a very swift cut off very dificult of entrance.—We have ran 82 Miles and encamped at the Mouth of the cut off of the Old Bluffs.—Abundance of Geese and Ducks but killed only 1 Malard— Bragg, Harris Dog hid all the meat that had been cooked for us.—

3ᵈ, TUESDAY.

Beautiful and calm, saw 3 Deer on the shore, a Prairie Wolf that was a long time runing under a high bank before it found a place whereby to climb, shot at by Michaux but missed. Geese, Sand Hill Cranes and another wolf. We passed the little Sioux River; saw 3 more Deer and another Prairie Wolf, swans, Pelicans and a great number of Geese, passed Soldier River at 2 o'clock. We were caught by a snag that scraped the roof of our Boat and torn it a little.—We have ran 82 miles and encamped at the mouth of the cut off of the old Bluffs, Killed only 1 Malard.—Bragg, Harris' Dog, hid all the meat that had been cooked for our suppers.

4ᵗʰ Wednesday.

*Cloudy and Coldish.—left early, canᵗ find my pocket Knife & fear that I have lost it.—My Knees very sore yet.—put on my {J****}. found my*

Knife—Stopped by the Wind at 10 at Cabané Bluffs—About 20 miles above Fort Croghan attempted to Hunt but {l****s} are too bad.— Saw Asle bushes and some black Walnuts—Saw the first Cycamores coming down and of course the last going upwards—Wind bound until night—nothing done.—

4ᵗʰ, WEDNESDAY.

Cloudy and coldish, left early, mislaid my pocket Knif[e,] my Knee very sore yet. Stopped by the wind at 10 at Cabane's Bluffs about 20 miles above Fort Croghan. Attempted to hunt but the country was too bad. Saw Hazle bushes and black walnuts, saw the first Sycamore coming down, and of course the last going upward, wind bound until night, nothing done.—

5ᵗʰ Thursday.

Blowed all night and clear.—beautiful Sun rise. Started early but stopped by the Wind at 8. B., H., & Squire have started for Fort Croghan— Appearance of rain—started at 3 & reached the Fort at ½ past 4. found all well, and kindly received. procured Green Corn for nothing and had abundance of bread made. bought 13 Eggs for 25 cents.—

Honey Bees are found here and do well, but none are seen above this place. I had a Slide on the bank as it had rained in the interim. Squire had also one at 12 this night when he and Harris & Sprague came to the boat after having played Whist up to that hour.—

5ᵗʰ, THURSDAY.

Blowed all night and clear, beautiful sun rise. Started early, but stopped by the wind at 8. Harris, Bell and Squire started for Fort Croghan, appearances of Rain. Started at 3 and reached the Fort at ½ past 4. found all well, and extremely Kindly received. Procured Green Corn for nothing, and had an abundance of bread made expressly for us. Bought 13 eggs for 25ᶜᵗˢ Honey Bees are found here and do well, but none are found above this place. I had a slide down the Bank as it had rained in the interim. Squires had also one when he and the rest returned to the Boat at 12, having been playing Whist up to that hour.

6ᵗʰ *Friday.*

Some rain and Thunder last night. tolerable day, breakfast at the camp and left at ½ past 8. Our Man Michaux was passed over to the officers to steer their barge down to Leavensworth but we are to keep in Company and he is to cook for us at night.—The whole is broke up and Captain Burgoin leaves in a few hours by Land with the Dragoons Horses &ᶜ—

Stopped at Belle Vüe at 9. good reception bought 6ˡᵇ Coffee, 13 Eggs, 2ˡᵇ Butter and some black pepper—Abundance of Indians of 4 different Nations. Major Miller the Agent a good Man—Left again at 11, fine day— Passed the Plate and its hundreds of snags at ¼ past 1, and stopped for the Men to dine! This stream quite full. Squaw on the Bar—The Village in sight. Killed 2 Pelicans but got only 1. encamped about 30 miles below Fort Croghan—Lᵗ Carleton supped with us and we played a game of Whist.—

6ᵗʰ, FRIDAY.

Some rain and thunder in the night. Tolerable day however. We all breakfasted at the officers' mess. We left at ½ past 8. Our man Michaux was passed over to the officer commanding the party of dragoons as far as Fort Leavensworth, but we were to keep company and the soldiers or Dragoons were to cook for us at night. This Fort Croghan is now broken up, and all the [scribal omission] and Capⁿ Burgoin leaves in a few hours by land with the rest of the troops, waggons &ᶜ &ᶜ. Stopped at Belle View at 9. Good reception, bought 6ˡᵇ Coffee, 13 eggs, 2ˡᵇ Butter and some black pepper. Abundance of Indians of 4 different Nations. Major Miller, the U. S. agent is a good man. We left again at 11. Fine day. Passed the Plate River and its hundreds of Snags at ¼ past 1, and stopped for the Soldiers to dine. The Plate River quite full, saw Squaws on the bar getting drinking water. The Village was in Sight. Killed 2 pelicans but got one only, encamped about 30 miles below Fort Croghan. Lieut. Carleton supped with us, and we played a Game of whist.

7ᵗʰ *Saturday.*

fine night and fine morn. Started too early and got on a bar for a few minutes—passed McPherson the first House in the State of Missouri at 8 o'clock.—Bell skinned the young of Fringilla Harrisi—Lᵗ Carleton came on board to breakfast with us, fine companion and a perfect gentleman— Indian Hoops were heard by his Men whilt embarking this morning—We encamped at the Mouth of Nishnebottana a fine clear stream next to the house of Mᵣ Beaumon pretty Wife & a fine run of 60 or 70 miles—

7ᵗʰ, SATURDAY.

Fine night and a fine morning, started too early and got lodged on a bar for a few minutes. Passed McPherson, the 1ˢᵗ House in the State of Missouri, at 8. Bell skinned the young of Fringilla Harrisii.[5]—Lieut C. came on board and breakfasted with us, a fine companion and a perfect Gentleman. Indian Hoops and Yells were heard by his men, whilst embarking this morning; we all encamped at the mouth of Nishnebottoma, a fine clear stream. Went to the House of Mᵣ Beaumont, a pretty wife and a fine run for us of 60 or 70 miles.

8 *Sunday.*

Started early—Cloudy and rain by 8 o'clock.—Stoped Twice and played Cards—Started at 12 and ran until ½ past four.—The Wind blowing hard stoped at a good encampment—Presented a Plate of the Quadrupeds to Lieutenant James Henry Carleton—and he has promised me a fine black bear skin and a set of Elk Horns.—Stopped on the East side of the shore in the evening. Saw a remarkably large flock of Geese passing southward.—

8ᵗʰ, SUNDAY

Started early. Cloudy and rain by 8 o'clock. Stopped twice. Started at 12 and ran until ½ past 4. The wind blowing hard we stopped at a good encampment, presented a plate to James Henry Carleton, and he has promised me a black bear skin and a pair of Elk Horns. Stopped on the East side of the River. Saw a remarkably large flock of Geese flying South.

---

5   Harris's sparrow, or Harris's finch in the Octavo *Birds of America,* plate 484.

9th Monday.

beautiful and calm, started early—Bell shot a grey squirrel which was divided and given to the fox and badger. S[****], H, B, & Sprague walked across the bend to the Black Snake Hills and Killed 6 Grey Squirrels, 4 Parokeets and 2 Partridges—Bought Butter, eggs & Whiskey for the Men— exchanged Knives with the lieutenant, and he gave me a damascus Double Barrelled Gun, and I presented him with the plates of the Quadrupeds. Started and ran 12 miles to a good camp on the Indian side—

9th, MONDAY.

Beautiful and calm. Started early. Bell shot a Grey Squirrel which was devided and given to the Fox and Badger. Squire, Bell, Harris and Sprague walked accross the bend to the Black-Snake Hills, and killed 6 Grey Squirrels, 4 Parokeets and 2 Partridges. Bought Butter, eggs and whiskey for the men. Exchanged Knives with the lieutenant and he gave me a damascus double barrelled Gun, and I gave him all the plates I had with the Port Folio. Started and ran 12 miles to a good camp on the Indian side.

10th Tuesday.

beautiful morning rather windy. Started early—Great flocks and Geese & Pelicans wounded 2 of the latter.—reached Fort Leavensworth at 4. Kind treatment and reception of Major Morton—Lieutenant James Henry Carleton gave me 1 Black bear skin fine 1 Damascus double barrelled Gun and a fine pair of Elk Horns—Wrote home, to John Bachman and Gideon B. Smith.—

10th, TUESDAY.

Beautiful morning, rather windy. Started early, great flocks of Geese and Pelicans; wounded 2 of the latter, reached Fort Leavenworth at 4. Kindly treated and received by Major Morton. Lieut James Hry Carleton gave me a black Bear Skin, the double barrelled Gun and a pair of Elk horns. I wrote home, to John Bachman and Gideon B. Smith.

11th Wednesday.

Sent our letters to the Lieut!—received a present of Fruit, Melons, Chickens & Bread & Butter from the Major!—Lieut. C. came to see us off. left at ½ past 6.—Weather Calm & beautiful—Game scarce. Pawpaws plenty—Stopped at Madame Chouteau many sick, got 3 pumpkins—Stopped at Liberty landing, and delivered the letters of Laidlaw to Black Harris.—reached Independence landing at Sun down—have run 60 miles.—No letters sent for us—Steamer Lebanon passed upwards at ½ past 8.—

11, WEDNESDAY.

Sent our letters to the Lieutenant, received a fine present from the Major of Fruits, Melons, Chickens, Bread & Butter. Lieut! C. came to see us off. Left at ½ past 6. Weather calm and beautiful. Game scarce. Papaws plenty. Stopped at Madame Chouteau, very sick. Got 3 Pumpkins for my Deer. Stopped at Liberty landing and gave M! Laidlaw's letters to Black Harris. Reached Independance at sundown, having ran 60 miles. No letters for any of us, Steamer Lebanon passed on her way upward at ½ past 8.—

12th Thursday.

beautiful and Calm stopped to buy eggs &c—none at a M! Shivers from Kentucky.—Ran well to Lexington where we got eggs and meat—and started ran 60 miles to day.—

12th OCT! 1843, THURSDAY.

Beautiful and Calm. Ran well to Lexington where we bought Eggs and meat. Ran 60 miles this day.

13th Friday.

heavy white frost & foggy. Started early and ran well—tried to buy butter at several places but in Vain, at Greenville bought Coffee.—Abundance of Geese & White Pelicans, many Sand Hill Cranes.—Harris killed a Wood Duck, passed Grand River. Stopped at New Brunswick, bought Beef &

*Butter. Beef 2½*cts *Butter 12½ Camped at a deserted Wood yard—ran between 60 & 70 miles—*

13th, FRIDAY.

Heavy white frost and foggy. Started early and ran well, Bought Coffee at Greenville. Ran betwen 60 and 70 miles, camped at a deserted woodyard.—

14th Saturday.

Windy night—& after 8 days good run I fear we will be stopped to day—Stopped by a high wind at 12, we were run ashore, and I undressed to push the boat, after which I got so deep in the mud that I had drag backwards—Visited 2 farm houses—bought Eggs, Chickens and corn bread—The Squaters visited our boat—Camped here, fine familly and a good Man from North Carolina.

Michaux Killed 2 Hutchins Geese, the fattest {I} ever saw in the flesh—ran about 20 miles—steamer Lebanon passed us going downwards about 1 hour before sun set—Turkeys & long tailed squirrels in abundance—

14th, SATURDAY.

Windy night, and after 8 days good run fear we will be disappointed to day. Stopped at 12 by a gale, which ran us ashore, I undressed to push off the Boat, after which I got so deep in the mud that I had to dragg myself backwards. Visited 2 famillies, bought Eggs Chickens and Corn bread. Gathered many remarkably large acorns of the finest white oaks I had ever seen. The farmers visited our Boat and we found that they were from North Carolina. Michaux killed 2 of Hutchins Geese, called by the French Boatmen "Babillarde" the fattest I ever saw and good eating. Turkeys and long-tailed squirrels very abundant.

15th Sunday.

Cold, foggy & Cloudy. Started early, passed Chariton River & Village and Glasgow—bought bread & Oats for my Deer. Abundance of Geese

& Ducks passed Arrow Rock at 11, passed Boonsville, the finest country
on this River. Rochefort, high rocky Cliffs—encamped 6 miles below
Rochefort, having ran 60 miles

15$\underline{\text{th}}$, SUNDAY.

Cold, foggy and Cloudy. Started early, passed Chariton River and Village
and at Glasgow bought bread and oats for my Deer, abundance of Geese
and Ducks. Passed Arrow Rock at 11. Passed Boonville, the finest country on
the River. Passed Rochefort also. High rocky Hills, camped 6 miles below
the latter Village having ran 60 [miles.]

16$\underline{\text{th}}$ *Monday.*

beautiful Autumnal morning. White frost & Calm left at 6. Strong current—
past Nashville, Mariane, Steamer Lexington going up. Jefferson City at 12,
bought Beef, Apples, Butter, bread & Coffee, passed the Osage & 4 Deer
opposite Smith Landing—camped at sun down—found Giraud the strong
Man—ran 61 miles—Met the Steamer Satan, badly steered—Abundance
of Geese & Ducks.

16$\underline{\text{th}}$, MONDAY.

Beautiful autumnal morning. White frost and calm, left at 6. Strong current, passed Nashville, Marian, met Steamer Lexington going up Jefferson
City at 12. Bought beef, apples, butter, bread and Coffee. Past the Osage
and 4 Deer opposite Smith Landing, camped at sun down, found Giraud
the strong man, ran 61 miles, met steamer Satan, badly steered. Geese and
Ducks in abundance.—

17$\underline{\text{th}}$ *Tuesday.*

Calm & very foggy—Started early & floated a good deal. Saw 2 Deer.—The
fog all off by 9 o'clock. passed the Gasconnade River at ½ past 9.—Landed
at Pinkney to buy bread &$\underline{c}$.—

Buffaloes have been seen mired and unable to defend themselves and
the wolves actually eating their nose during the while and killing them

after all—passed Washington & encamped below it at Sun down—a
good run. The Drunken green Man—

17ᵗʰ, TUESDAY.

Calm but very foggy. Started early and floated a good while. Saw 2 Deer. The fog all dispersed by 9 o'clock, passed the Gasconade River ½ an hour after. Landed at Pinkney to buy bread &ᶜ &ᶜ.

Buffaloes have been seen mired and so unable to defen[d] themselves, that the wolves actualy eat their nose and killing them after all. Passed Washington and encamped below it. Have made a good run. The Drunken Green Man.

18ᵗʰ Wednesday.

fine & Calm—Started early—passed Mount Pleasant—Landed at Sᵗ
Charles to purchase bread &ᶜ. Provost was drunk this day & went off by
land to Sᵗ Louis we gave him $1.50, passed the Charboniere and encamped
about 1 mile below. Steamer Tobacco plant landed on the Island opposite—
Killed 6 Grey Squirrels, B. & H.

18ᵗʰ, WEDNESDAY.

Fine and Calm, started early. Passed Mount Pleasant. Landed at Sᵗ Charles to buy bread &ᶜ. Provost was drunk this day and went accross the Country to Sᵗ Louis. Gave him $1.50. Passed the Charboniere and encamped about 1 mile below the steamer Tobacco Plant. On the Island opposite, Bell and Harris Killed 6 Grey Squirrels.

19ᵗʰ Thursday.

fine & Calm—but very foggy & white frost—Started early. The Steamer
after us—forced to stop on a Bar for the fog and lost 1 hour—reached Sᵗ
Louis at 3 of the afternoon—unloaded & sent all things to N. B.'s Ware
house.—Wrote home—

19ᵗʰ, THURSDAY.

Fine and calm but very foggy with white frost. Forced to stop on a bar for fog, and left in 1 hour. Reached Sᵗ Louis at 3 of this afternoon. Unloaded and sent all our things to N. Berthoud's warehouse, wrote home, on the 20<u>th</u> and to Maria Martin on 21<u>st</u>.

Left S<u>t</u> Louis on the 22 Oct<u>r</u> 1843 in Steamer Nautilus, for Cincinati.

Reached Home on Sunday afternoon at 3 o'clock P.M. 6<u>th</u> Nov<u>r</u> and thank God found all my familly quite well.—

BACK MATTER THAT OCCURS ONLY
IN THE FIELD NOTEBOOK

Wrote home on the 20<u>th</u>—

Wrote to Maria Martin on the 21<u>st</u>—

Left S<u>t</u> Louis 22<u>d</u> Oct<u>r</u> in Steamer Nautiluz for Cincinnati.—

Reached home at 3 o'clock P.M. 6<u>th</u> Nov<u>r</u> 1843 and thank God found all my familly quite well!—

[Other ms material:]

[On the inside back cover:]

Tho<u>s</u> W. Bakewell
55 Columbia S<u>t</u>
John J. Bowen—
~~Corner of~~ Broad &
Chestnut—Phil<u>a</u>—

[Top half of next page moving inward, and written upside down:]

Bell's Vireo
Harris Finch
A new finch allied to Henslows,
Meadow lark—
Sprague's Lark—
New Grackle—
New Pallida—
C W Webber Princeton N J—[6]

6   The signature of Charles Wilkins Webber in the back of this notebook corroborates Webber's claim to have met Audubon on this return trip.

[On the next right-hand leaf:]

Indian Nations through which we have passed—

1 Assiniboins—

2 Gros Ventre—

3 Mandans—

4 Ricarees—

5 Poncas—

6 Omahas—[a French bracket joins Poncas and Omahas with notation: "now together"]

7 Santees & Yanctons (Sioux)

8 Potowatamies—

9 Ioways—

10 Sacs—

11 Kickapoos—

12 Ottos—

13 Chipeways *seen only*

15 Crees

16 Black feet (seen)

17 Crows (seen)

18 Pawny Loups—

19 Grand Ponies—

20 Pawnies republican—

21 Tapaye

22 Delaware

23 Shonies—

PART IV

AUDUBON'S CONSERVATION
ETHIC RECONSIDERED

Audubon's Hunting and Conservation
Ethic as Represented in the Biographies

In writing about Audubon, biographers have had to navigate a course somewhere between praise and condemnation. Was he a great naturalist and illustrator inspiring a nation to love and protect its birds, or was he a heartless butcher of birds? When Audubon began to publish his "bird biographies" and "episodes" in 1831, he knew that many readers would judge him for killing as many birds and other animals as he rather freely narrates having done. He was right, of course. Anyone who goes public on so vast a scale as the 3,170-page *Ornithological Biography* invites the scrutiny of the reading public. This concern about the killing was already voiced in the very first biographical sketch of Audubon, published in 1834 ("Biographical Sketch"). While the purpose of offering high praise for the man and his work is clear throughout the sketch, the anonymous writer also expresses the anticipated reservation and chagrin. Speaking in the plural, as if representing Audubon's readers generally, he explains,

> We like him, when speaking of [his] early years, and confessing, that the moment a bird was killed for the sake of forwarding his researches, however beautiful it had been in life, the pleasure arising from the possession of it became blunted, for he felt that its vesture was sullied, and that it no longer was fresh from the hands of its Maker. He wished to obtain all the productions of Nature, but he wished life with them. To the present day, we find him speaking of the necessity of resorting to deadly means to secure the objects of his study, as costing him pain; and

this tender feeling is even more apparent from the affecting manner in which he describes the specimen than from any direct attestation. (198)

This writer then cites the example of the "Golden Eagle" essay (just published in volume 2 of *Ornithological Biography*), where Audubon narrates at deliberately agonizing length his awkward, failed attempts to asphyxiate with burning charcoal the living eagle he had just purchased and his subsequent method of killing: "I thrust a long pointed piece of steel through his heart, when my proud prisoner instantly fell dead, without even ruffling a feather" (OB 2:465). This is harsh, and the writer of this first full view of Audubon feels compelled to apologize for it, so he claims that Audubon expresses "by the manner of description" "how it went to his heart to have been obliged thus to treat his precious victim" ("Biographical Sketch" 198). But this is a biased reading. What Audubon actually wrote shows no regret: "at last I was compelled to resort to a method always used as the last expedient, and a most effectual one" (OB 2:465).

Several years later, when Audubon and his sons were producing the more affordable Royal Octavo edition of *The Birds of America*, with a much broader readership in mind, either Victor or John apparently persuaded the more freely expressive father to delete the entire scene, cutting the essay in half. There is killing enough to go around in the bird biographies, but this particularly heartless passage was deemed to run an unnecessary risk of alienating readers at a time when profit was the motive.

Audubon's voice throughout the bird biographies is a piece of rhetoric designed to keep his readers willing to forgive him for his sins.[1] "Dear Reader" and "Kind Reader" regularly announce his desire for friendly relations. Even though he grew increasingly confident in his relationship with the "Kind Reader" through the years of writing the bird biographies, Audubon remained mindful of the likelihood that some readers would be repulsed by accounts of the gunning. Thus the tension between adulation and disapproval looming in the biographies was in place from the beginning.

1   I discuss this point at length in "The Written Ethic."

In the rest of this chapter, I survey and analyze the full range of biographi-cal representations of Audubon published from the end of the nineteenth century to 2004, a year in which three new biographies appeared.

If Audubon's actual thought about conservation and how humans should live in relation to nature is difficult to discern, it's partly his own fault: he wrote so much over so long a period of time through all the stages of his life that anyone would be challenged to pin him down. Further obscuring the truth are the facts that some editors have misreported what he wrote and that quite a few people who have written about Audubon have deliberately manipulated the evidence to satisfy a particular agenda. While complete objectivity is, of course, impossible, the range of disparate representations of Audubon's likely position in conservation debates is so wide as almost to defy belief. Could all those authors be writing about the same man?

Despite Audubon's awareness that some of his readers would occasionally object to so much killing in his published work, and despite the faint glim-mer of this view in the sketch of 1834 noted earlier, the nineteenth century was largely silent about his gunning. Accounts of him generally emphasize the romantic, indefatigable hunter-naturalist who creates an American natural history that surpasses that of the European closet naturalists. In association with Grinnell's Audubon Society (founded 1887), Audubon is the protector of birds and an American sportsman. Maria Audubon's 1897 publication of her version of her grandfather's journals becomes the crowning moment of the nineteenth-century adoration by making Audubon—even if quite awkwardly and unevenly—into a visionary conservationist. Maria's false apotheosis, however, marks the turning point in representations of Audubon's killing of birds. The 1880s and 1890s were a time of rapidly increasing awareness of major declines in wild animal populations in the United States. Among the most charismatic of these declines were those of the bison, passenger pigeons, Carolina parakeets, and alligators. The 1885 publication of Symington Grieve's *The Great Auk, or Garefowl* had made the great auk one of the best known examples of recent extinctions caused by human depredations. There was, then, culturally speaking, an elephant in the room, and the biographies register this development in the public view of so much killing.

Hezekiah Butterworth's 1901 biography intended for young male readers, *In the Days of Audubon: A Tale of the "Protector of Birds,"* reveals the newly felt need to remove Audubon from the degree of killing reported in *Ornithological Biography*. Butterworth was well known for his books for young readers that promoted patriotism, nature study, and Protestant piety, and he claimed that he blended fiction with facts only in order to create a "fiction true to the spirit of fact" (218) and that he occasionally altered Audubon's "words for the sake of a free, interpretative narrative" (128). Nevertheless, this book labors to minimize the numbers of birds Audubon killed and to assure the impressionable readers that every death was justified and caused Audubon pain: "He never destroyed life if he could help it. He studied living specimens, and when he could set them free he gave them again to the fields, air, and sky" (115). Butterworth's readers should also consider that Audubon the artist and naturalist could justify his killing of birds on the belief that "it will never need to be done again" (114) once he had immortalized each species in his art and natural history essay. Typical of Butterworth's method is his treatment of Audubon's golden eagle story. Butterworth presents the narrative from the *Ornithological Biography* of Audubon's failed attempt to kill the eagle he purchased, including the thrusting of the "long pointed piece of steel through his heart" (OB 2:465), but for the golden eagle, Butterworth substitutes "a caracara eagle, the Brazilian bird," and he moves the setting from Boston to St. Augustine, Florida. By these substitutions, Butterworth sends his hero to an alien place to kill an alien species, thereby diminishing his readers' emotional connection to the bird and removing the undesirable fact that Audubon killed so heartlessly a bird represented in one of the most noble and powerful of all the images in *The Birds of America* (plate 181).

The Audubon of John Burroughs's biography of 1902 is essentially a poet in paint, a passionate lover of nature, "a natural hunter, roamer, woodsman;

as unworldly as a child, and as simple and transparent" (x). Burroughs remains so focused on the personality of the man and his adventures that he raises no concerns about diminishing animal populations, not even the popular need for bird protection. From the Labrador journals, Burroughs makes no reference to the eggers whom Audubon condemned in an episode. His Audubon in Labrador is simply a lover of beauty and birds; Burroughs cites a moment of pity for devoted parent ring plovers as representative of the man's treatment of birds: "'I would not have shot one of the old ones, or taken one of the young for any consideration, and I was glad my young men were as forbearing'" (104). From so prominent a writer as Burroughs was in 1902, this book is remarkable for being so much pablum. It is an Audubon cleansed of gunning and who never voiced his objections to human depredations. Readers would have sensed vaguely a reason to admire the famous naturalist and artist, but Burroughs so reduced Audubon to a "picturesque" character (in the book's final paragraph) that they would not have known that Audubon ever did or said anything relative to conservation. It's a surprisingly innocuous image from an important nature writer.

The first attempt to complicate the character of Audubon occurs in the revisionist biography by Francis Hobart Herrick, *Audubon the Naturalist: A History of his Life and Time* (1917). Herrick announces that he had uncovered "new documentary evidence in surprising abundance" that "obliged" him "to draw conclusions contrary to those which have hitherto been accepted, and the new light thus obtained enables us to form a more accurate and just judgment of Audubon the man, and of his work" (1:22–23). Herrick's Audubon, then, because based on more primary evidence, has visible flaws, which should surprise no one, he notes, "for the most admirable of men have possessed faults" (1:6).

Even though Herrick wrote this in the wake of the extinctions of the Carolina parakeet and the passenger pigeon, the world was at war, and concerns about the preservation of wildlife species and environmental ethics were largely muted. Herrick generally reports Audubon's killing of birds without comment, tacitly accepting it as simply necessary for his work. He does not shrink, however, from quoting a statement from a letter Audubon wrote from Florida, dated December 31, 1831, that many people find one of the most alarming in all that he wrote: "The birds, generally speaking, appeared wild and few—you must be aware that I call birds few, when I shoot less than one hundred per day." Herrick was the first to quote this line since its initial appearance in an obscure natural history journal in 1832, so it was here that many people first saw this dimension of Audubon's gunning. Herrick, however, immediately offers a thorough explanation and defense, arguing that it would be wrong to conclude that Audubon "was a reckless destroyer of all bird life": "It must be remembered that this was over eighty years ago, when the unrivaled abundance of our birds was such that the necessity of their conservation had hardly entered the dreams of the most discerning." Readers also needed to consider that many specimens of a species were needed to determine all the features and variations of that species. Audubon was also supplying his assistant in Edinburgh with numerous birds for his anatomical drawings that were appearing in the *Ornithological Biography*. Finally, Audubon needed money to fund his work, and European museums and collectors were eager to purchase bird skins (2:17). Thus it was that in 1917 the public learned of a particularly disturbing remark from the supposed "protector of birds" and was given a rather thorough defense to consider.

After becoming the first biographer after Maria Audubon to complicate Audubon's image of innocence, Herrick also became the first to give Audubon the prairie epiphany created by Maria in her partially forged journal entry for August 5, 1843. Herrick does not use the word "extinction," but he gives Audubon the ability to foresee "the departure" of the bison as he quotes the problematic prediction that "before many years the Buffalo, like the Great Auk, will have disappeared" (2:255–56). With the publication of Herrick's groundbreaking, two-volume biography (and its subsequent reissue twenty-one years later), Maria's fiction had its desired effect. Audubon the visionary environmental saint was born.

But he didn't live long. In 1929 Howard Corning published his edition of

Audubon's manuscript journal from his Ohio and Mississippi River travels of 1820 and 1821. Audubon's own words couple accounts of astounding abundance of avian life with a complete lack of compunction about killing as many birds as possible—both as food for the passengers and crew of the boat he traveled on and to advance his collection of drawings. Here's a typical example of the previously obscured Audubon that many readers from 1929 onward have found disturbing: "Shortly after Breakfast I saw some *Terns Winowing* in the Eddy below us. Killed Two on the Wing—on the Falling of the first, the Second approach[d] as if to see What was the Matter. I shot it dead" (107). The publication of this journal made it all the more difficult for readers interested in Audubon to deny the symbiotic relationship between his gun and his paintbrush. Audubon himself points to the common origin of both his hunting and his art: "I believe few men can boast of having killed many of them [i.e., peregrine falcons], for 15 years, that I have hunted and seen probably one hundred I never had the satisfaction of bringing one to the ground" (57–58). Looking back, then, with Audubon from 1820 to 1805, his readers see that there never was a time when Audubon saw a bird and did not think of his gun. In the entry for March 16, 1821, Audubon's own syntax emphasizes the proximity of gun and art: "I took a Walk with my Gun this afternoon to see the Passage of Millions of Golden Plovers." While there is evidence of a hunting ethic in this journal (which I discuss later in a slightly different context), it's hard to hear above the gunfire.

In 1935, following the simultaneous publication of Aldo Leopold's *Game Management* and the creation of the Civilian Conservation Corps (CCC) early in 1933, Donald Culross Peattie published his song of praise for what Audubon represented to him: dismissing the "noble fiction" of Audubon the protector of birds, Peattie shapes an Audubon passionately focused on the study of natural life in the then still living wilderness. Leopold's *Game Management* shows the way to sustainable agricultural methods alongside healthy wild animal populations. The CCC was devoted to putting the Depression-era unemployed to work restoring America's devastated forests and eroded croplands. And Peattie's *Singing in the Wilderness* praises

Audubon as one who—at a time when we could not kill the bison or take down the forests fast enough—reminds Americans of where we went wrong. Even amid the economic concerns of the Great Depression, the 1930s were a time of increasing awareness of environmental loss. Peattie disparages the biographers who focus on the social, cultural, and family history of Audubon as "detailed biographers" (160). In the depths of the Great Depression, Peattie presents Audubon as a "national hero" (210) for his ability to see past the limits of his fellows' economic vision to the truer life of our natural environment. This is the Audubon who "gave everything he had for the most beautiful thing he could see" (164). To achieve this Audubon, however, Peattie hardly alludes to the gunning or killing. While admitting that the recently published 1820–21 journal presents "sometimes quite painful reading" (189), Peattie becomes the first author to find more meaning elsewhere.

Five years later, when Peattie published his lavish edition of Audubon's published prose, *Audubon's America: The Narratives and Experiences of John James Audubon*, his selections and introductions reflected the sentiments at work in *Singing in the Wilderness*. Peattie's idea was to let the breadth and tenacity of Audubon's testimony about a momentous time of unimaginable natural abundance and rapid change in "our heroic age" (3) persuade Americans to emulate his love of life and to adopt a love for the continent that had given life to American history and culture. For the first time in the twentieth century, however, Audubon's identity as hunter is emphasized. Peattie's young Audubon viewed "nature as a hunting preserve" (8). But if he was "an ardent sportsman," he was one "in the lavish style that wilderness conditions then made no crime" (253). Peattie contextualizes Audubon's killing in a way that excuses it: The Mississippi Valley "would have been unrecognizable for us. It was then still a wilderness, with no more than farmed clearings here and there." Many lived by hunting. "Men hunted the bear and the deer and the coon for a living, as they hunted cougars and Indians out of enmity for natural foes. God had put robins and thrushes in the forest to be eaten as surely as He had passenger pigeons and wild ducks" (14). In the final five pages of his 319-page narrative, Peattie brings Audubon onstage as an early—if not the first—voice for restraint in the destruction of the American bison. Over the fifty or sixty years preceding the publication of *Audubon's America*, the fate of the bison had been debated nationally from many differing perspectives. Once on the brink of extinction, "the dominant animal species of the largest biome on the continent" (Isenberg 196) had been allowed to survive amid the tangled motives of several government agencies, private ranchers, and the American Bison Society, which was founded in 1905 (Isenberg 166–86). The burly icon of America's wild western prairies appeared on the newly designed "Buffalo" or "Indian Head" nickel in 1913. The salvation of the bison was a popular story, and Peattie chose to close his narrative of Audubon's life by depicting the naturalist amid the legendarily vast bison herds in 1843, not at his home in New York where he died in 1851. Relying upon Maria Audubon's fabricated journal entries for August 5 and 8, Peattie allows his national hero to become a visionary conservationist by foreseeing the likely extinction of the bison and calling for the necessary restraint: "before many years the Buffalo, like the Great Auk, will have disappeared; surely this should not be permitted" (315).

In 1937 Stanley Clisby Arthur became the first biographer to be perfectly willing to have his readers be appalled by some of Audubon's gunning practices and to expose the rampant misrepresentations of Maria Audubon's edition of the journals as well as her destruction of many of the original manuscripts. His work for the state of Louisiana in the Department of Conservation as director of the Division of Wildlife may have left him disinclined to apologize for the behavior of a fellow sportsman. Arthur's close study reveals Audubon "as sportsman rather than student, as hunter-naturalist rather than ornithologist, one who loved nature more than he loved science, yet whose tenderness and love of bird life were coupled with the lust to kill the objects of his admiration . . . the wonder subjects of his facile pencil" (15). Having introduced him as hunter-naturalist and sportsman, Arthur does not withhold any of the accounts of shooting and killing from Audubon's 1820–21 journal (104–13). Arthur wants his readers to know that previous editors, and even Audubon himself, drew fictitious veils over reality when that reality included disturbing incidents. One very clear case is that of an opossum that was brought aboard the flatboat on which Audubon was descending the Mississippi River on November 14, 1820. Arthur explains that the original uncensored journal entry rather nonchalantly recounts the brutal conclusion of the small mammal's life: after Audubon's dog had mauled it apparently to death and it was thrown overboard, "the moment he toucht the Watter he swam for the Boats—so tenacious of Life are these animals that it took a heavy blow of the Axe to finish him." Arthur exposes the censoring by quoting the version of the same incident that Audubon published in 1835 entitled "The Opposum." In this episode a pair of opossums are thrown overboard because they pretend to be dead when crew members harass them, but when they then swim back to the boat, the "poor things" "were taken up, and afterwards let loose in their native woods" (111). Arthur's Audubon is a "many-sided human being," and because Arthur believed the man himself "is far more interesting than aught he accomplished" (13), he wanted his readers to know that much of what had been reported about Audubon had been censored or bowdlerized by editors, especially Maria R. Audubon. Although this granddaughter claimed to present "'only Audubon the man' before her readers, 'and in his own words so far as possible,'" Arthur explains, she "then suppressed the very passages in those documents that would more clearly illumine his true character" and burned the originals (14). With regard specifically to Audubon's journal for 1822, Arthur quotes from a letter that Maria wrote in 1905: "*I burned it myself* in 1895" (243). Thus it is that since the publication of *Audubon: An Intimate Life of the American Woodsman* in 1937, we have known that Audubon's actual beliefs about how humans should live in relation to animals and the natural environment have not been faithfully reported and that the truth has been deliberately obscured.

Framing and imbuing the decade of the 1960s—the decade of Rachel Carson's *Silent Spring* (1962) and the birth of the modern environmental movement—is the interpretation of John James Audubon as an emerging conservationist created by Alice Ford. In stark contrast to Arthur's frank recounting of Audubon's passion for hunting, Ford opens her 1957 *The Bird Biographies of John James Audubon* with assurances to her readers that she will include no passages that "would offend the squeamish and outrage the conservationist" (ix–x). Throughout this handsome volume (including color reproductions of twelve of the original watercolor paintings for *The Birds of America*), as she selects and freely edits and rewrites bird biographies from *Ornithological Biography*, Ford tones down the killing. Her readers will not see the figurehead of the National Audubon Society shouting "Oh that we had more guns!" in the midst of a gunning frenzy in the Canada goose biography (OB 3:1–18). Nor will they witness the valiant, wounded whooping crane's death by boat oar in the biography of the only species ever to set Audubon to flight (OB 3:202–13). And that hour Audubon spent on a Newfoundland island shooting puffins on the wing, assisted by two sailors reloading his guns, will trouble the dreams of none of her readers (OB 3:105–11). Similarly, the full biography of Audubon that Ford published in 1964 avoids discussion of the gunning and focuses instead on the challenges he overcame on his road to success and on his relationship with his beset wife. In the latter portions of the biography, however, Ford does raise the issue of conservation by showing that a sensitivity to human depredations was awakening in Audubon beginning in 1833 on the expedition to Labrador. There, when Audubon witnessed the objectionable overharvesting of eggs, Ford insinuates (without evidence, for there is none) that Audubon reflected on "his own past excesses; as recently as the Floridas cruise he had called the day that yielded no more than a hundred birds a poor one" (*John James Audubon* 303). In her treatment of the Missouri River expedition, Ford distinguishes Audubon from his companions by claiming that he "was not guilty himself of wanton killing" and foresaw with sorrow the extinction of the bison (401).

In *Audubon, by Himself* (1969) Ford works to make her Audubon more relevant to readers' heightened awareness of ever-worsening environmental crises in the late 1960s. (The year after the publication of Ford's book saw the passage of the Clean Air Act, the founding of the Environmental Protection Agency, and the first observation of Earth Day.) Drawing upon Audubon's writings to create a first-person narrative of his life in his own voice, the editor arranges for this version of Audubon to take the stage as an enthusiastic sportsman or hunter-naturalist but presents him as maturing into a thorough conservationist. The younger Audubon in this narrative is a hunter passionately driven by desire to possess a prized bird, the Bird of Washington (38), and he is an enthusiastic sportsman hunting swans with Shawnees (45). Later, however, Ford relies upon several of Audubon's published "episodes" to show her more mature Audubon providing critical testimony about heedless, unsustainable depredations by humans on the live oak trees of Florida (183–85), on sea turtle eggs in the Florida Keys (200–205), in the forests of Maine (212–15), and on ocean bird eggs in Labrador (231–35). She completes the conversion of the nineteenth-century hunter-naturalist into a 1960s environmentalist by dramatically concluding the narrative at the moment in Maria Audubon's edition of the Missouri River journals when Audubon, profoundly moved by the slaughter of bison he has witnessed, utters the "Great Auk Speech": "But this cannot last. Even now the herds are smaller. Before many years the Buffalo, like the Great Auk, will have disappeared. Surely this should not be allowed to happen" (258). Over the course of twelve years, through three books, Alice Ford composed a saint's life.

However, Saint Audubon had a more extreme and slightly earlier apotheosis in Alexander B. Adams's biography, published in 1966. The five-page prologue comprises a belabored expiation for the environmental sins committed by nineteenth-century Euro-Americans, while making no direct mention of Audubon. The Americans of Audubon's day and before cut down the forests and wiped out the wild animal populations because the supply seemed inexhaustible: "Shooting a hundred birds, or five hundred birds, or a thousand birds, was like taking a cup of water out of the ocean" (9). In the opening of his biography of Audubon, Adams depicts a young nation amid incredible natural bounty in need of a visionary leader: "What they needed were men to do their seeing for them, whose vision reached a little further than theirs, who did not have to wait until a bird or an animal or a tree disappeared to know it could disappear." This visionary would "make them think of golden plovers as something more than a crop that could be harvested free except for the cost of shot" (12–13). Adams himself was a devoted conservationist, which the dust jacket copy makes clear, and he seems to assume a readership drawn from the post–*Silent Spring* 1960s who will be predisposed to condemn Audubon for all the killing; thus he labors to explain why his readers have no right to judge a man of Audubon's visionary powers. Adams goes so far as to argue that Audubon's father, Jean Audubon, was not culpable for owning and profiting from a plantation on San Domingo worked by slaves, arguing that the slave economy and French imperialism saved the "deserted, poverty-ridden island" from "reverting slowly to the semijungle Columbus had discovered" (15–16). The chief weakness of Adams's Audubon was that he "misused so many years" trying to become a merchant (471). Once the great American naturalist realized that he must devote his life to the completion of the images in *The Birds of America*, almost anything could be justified, Adams suggests. Adams depicts Audubon condemning various abuses of animal populations, including bison on the 1843 expedition (in his treatment of which Adams cites the "Great Auk Speech" [461]), but what redeems Audubon is the power of his art to inspire people, to teach them to "see a bird for what a bird is" (472). Adams's anger at his fellow Americans for the environmental corruptions and losses that accelerated dramatically after World War II seems manifest in his image of Minnie's Land, the Audubon family home in New York, as having "been encircled by steel-framed, crime-filled buildings" following Audubon's death, "which engulfed the stream and its two ponds and made the edge of the river an unsafe place for a grandfather to sit and dream in the twilight" (471). Adams's strategy was to offer Americans an Audubon whose environmental vision could "give the rest of us one last, irretrievable chance to save what is left" (472). Adams's biography seems an act of desperation by a 1960s environmentalist—and it incited the following reaction.

The removal of all saintly associations from Audubon's reputation is the announced intent of the British bookseller and sometime author John Chancellor's *Audubon: A Biography* (1978): "Audubon was not a latter-day St Francis of Assisi and he indulged in what appears to us today to be revolting and wantonly indiscriminate slaughter of birds" (7). While it is fitting that the National Audubon Society is named for a sportsman because hunters, "we are told, are the best conservationists," the society has, Chancellor insinuates, "perhaps deliberately" detached its name "from the man, whose record as a protector of wildlife was not a good one" (8). Chancellor's portrait presents "a curious character, with good and bad points," who "becomes steadily more unattractive as the book progresses"—until, that is, "the year 1826 when he landed in Liverpool and went on to Scotland" (7). This Briton is certainly correct to emphasize the crucial role England played in Audubon's achievement by supplying both his wife and his engraver, but Chancellor's willingness to employ clearly underhanded methods to achieve what amounts to an attempt at character assassination seriously compromises his credibility. He is also justified in his emphasis on Audubon's own accounts of his killing of many birds and other animals. But his presentation of this material is clearly designed to incriminate. When he quotes the account of birds and squirrels killed on the first day of gunning in the journal of 1820–21, for example, Chancellor does not mention that most of these were shot to feed the people on the flatboat (88). When reporting Audubon's intent in Florida to shoot twenty-five brown pelicans, Chancellor pretends to naïveté: "Why he should have needed so many birds for a single drawing is curious" (178). Yet he omits the overt explanation that Audubon provides: to sell the valuable skins. But Chancellor's contempt for Audubon takes him further than elementary biased reporting. While Audubon's credibility is easy to attack on some points, a chief emphasis in Chancellor's portrait is on Audubon as a liar. When he recounts the story Audubon wrote about having to redraw some of his early drawings after they were destroyed by Norway rats (OB 1:13–14), Chancellor casts the general aspersion—with no hint of probable cause—that the story may be "perhaps—knowing the author—untrue" (90). After relating the story from the journal of 1820–21 of Audubon's having left his portfolio of drawings behind at Natchez, Chancellor resorts to gossipmongering: "His spirits were low because he had lost somewhere in Natchez—perhaps in a brothel or gambling den—a portfolio containing some of the drawings which he had done since leaving Cincinnati" (96). Whether Audubon left the portfolio behind in "a brothel or gambling den" is unknown, but there is utterly no evidence to support either suggestion, whether this biographer qualified them with "perhaps" or not. In a way, one can see the basic theme and method of Chancellor's book emblemized on the first two pages of the first chapter, where the story of Audubon's birth faces a full-page illustration of Audubon's great black-backed gull. Of all the 435 plates in *The Birds of America*, this one most emphatically depicts the agony of a bird wounded by a human. The gull has crashed to the ground, its right wing is raised high, its uplifted, gaping beak and extended tongue suggest a cry of pain, and blood flows from the shoulder wound and puddles beneath the bird. Audubon shot the gull and painted it himself, but this image has nothing to do with the story of his birth. Chancellor has placed it here to establish that this biography will portray a butcher of birds.

The roller coaster of representations took another dramatic twist in 1980 with *On the Road with John James Audubon* by the married team of Mary Durant and Michael Harwood. Without mentioning Chancellor by name, they dismiss his book as not "of the first rank" (619). Their Audubon is a romantic and "heroic figure in American lore" (xii) who is largely responsible for our ongoing efforts to preserve bird species: "He is the focus, the wellspring of our bird-protection" (211). While acknowledging that it has "become common to see Audubon assailed as a bird-butcher" (209), Durant and Harwood want to convince their readers that this is a ridiculous charge because it is anachronistic and ignores the scientific need to gather many specimens at the beginning of American ornithology. Audubon could not have completed his great work without killing many birds (210). The determination of boundaries between species depends upon the examination of many specimens of each species, as does the determination of the plumage of females and of young and immature stages of growth. Audubon also advanced his work and raised money by shooting many specimens to give or sell to European "bird-men" (211). When they anticipate their readers objecting that Audubon took part in many slaughters with no scientific purpose, Durant and Harwood respond by acknowledging that he did indeed enjoy the sport of gunning but that "*everyone* thought wildlife on our fecund continent was too numerous to be destroyed." "Going gunning then, and being good at it, was a legitimate sport—just as golf or bowling today" (211). Underpinning their argument is the proposition that most of the lives of the birds Audubon killed were necessary sacrifices for the work that he would ultimately produce and for the subsequent inspiration of others to organize societies and pass legislation for the protection of birds and other wildlife. Our hero definitely was excessive at times, but taken all in all, his gunning had constructive results for our culture and our wildlife. Durant and Harwood also present Audubon as one of the first to recognize that humans were diminishing some bird populations: "Consequently, he changed his behavior, and tried to get others to change theirs" (212). To bolster their interpretation, these authors emphasize those moments in Audubon's narratives when he was moved to peaceful coexistence with the birds, such as this moment in the Florida Keys: "Although desirous of obtaining the birds before me, I had no wish to shoot them at that moment. My gun lay loosely on my arms, my eyes were rivetted on the Flycatcher, my ears open to the soft notes of the Dove. Reader, such are the moments, amid days of toil and discomfort, that compensate for every privation" (qtd. at 375). They also omit evidence that weakens their case, such as Audubon's letter from Florida in December 1831 in which he defines shooting "few" birds as fewer than one hundred each day. The Missouri River journals tell a truly bloody story of many hunting and gunning experiences, and Durant and Harwood expunge practically all of that detail, thereby presenting a gentle and sanitized version of Audubon's last expedition. Despite all the excitement and high adventure of his time at Fort Union, this Audubon now longed romantically for "his civilized hearth, surrounded by grandchildren" (566). Thus the authors produced a romanticized image of the grandfather of American conservationism sitting by the warm glow of his hearth surrounded by grandkids—for which there is no evidence.

Just five years later, in a superb narrative essay (richly illustrated) about the Missouri River expedition, "Mr. Audubon's Last Hurrah," Harwood revised the final impression of Audubon left by *On the Road with John James Audubon*. This revised Audubon does not long, toward the end of the trip, for "his civilized hearth, surrounded by grandchildren"; he rather "regrets having promised his family that he would be home in the fall" (110). Arranged like a journal, this essay treats fully the gunning and hunting behavior of all involved, including the aggressively competitive spirit of Audubon's companions that emerged during the organized bison hunts. In this version of the story, Harwood does not shrink from acknowledging how bloody, violent, and dangerous these hunts sometimes became. He cites Audubon's friend Edward Harris's prescient statement of regret that they had "destroyed these noble beasts for no earthly reason but to gratify a sanguinary disposition which appears to be inherent in our natures" and their pang of conscience that left them "completely disgusted with ourselves and with the conduct of all white men who come to this country" (qtd. at 110). Thus Harwood presents the killing glossed over in the previous book and raises the difficult ethical questions about that killing but shows Audubon to be among the first Americans to yearn for a restraint not yet seen on this continent. Harwood is the first to see the importance of Audubon's journal entry for December 12, 1826, where, from Edinburgh, he calls for Walter Scott to come to America to describe the wilderness and thereby "[w]restle with mankind and stop their increasing ravages on Nature." Harwood suggests that in Audubon's later writings, he himself supplied "the Scott-like portrait of his adopted country" (95). On the 1843 trip, this Audubon is complicit in much of the killing but also an early voice for restraint; Harwood cites a line from Maria Audubon's fictional August 8 entry: "My remonstrances about useless slaughter have not been wholly unheeded" (qtd. at 110). Just as Durant and Harwood avoided the "Great Auk Speech" because of their suspicions about Maria Audubon's edition, so does Harwood here because, as he writes, "Her version of what Audubon wrote remains a useful source of information but must be quoted cautiously and double-checked against other sources whenever possible" (116). For a writer so immersed in the life and mind of Audubon to resist the temptation to grant his Audubon the "Great Auk Speech" required strong suspicions and admirable restraint.

After complimenting Durant and Harwood for *On the Road with John James Audubon* ("it set a tone that I found exhilarating"), Shirley Streshinsky explains in *Audubon: Life and Art in the American Wilderness* (1993) that she "did not set out to break any new ground; I wanted simply to tell the full, fascinating story of an extraordinary life against the background of a young, vibrant America" (xvii). This version of Audubon's "fascinating story" has the structure of a conversion narrative, not from sinner to saint, but from a man with a more or less typical early nineteenth-century gunning ethic to a visionary artist whose graphic representations of birds justify his name's association with today's conservation movement. Streshinsky acknowledges that Audubon did some killing that many of her readers will condemn, and she wants to teach them to see that killing in the context of Audubon's day: "[H]is was an era with a different ethos, and [...] he reflected the attitudes of his times" (xviii). This biographer proceeds, however, to diminish the impact of the gunning by recounting very little of it and by carefully contextualizing the little that she mentions. She renders the abundance of game tallies from the 1820 descent of the Mississippi River ethically neutral by ensconcing them in Audubon's dedication and hard work: "Finding, shooting, and drawing the birds was his all-consuming mission. He was focused now; the goal was set; to reach it he would have to work as hard as he ever had. He hoped to draw one bird each day" (109). In her presentation of one of the earliest tallies of animals killed, from the journal of 1820–21, which some have used to condemn Audubon, Streshinsky deftly directs attention away from the tally to young Joseph Mason's early success: "Audubon brought back 'thirty Partridges—1 Wood Cock—27 Grey Squirrles—a Barn Owl—a Young Turkey Buzzard.' Audubon taught Joseph how to shoot, too; managing to kill three wild turkeys, the boy 'was not a little proud when he heard 3 Chears given him from the Boats.' It was one of the rare bright spots in an otherwise dismal journey" (108). Like Michael Harwood before her, Streshinsky cites the important journal entry of December 12, 1826, as evidence that Audubon was among the first Americans to express concern about environmental degradation and changes to the landscape brought on by westward expansionism and commerce (191).

In her treatment of the Missouri River expedition, Streshinsky nominally admits that much killing occurred, but she distances Audubon from it by emphasizing his role as spectator (355). She also distinguishes Audubon from the others, like Harris, who might have regretted some of the excessive killing but still pursued the hunt at every opportunity: "Audubon made a few short forays out of the fort, usually riding in a cart or carryall. Sometimes they camped out overnight. While the others were hunting, Audubon stayed behind to fish, and if the hunters came back without any fresh meat for their dinner, three or four catfish would be waiting for them. On these one- and two-day trips, Audubon could see the countryside, glimpse its wildlife, live the rough life demanded in the wilderness" (355–56). By 1843, Streshinsky's Audubon has acquired or shaped a humbler environmental ethic than that of his contemporaries, but, again like Harwood, Streshinsky rejects Maria Audubon's edition of the journals as unreliable and does not cite the "Great Auk Speech" (xiv). Instead, she discusses a statement published in the essay on the American bison in *The Quadrupeds of North America* to leave her readers with a visionary Audubon: "If others thought the great herds would last forever, Audubon did not. He considered the bison to be a link to other huge American animals, now extinct, and he predicted its demise" (355).[2] In Streshinsky's biography, Audubon's ethic and his love for the wild emerge most apparently from the affective art of the plates of *The Birds of America*. His writings, however, also echo "with warnings; he saw the changes civilization wrought, and he was among the first to sound the alarm" (xix).

The three book-length biographical studies of Audubon that appeared in 2004 all agree that we should view his gunning practice as consistent with a spirit of conservationism and the needs of science, but each does so from its own perspective.

While highlighting the crucial support Audubon received in Liverpool in 1826, especially from the Rathbones at Greenbank, their family seat, the British writer and naturalist Duff Hart-Davis casts his account of any of Audubon's gunning as his lust for knowledge, not blood. Hart-Davis presents Audubon's infamous definition of "few" from the Florida expedition in 1831 in the forgiving context of scientific and financial necessity (190). Contrary to the usual interpretation of Audubon's bird biography of the golden eagle as evincing his heartlessness, Hart-Davis builds pathos for his Audubon by attributing the illness he experienced after he killed the bird to overwork and to his compassion for the bird (195).

Like many before him, in his presentation of the Missouri River expedition Hart-Davis renders an Audubon who undergoes a late conversion to a much humbler ethic. At the center of this brief (three-page) treatment of Audubon's last expedition, Hart-Davis quotes the "Great Auk Speech" (264). This author stresses that behind the killing was a man with compassion for animals and one who, at least late in his career, exercised restraint.

---

2   Streshinsky concludes that Audubon predicted the extinction of the bison from the following statement apparently added by Bachman to the American bison essay: "perhaps sooner to be forever lost than is generally supposed." This statement was first published in 1851 (the year of Audubon's death) in volume 2 of the three-volume edition of the essays to accompany the two-volume set of the lithographic plates, both bearing the same title, *The Viviparous Quadrupeds of North America*. The original manuscript of the bison essay at the Beinecke, which Audubon prepared to send to Bachman, does not include the reference to the fear of extinction. See my discussion of this manuscript, pages 21–22.

In *Under a Wild Sky: John James Audubon and the Making of "The Birds of America,"* William Souder hopes not only to persuade his readers not to condemn Audubon for his killing of so many birds but even to teach them to admire Audubon for being such a good shot, especially considering the conditions in which he hunted and the kinds of guns he used. More so than any other biographer, Souder views Audubon's gunning ethic from the perspective of the hunter. "To the uninitiated it seems a loud and cruel pastime. To the hunter, there is something ineffable yet almost physical in the pleasure of taking a bird from the air. The tang of fall on the wind, the swing of the gun, its powerful slam against the shoulder, the long, slanting parabola of the stricken bird falling to earth—all of it is an experience that, for some, begs repeating" (98–99). Unlike all earlier biographers, who step gingerly around the fact of Audubon's killing of so many thousands of birds, Souder asserts that it "is wrong" to excuse Audubon's gunning as "a function of having no camera, no modern optical device with which to 'capture' his subjects unharmed." We need make no apology for Audubon's gunning: "Many things *were* different in Audubon's time, but field ornithology has in truth changed little since then. Modern ornithologists still collect bird specimens all over the world. They still shoot them with shotguns. Many of these same scientists are both conservationists and avid bird hunters" (99). Souder engages his readers so confidently on this point that he is willing even to exaggerate Audubon's 1831 definition of "few" birds: "He sometimes said a day in which he killed fewer than a hundred birds was a day wasted" (97). Audubon said this once, not "sometimes," and he said it was a day on which he shot few birds, not a day wasted; nevertheless, Souder suggests that it is time for us to get past our collective need to be troubled by the gunning conducted by one of our icons of conservationism (138).

Like William Souder, Richard Rhodes sets Audubon's gunning in the context of early nineteenth-century hunting codes: "Audubon engaged birds with the intensity (and sometimes the ferocity) of a hunter because hunting was the cultural frame out of which his encounter with birds emerged" (74–75). In Audubon's day, Rhodes argues, observing wild animals for almost any purpose involved a gun: "Much of what seems contradictory in his narratives—learning birds, studying birds, concerning himself with their population dynamics and stresses but also killing birds for food and for sport, sometimes in great numbers—follows from this fact. To argue that he should have known better is anachronistic and nostalgic." Beyond the anachronism, Rhodes wants his readers to consider how removed we are today from the killing of all the domesticated species of animals consumed as food (75). A scent of hypocrisy attends the condemnation today of Audubon as cruel and heartless. With this idea in place, Rhodes frankly acknowledges the extent of the killing that Audubon reports in the journal of 1820–21 (157). Audubon's killing is not driven by bloodlust; it all serves the practical purposes of feeding the people on the boat, advancing Audubon's nascent ornithology, and finding new birds for him to draw at a time when the killing of "animals and birds was unremarkable" (375). Rhodes presents a similar argument in his treatment of Audubon's awkward and bungled attempt to asphyxiate the golden eagle in Boston in 1833. Readers should consider that Audubon "believed he was performing a service, conducting an experiment, hoping to devise a less painful way than shooting or throttling to kill and to preserve" (375). Rhodes then establishes for his Audubon a clear conservationist leaning, evident in the Labrador expedition of the same year. When men conducted the "prodigal expropriation and slaughter" for commercial markets on the unsustainable scale that Audubon observed on the gannet rocks and elsewhere on this expedition, Audubon was moved to write eloquently against the unregulated practices, and Rhodes emphasizes the conservationist implications of these writings (385–86). While giving the Missouri River expedition short shrift (as does every other biographer), Rhodes stresses the occurrence of a change in Audubon on this trip. Even at the arrival at Fort Union, where the "first order of business" was "hunting buffalo," according to Rhodes, "[s]port hunting no longer appealed much to Audubon" (427). And after witnessing "the continual slaughter of buffalo and wolves in the neighborhood of the fort," Audubon began to protest: "He was impressed on August 8 when a hunter 'and two men had been dispatched with a cart to kill three fat cows but *no more*'; he felt his 'remonstrances about useless slaughter have not been wholly unheeded'" (428). In the next paragraph Rhodes sets Audubon's prescient ability to imagine the future near-extinction of the bison in dramatic relief by juxtaposing it with a report Audubon recorded of bison herds so extensive that they blackened the snow-covered hills, "so crowded was it with these animals [. . .]. In fact it is *impossible to describe or even conceive* the vast multitudes of these animals that exist even now and feed on these ocean-like prairies" (qtd. at 428). With only ten pages remaining in this biography, Audubon the visionary conservationist on the northern plains looms into view.

In the following chapter, "The Lived Ethic," I will examine the available evidence of what Audubon believed about how humans should live in relation to nature. This evidence spans the time from 1820 to 1844, which is from the earliest surviving journal until shortly after his return from the Missouri River expedition. I'm tempted to comment that collectively the evidence comprises an incomplete tangled mess and that Audubon in this regard is not different from most people I know. He was not a systematic thinker but my goal is to discover which beliefs were least subject to change as his circumstances or motives changed and which beliefs or behavior were more the product of passing or fleeting or anomalous vagaries.

The Lived Ethic

If the conditions of life that John James and Lucy Audubon faced in October 1820 were not truly desperate, they were very nearly so. After their marriage in 1808 and migration down the Ohio River to Louisville and then on to Henderson, Kentucky, the Audubons lost, after some initial success, their family wealth and possessions. Practically all American businesses dependent upon trade with Europe suffered when the War of 1812 broke out in June of that year. A few months prior, Audubon had invested heavily with his brother-in-law Thomas W. Bakewell in a trade venture based in New Orleans. That money was soon lost. Audubon's former business partner Ferdinand Rozier owed him money but was unable to pay. The steam-powered grist and lumber mill that Audubon built with Thomas Bakewell and a third partner opened in 1817, but the demand for its services was not what they had hoped. Early that winter, the Audubons faced the sad loss of the two-year-old daughter they had named for Lucy's mother, Lucy Green Bakewell. When his two partners withdrew from the mill enterprise, the full economic burden fell onto Audubon's shoulders. A new emotional stress came when he learned in the spring of 1818 that his father in France had died on February 19. Because relatives were contesting the will that left property to Audubon, he became worried that news of his carefully concealed illegitimate birth might reach Kentucky if they sought to locate him (Herrick 1:248–57; Ford, *John James Audubon* 94–102).

The next year brought his struggles as a businessman to an end. In May 1819 a Mr. Samuel Bowen, who had purchased a small steamboat from

Audubon but with "worthless paper," absconded with the boat to New Orleans, where he then passed the boat over to someone to whom he was indebted. Although Audubon pursued Bowen down the river, he failed to capture him or secure the boat. When Audubon returned to Henderson, he was told that Bowen had also returned and had threatened to kill Audubon. Bowen attacked Audubon with a club of some sort, but Audubon managed to stab him with the dagger he had deliberately carried for protection. This he did with his right hand in a sling because he had injured it in the mill. Even though he was found innocent in court, this was a truly distressing time for both Audubon and Lucy. The final blow, however, came with the closure of the mill that year, following which Audubon was arrested for debt and imprisoned in Louisville. Bankruptcy and the loss of possessions ensued. Around this time, Lucy would give birth to her fourth child, Rosa (Herrick 1:257–60; Ford, *John James Audubon* 112).

Audubon then found what appeared as promising employment as a taxidermist and artist in Dr. Daniel Drake's new Western Museum in Cincinnati. Audubon went alone at first, leaving Lucy in Henderson with their two sons and infant daughter. Before Lucy could move her family to Cincinnati, the seven-month-old Rosa died and was buried in Louisville (Ford, *John James Audubon* 112). After several months of work and irregular payments of his otherwise respectable salary, and even after some small success offering drawing and painting classes, Audubon and Lucy saw that the museum would not sustain their family.

Clearly, that was a rough period for the young couple, but these were the challenging conditions that set the stage for Audubon's departure from Cincinnati on a flatboat on October 12, 1820, with practically no money and only his artistic talents to support him and his young assistant, Joseph Mason, on their descent of the Ohio and Lower Mississippi Rivers. Audubon was profoundly beholden to Lucy, not only because he had uprooted her from her comfortable Bakewell family home some twenty miles from Philadelphia and lost in bankruptcy the money, furnishings, and other possessions she had brought to their partnership, but also for her having born four children and endured the loss of the two little daughters. Together the couple decided that their best chance for a future prosperity lay in making a serious, frightening gamble. The watercolor drawings of the birds of America that Audubon had developed by this time seemed promising enough to them to send him off to complete and publish them. It was a long shot, to say the least, but no other viable options appeared.

Also affecting Audubon's mood on this trip was the embarrassment of his financial dependence. While the terms of Audubon's agreement with Captain Aumack for his and Mason's passage are only cryptically indicated in Audubon's journal from this trip, in words of caution addressed directly to his two sons he reveals that Aumack offered them free passage in exchange, presumably, for the hunting they would do to feed passengers and crew on board. But on the evening of November 16, 1820, a misunderstanding or disagreement with Aumack that troubled the waters moved Audubon to advise his sons: "Never take a Passage in any Stage or Vessel without a well understood agreement betwen You & the owners or Clerks & of all things Never go for Nothing if You Wish to save Mental Troubles & Body Viscissitude."[3] The next day, when the boat passed from the Ohio into the Mississippi River, Audubon's embarrassment hurt him again when he recalled the last time he had traveled through this confluence, eleven years earlier, with his business partner and a freight of goods to sell: "Now I enter it *poor* in fact *Destitute* of all things and reliing only on that providential Hope the Comforter of this Wearied Mind—in a flat Boat a Passenger." In these circumstances and in this frame of mind, as one can easily imagine, Audubon would not be considering much the individual lives of birds and other animals it was his job to shoot.

Another major factor in this context is the truly amazing abundance of avian life reported in this journal. On the simplest level are the daily game tallies, most of which would be eaten by those on the boat. In the very first

3   This and the following quotations are taken from Audubon's manuscript journal, 1820–21, at the Ernst Mayr Library, Museum of Comparative Zoology Archives, Harvard University.

entry of the journal, Audubon reports what becomes a completely typical tally: "We shot thirty Partridges—1 Wood Cock—27 Grey Squirels—a Bared Owl—a Young Turkey Buzard and an Autumnal Warbler." On November 14 one of their party "[k]illed 26 Starlings Sturnus Pradetorius, all Young— eat them at our Supper, good & Delicate." Flying over the river and above the endless forests on each bank are flocks of ducks, geese, mergansers, grackles, finches, and numerous others. As the flatboat drifts ever farther down the river, ivory-billed woodpeckers occur in greater numbers and frequency, along with the immense flocks of Carolina parakeets and red-winged blackbirds. He gains an even vaster perspective on the incredible bird populations when he reaches New Orleans and visits the markets: "[T]he Market is regularly furnished with the *English Snipe* Which the french Call *Cache cache*, Robins Blue Wing^d Teals Common Teals, Spoon Bill Ducks, Malards, Snow Geese, Canada Geese, Many Cormorants, Coots, Watter Hens, Tell Tale (Godwits, Call^d here *Clou Clou*) Yellow Shank Snipes, some Sand Hills Cranes, Strings of Blew Warblers, Cardinal Grosbeaks, Common Turtle Doves, Golden Wing^d Wood Peckers &^{c}." The most extreme evidence of avian abundance in this journal occurs in the entry for March 16, 1821: "I took a walk with my Gun this afternoon to see the Passage of Millions of *Golden Plovers* Coming from the North Est and going Nearly Ouest." Of the "Millions," according to Audubon's reckoning, four hundred gunners that day killed 144,000 plovers. When Audubon asked whether such passages occurred frequently, he was told that six years before the same thing had happened.

Although a complex of motives and conditions drove Audubon to kill whatever he needed or wanted on this trip without compunction, this journal does show that he saw the need for restraint in some circumstances. He clearly understood the cause of and correction for the local extinction of game birds. After having regularly shot numbers of the bobwhite quail on his descent of the Mississippi River, he learned of their scarcity near New Orleans: "[T]hese Birds are here much sought and hunted down without Mercy, not even do the Sportsmen permit a few Paires to remain untouch &

thereby the race is nearly extinguished Near the City" (February 25, 1821). He's no Aldo Leopold, but he does show the ability to imagine a situation in which the sportsmen of a region would cooperate to allow the recovery of a species. Another instance shows that if he could do his work without killing so many, he would: "This afternoon having finished My Drawing of the Red Cockaded Woodpecker and satisfied of its Correctness by a Close comparison to the living original I gave it its Liberty, and was glad to think that it most Likely would do well as it flew 40 to 50 Yards at times and seemed Much refreshed by its return to Liberty" (July 29, 1821).

The entry for August 12, 1821, might serve to epitomize Audubon's gunning ethic at this point in his career. First, the passage reveals three principles of restraint that have at least occasionally guided him all along. He will not shoot a bird if he cannot retrieve it: "Saw a White Ibis on a Log where it sat a Long time arranging its feathers using its scythe Shape^d bill very dexterously; Could have killed it but having No boat and afraid of sending a Dog in the Lake Left it setting peaceably." On this trip at least (he will suspend this principle later when he begins to sell more bird skins), he will not shoot a bird if he has already drawn its species. In this instance, he has already drawn the "White & Black Creeper" and momentarily believes that a "Yellow Throated Warbler" (which he has not yet drawn) that is flying toward him is another "White & Black Creeper." He realizes his error and explains, "[H]ad I not seen one fly directly toward me and discovering then the beautiful & rich Yellow Throat I would not have shot one." Similarly, he will not shoot a bird if he will not be able to draw it before it begins to decay: "[S]aw only the one I shot to day and having as much as I Knew I could Well draw before they would be Spoiled by the heat of the Wheather *returned* to the House." In this, the earliest of Audubon's journals to have survived, it is clear that, although he kills many birds, he is often guided by an ethical impulse, rudimentary and inconsistently applied as it may be. Later in the same, lengthy entry, Audubon writes of having shot two Mississippi kites, a young one and its mother, in a way that might have elicited pathos for the birds, but he shows none: "[T]he young was too far gone to relish food—the Mother exibited much distress and after several trials to Make the young Bird take it [a grasshopper] it dropt it and taking [h]old of her Offspring by the feathers of the back carried it off with ease for about 25 yards to another tree where I follow^d and killed both at One Shot." Intending to draw both birds later, he "purposely hided them under a Log, but on my return some quadrupedes had discovered them and eat both—I regret much the Loss." When he later writes of this mother and her young one in *Ornithological Biography* (2:108–13), he expresses sympathy for his victims, but in his journal in 1821, he is all business because this species is hard to procure and he has not drawn it before. When he loses them to another predator, he expresses regret over his loss, not the waste of life. Thus, while there are traces of a gunning ethic and some impulse toward restraint in the journal of 1820–21, the most powerful motive is his need to advance his project of drawing his country's birds for his family's future prosperity. The power of this motive surprises even himself: "Nothing but the astonishing desire I have of Compleating my work Keeps my spirits at par" (January 6, 1821). Yes, and his ethic at bay.

Because none of Audubon's journals from 1821 to 1826 has survived, the next important document that sheds light on the history of what we might loosely refer to as his conservationist thought is his journal of 1826. Having failed to find an engraver and publisher for his bird paintings in Philadelphia or New York City, he sailed to Liverpool to seek one in England or Scotland. The contrast between the tamed environment he saw there and his beloved American wilds gave him a perspective that was new to him. Ever since late summer 1803, when he was an eighteen-year-old French emigrant, Audubon had enjoyed the freedom to roam the American countryside, and ever since his move to Kentucky in the spring of 1808, he had enjoyed the freedom to explore the wilder, unsettled regions. Even before the ship *Delos* reached the Liverpool docks, the landscape in view disappointed him: "[B]ut what nakedness the Country exibits—scarce a patch of Timber to be seen—our fine Forests of Pines, of Oaks, of heavy Walnut Trees, of magnific magnolias, of Hickories, or ash, or Sugar trees; and represented here by a {diminute} growth—named *Furze*" (49). In America over the preceding twenty-three years, Audubon had become a prodigious walker, once claiming that he could walk five miles per hour practically all day long (AJ 1:270). Just three years earlier, with his elder son Victor, he claimed to have walked the 250 miles from the mouth of the Ohio River to Henderson, Kentucky, in seven days (OB 3:371–75). It is easy to understand, then, that he would disparage the confining walks offered in tame England, even from the home of his favorite British family, the Rathbones: "[W]e walk^d between dreary Walls contenting ourselves with the distant objects without the sweet privileges of moving freely to & fro to right or Left or to advance towards any particular object that might with a wish attract the Eye." He wished he could have shown his friends what he was missing just then: "our Wild Scenery of the Woods of America." This contrast between American wilds and the "dreary Walls" of Britain moved him to complain to Lucy that "in England, all is Hospitality within, all is Aristocratic without their Dwellings—No one dare *trespass*, as it is called, one foot on the Grass" (*Journal of 1826* 122).[4]

If he was disgusted by the aristocratic control of the landscape, he was even more disgusted by the raising of game animals in enclosures, to be released and shot at the leisure of the landed classes. In reaction against what he perceived as a cruel practice (on a British lord's estate near Manchester),

4 A letter by Mrs. Hannah Mary Rathbone, the matriarch of the Rathbone family whom Audubon admiringly referred to as "the queen Bee of this honeyed mansion" (*Journal of 1826* 121), provides interesting and helpful context for Audubon's time with the Rathbones. On August 20, 1826, Mrs. Rathbone wrote to her son Theodore about their American guest: "[W]e have the prince of bird killers now, raising them, (almost) from the dead, in thy little room, — Mr Audubon — he has brought from North America a most astonishing collection of birds which he painted there (after shooting them) in a way of his own, quite singular, and beautiful." She goes on to shed light on Audubon's uneasiness in the confined walks of Liverpool: "[H]e fixed himself, at an Inn in Dale street; shut up in a Town and longing for the Society of his beloved family; he seemed so miserable, we desired he would come to Greenbank, while he is waiting for letters from home, which unfortunately are some how detained from him, — he rises before the Sun, and wanders through our inhospitable country, imprisoned by stone walls, and warned that, steel traps, spring guns, and great dogs surround him, comes back, kills a bird, and paints it, for some of the family."

Audubon is moved for the first time to articulate his hunting ethic, which had remained unspoken before: "I thought it more Cruel to permit them to grow gentle, nay, quite tame and sudenly and by Frisks murder them by Thousands than to give them the Fair Play that our Game [h]as with us in our Forests of being Free, ah, yes, Free and as Wild as Nature made them to Excite the active healthfull pursuer to search after it and pay for it thro the pleasure of Hunting it downe against all dificulties" (152). While he is also bragging or posturing a bit in this semiprivate journal, for this American sportsman, the hunter became worthy of his game or prey by his willingness and ability to pursue the animal through all the challenges and discomforts presented by that animal's natural habitat. Thus it was that in England, Audubon learned to associate his hunting ethic with the freedom to hunt that he was accustomed to in the American wilds. The image of the worthy hunter, newly realized, became for Audubon an emblem of American freedom, as opposed to the British and European aristocratic control of all hunting—and even of traipsing through the countryside.

This causal association in Audubon's mind between hunting and freedom in America briefly explodes into view (even though he and Lucy promised one another never to talk about politics [37]) in his account of a pleasant dinner party one evening in Liverpool. Despite the general cordiality of the gathering, the fact that game birds shot that day on a moor were served made it impossible for Audubon to fully enjoy them: "The *Moore Game*, however, was highly *tainted*, the True Flavour for the Lords of England == Common people, or persons who have no title _Hereditary_: Those who are not *Heritics by Birth* have to write a very particular note of thanks for every paire of Rotten Grous they receive from a Fatten^d Friend—Now in America *Freedom is Hereditary*!!!—Grous and Turkey, the Elk, the Bufaloe or the *Venaison* reach^d the palate of all Individuals without a Sigh of Oppression" (132). Such an outburst of political philosophy is rare in Audubon's surviving writings, but the intensity of this condemnation of political oppression reflects the tenacity with which Audubon clung to his American identity. He had been naturalized as an American citizen in Philadelphia on July 3,

1812, just fifteen days after President Madison signed the declaration of war with England on June 18. As Alan Taylor explains, the three years of this war set in motion a rising tide of confidence in the United States that its freedom from British rule was assured (439). Amid this rising confidence, Audubon, as a new citizen, was trying to make his way into commerce at Henderson, Kentucky. And amid all the French émigrés throughout the Ohio River Valley, from Louisville to St. Louis, Audubon's valuation of the freedoms he enjoyed was likely also enhanced by his awareness of the fearful changing conditions in France from the time of the revolution of 1789, the subsequent Reign of Terror (some of which he experienced in his youth in Nantes), and Napoleon's rise to imperial power. Audubon nurtured a sense of satisfaction for having escaped France without having been conscripted into one of Napoleon's armies. As the twenty-year-old Audubon himself wrote on May 20, 1805, perhaps somewhat satirically, to his future father-in-law while visiting his father in Nantes, "I am here in the Snears of the eagle," meaning Napoleon, whose grasp he would evade (Princeton). He always maintained a sense of the uncertainty of the human world, but through his first twenty-three years in America he lived as if political liberty were assured. The relatively oppressive conditions he observed in England in 1826 to some degree politicized the pleasures he experienced as an American sportsman and naturalist.

These new insights, made possible by the contrasting ways of England, when later combined with his assurance that *The Birds of America* would indeed be published and his fame and his family's welfare secured, led him temporarily into unexplored regions of conservationist and even preservationist thought. His journal entry for December 12, 1826, stands out as the most extreme call for human restraint on the American frontier that anyone at that time had yet uttered. On this evening, his mood was expansive because his mind had recently been liberated from all his doubts and fears. The first two plates of *The Birds of America* had been produced by Edinburgh's chief engraver and were deemed "truly beautiful" (*Journal of 1826* 339). His portrait was to be painted and engraved for publication.

And just the day before he had received two long-delayed letters from Lucy reassuring him of his family's health and of Lucy's continued commitment to their life together—and that after just two days earlier having written Lucy an ultimatum about their marriage because he had heard so little from her in the past months. In the midst of this long-pursued happiness, and in anticipation of soon being introduced to Scotland's most revered author, Walter Scott, Audubon moved into an intense reverie in which he calls upon Scott to save America's wild natural bounty from destruction by those building new settlements in ever farther westward locations: "Oh, Walter Scot, where art thou? Wilt thou not come to my Country? Wrestle with Mankind and stop their Increasing ravages on Nature & describe her Now for the sake of Future Ages" (346). This is radical. No one else was calling for a *stop* to the changes in the land Americans were then making.[5] And what he refers to here as "ravages" everyone else called "progress." Naturalists in Audubon's day, Richard Judd explains, had mixed feelings about the rapidly expanding settlements. They "celebrated the westward movement. A coming-together of evolutionary naturalism and manifest destiny, the change from forest to fruitful field would complete the logic of the integrated landscape. Yet [...] they displayed a subtle uncertainty about the pioneering family in its new wilderness home. These misgivings brought the first systematic expressions of conservation thought in America" (220). Thus, Audubon's misgivings coincided with the origins of conservationism, but his 1826 call to stop the "ravages" goes beyond mere misgiving and is therefore all the further advanced for its time.

One of the prevailing myths in America at that time, and the principle one that tended to justify unsustainable practices, was that of the

5   Daniel Drake, who founded the Western Museum Society in Cincinnati and for whom Audubon worked briefly in 1820, published a similar call for a portrait of the original American wilderness to be made in 1834: "Thus the teeming and beautiful landscape of nature, fades away like a dream of poetry, and is followed by the useful but awkward creations of art. Before this transformation is finished, a portrait should be taken, that our children may contemplate the primitive physiognomy of their native land, and feast their eyes on its virgin charms" (17).

inexhaustibility of the natural environment in North America. Seen in this context, Audubon's late-night, private utterance on December 12, 1826, stands out as one of the earliest arguments against that myth. Roderick Nash writes of Henry David Thoreau (1817–62) that he "was one of the first Americans to perceive inexhaustibility as a myth," citing as evidence Thoreau's journal entry for January 3, 1861, thirty-four years after Audubon's private call to stop the "ravages" (36).[6] As Audubon continues to express his concerns about the American natural environment in the December 12 journal entry, he develops the image of America as a pathetic, perishing daughter of a feminized nature:

> Neither this Little stream—this swamp, this Grand sheet of Flowing Watter, nor these Mountains will be seen in a Century hence, as I see them now. Nature will have been rob[d] of her brilliant charms—The currents will be tormented and turned astray from their primitive courses—The Hills will be levelled with the swamp and probably this very swamp have become a mound covered with a Fortress of a thousand Guns—Scarce a Magnolia will Louisiana possess—The Timid Deer will exist no more—Fishes will no longer bask on this surface, the Eagle scarce ever alight, and those millions of songsters will be drove away by Man—Oh, Walter Scot, come, Come to America! Sit thee here. Look upon her and see her Grandeur now. Nature still nurses her, cherishes her, but a Tear flows in her eye, her cheak has already changed from the Peach Blossom, to sallow hue—Her Frame Enclines to Emaciation, her step is arrested. Without thee, Walter Scot, unknown to the world she must Die. (*Journal of 1826* 348)

Audubon had been living amid the "Grandeur" for twenty-three years, but he had witnessed and participated in (by trying to operate a sawmill on the Ohio River) the increasingly rapid felling of the forests and the local extinctions of many species of wildlife, and he knew the economic pressures

---

6   Thoreau was reacting against the steady destruction of New England's forests: "Thank God, men cannot as yet fly, and lay waste the sky as well as the earth!" (qtd. in Nash 36).

and the determination of the people to transform America's interior into a wealthy nation. So he feared for the future of this grandeur, and this night, alone in his room in Edinburgh, he expresses a vain hope that his compatriots back home might be persuaded to stop.

They did not, and he knew they would not, could not. But considerably in advance of everyone else in the United States at this time, America's most ambitious naturalist put into words what Susan Fenimore Cooper, Henry David Thoreau, and George Perkins Marsh would begin to utter twenty or more years later. The night of December 12, 1826, marks Audubon's most visionary conservationist moment. He would not return to this extreme.

With the exception of signs of strains in his relationship with Lucy, Audubon's success in 1826 was unqualified. Between April and December, his status changed from unknown business failure to the most promising nature illustrator and naturalist in America and Europe. From January 1827 through March 1829 he worked feverishly to solicit subscribers for *The Birds of America* while also painting and writing. This was the time to make good on his promises to Lucy and their sons to secure their future prosperity. While monitoring as closely as he could manage the quality of the prints being produced by the London engraver Robert Havell Jr., he also traveled throughout England and made a trip to Paris to woo the wealthy with his lavish, extravagant images of American birds. He was now the head of a family business and had to act more pragmatically than he ever had before. This new pragmatism accompanied his sense of purpose as well upon his return to New York City in May 1829. He left Havell with enough paintings to engrave through the next seven or eight months and returned to America to procure and paint more of the needed bird species and to retrieve Lucy, if she would agree to accompany him back to England. He was understandably anxious about not being physically present in London while the printing of his plates was still underway, but he knew that a couple of hundred more bird species were needed to complete his *Birds*, and he did want Lucy with him, as he wrote to her in the third person on May 10, 1829, "to enliven his spirits and assist him with her kind advises" (*Letters* 1:83). But this head of the family business did not want to lose time by traveling to Lucy at Bayou Sarah in Louisiana. His focus was on getting the birds he needed: "[T]o accomplish the whole of this or as much as I can of it between now and my return to Europe I intend to remain as *stationary* as possible in such parts of the Country as will afford me most of the subjects, and these parts I know well" (1:81). He underlined the word "stationary" to stress to Lucy his determination to stay focused on his work even though he had been away from her for a full three years. This letter also contains his ultimatum to Lucy: she must choose now whether she will accompany him to London. If she agrees to come, "my heart will bound also, and it seems to me that It will give me nerve for further exertions!" But if she will not come, "*I never will put the question again* and *we probably* never will meet again" (1:83). Two months later, on July 18, having received from Lucy two letters "full of doubt & fear, nay quite the contrary of what I expected," Audubon wrote to their elder son Victor in hopes that he might be able to persuade his mother to conform to Audubon's wishes. His single-minded determination to advance the work he needed to do remained undiminished, even with his marriage and family unity at stake: "*I cannot go to her* because was I to lose my Summer by so doing I would miss the birds that I want" (1:89). Later in the same letter, Audubon reveals a rather jaundiced view of humankind that may also have increased his determination to complete his work with as few ethical delays as possible: "[U]pon my word it is wond[erful] to live a long life and see the movements, thoughts, and different actions of our poor degenerated species.—good only through interest; forgetfull, envious, or hateful through the same medium—well my Dear Boy so goes the world and with it we must move the best way we can" (1:92). That the family climate, so to speak, at this time was one in which practicalities held precedent over dreams and ideals is further confirmed by Lucy, writing from St. Francisville, Louisiana, two months later (September 29, 1829). She tells her husband that she knows he "will never be happy but in toiling for" fame, whereas about herself she asserts, "[W]hen it is positively known I am no longer of any service to the world, the world will take no interest in me! I have not lived so long not to know that" (Beinecke). The completion of *The Birds of America* had taken on paramount importance for Audubon during the preceding two years. Through the early and mid-1830s, while he traveled to hunt and draw the needed bird species, his gunning ethic would reflect his determination and a newly prescribed pragmatism.

The radical preservationist thoughts Audubon expressed on December 12, 1826, did not make their way into the more pragmatic 1830s. They were not realistic. There existed then insufficient context within which they might have found traction. Early in 1831, in "The Ohio," the first of the essays on frontier life that he included in the first volume of *Ornithological Biography*, Audubon went public with some vestiges of that evening's thoughts, but they were heavily qualified for publication. The earlier preservationist impulse had vanished behind a veil of ambiguity. In this lovely essay Audubon composes some of the description of the American wilderness he once called upon Walter Scott to write:

> Nature, in her varied arrangements, seems to have felt a partiality towards this portion of our country. As the traveller ascends or descends the Ohio, he cannot help remarking that alternately, nearly the whole length of the river, the margin, on one side, is bounded by lofty hills and a rolling surface, while on the other, extensive plains of the richest alluvial land are seen as far as the eye can command the view. Islands of varied size and form rise here and there from the bosom of the water, and the winding course of the stream frequently brings you to places where the idea of being on a river of great length changes to that of floating on a lake of moderate extent. Some of these islands are of considerable size and value; while others, small and insignificant, seem as if intended for contrast, and as serving to enhance the general interest of the scenery.

Following this glowing scene of undisturbed natural beauty, Audubon introduces a note of caution that he and Lucy felt on their descent of the Ohio from Shippingport (a former port at the Falls of the Ohio) to Henderson,

Kentucky, in October 1812: "We foresaw with great concern the alterations that cultivation would soon produce along those delightful banks" (OB 1:30). When Audubon wrote this essay, he had witnessed the "alterations" settlers were making along the Ohio and Lower Mississippi Rivers from April 1808 until April 1826, eighteen years. During this long period of observation, he would have noticed, beyond the mere changes themselves, the increasing rate of acceleration of those changes. One traveler on the upper Ohio River "found only thirty families" before 1795, but "when he retraced his route in 1802, he found farms every two or three miles down its entire length. The following year, Ohio, with some 45,000 residents, became a state and, by 1810, its population had grown to more than 230,000" (Judd 74). The momentum of this growth was unstoppable, of course, but still in his first published book, in which he is most solicitous of his readers (which I will discuss more fully in the next chapter), Audubon was willing to express some note of caution amid a tone of lamentation.

In the history of Audubon's thoughts about the need for human restraint on the American frontier, "The Ohio" is especially important for its perfect reflection of the conflicted relationship most Euro-Americans had with the land upon which they depended for their livelihoods at this time. The continent's beauty and bounty lured them in, and they loved it, it excited them, yet they had to compromise that beauty and diminish the bounty with every step they took toward survival and prosperity. As James Fenimore Cooper's 1823 The Pioneers makes clear, there was a broad spectrum of varying attitudes toward the land and its exploitation at this time, from those who wanted to leave it practically unaltered by humankind to those who would extract resources with no regard to future generations or the health of the environment, and Audubon deliberately crafted an ambiguous statement in his first published comment on the issue. In the final two paragraphs of "The Ohio," the by-then-famous American naturalist builds a powerful case for restraint but then strategically steps back from that position:

When I think of these times, and call back to my mind the grandeur and beauty of those almost uninhabited shores; when I picture to myself the dense and lofty summits of the forest, that everywhere spread along the hills, and overhung the margins of the stream, unmolested by the axe of the settler; when I know how dearly purchased the safe navigation of that river has been by the blood of many worthy Virginians; when I see that no longer any Aborigines are to be found there, and that the vast herds of elks, deer and buffaloes which once pastured on these hills and in these valleys, making for themselves great roads to the several salt-springs, have ceased to exist; when I reflect that all this grand portion of our Union, instead of being in a state of nature, is now more or less covered with villages, farms, and towns, where the din of hammers and machinery is constantly heard; that the woods are fast disappearing under the axe by day, and the fire by night; that hundreds of steam-boats are gliding to and fro, over the whole length of the majestic river, forcing commerce to take root and to prosper at every spot; when I see the surplus population of Europe coming to assist in the destruction of the forest, and transplanting civilization into its darkest recesses;—when I remember that these extraordinary changes have all taken place in the short period of twenty years, I pause, wonder, and, although I know all to be fact, can scarcely believe its reality. (OB 1:31–32)

The rhetorical momentum of this lament leads the reader to expect a call to "stop their Increasing ravages on Nature," as Audubon wrote in his journal on December 12, 1826, but here his pragmatism causes him to diffuse that momentum: "Whether these changes are for the better or for the worse, I shall not pretend to say" (OB 1:32). This rhetorical maneuver has the structure of a betrayal, an act of cowardice. The choice of words, the metaphors, and the images of the lamentation call for a stop to the destruction: the "grandeur and beauty" were once "unmolested by the axe"; the "vast herds" of wild game "have ceased to exist"; the "villages, farms, and towns" have brought "the din of hammers and machinery" to the former "state of

nature"; the forests are falling "under the axe by day, and the fire by night";
commerce is *forced* upon the "majestic river"; and the European immigrants
have come to "assist in the destruction of the forest." On the evening of
December 12, 1826, all this ceaseless violation resulted in an image of nature
as pallid, weeping, and starving: "Her Frame Enclines to Emaciation, her
step is arrested." She is dying. But now his public persona suggests that a
good picture of the former appearance of the Ohio's wilderness will suffice:
"[B]ut in whatever way my conclusions may incline, I feel with regret that
there are on record no satisfactory accounts of the state of that portion of
the country, from the time when our people first settled in it." And he calls
now on America's two most famous authors, Irving and Cooper, instead
of Scott, to write the description of America's original wilds so that future
generations will know: "I sincerely hope that either or both of them will
ere long furnish the generations to come with those delightful descriptions
which they are so well qualified to give, of the original state of a country that
has been so rapidly forced to change her form and attire under the influence
of increasing population." Whereas in 1826 the country was emaciated and
dying, now she is only losing a little weight and changing her clothes. In
his closing gesture, he hops down from the rhetorical fence he was sitting
on by adopting the most popular word used to characterize all the changes
to the land: "I hope to read, ere I close my earthly career, accounts from
those delightful writers of the *progress* of civilization in our western country" [emphasis mine]. In the space of a single paragraph, Audubon's public
persona distances himself from the implications of one of the most powerful
conservationist lamentations published in the first half of the nineteenth
century and praises the agents of that change as heroes: "They will speak of
the CLARKS, the CROGHANS, the BOONS, and many other men of great
and daring enterprise. They will analyze, as it were, into each component
part, the country as it once existed, and will render the picture, as it ought to
be, immortal" (OB 1:32). By accepting the preservation of the *picture*, rather
than insisting upon the preservation of the living environment, Audubon

enacts the compromise that most white Americans also accepted, but he does so only after expressing the lamentation fully.

After a year and a half in London, Edinburgh, and Paris writing the bird biographies, seeing to the quality of the plates, and always soliciting subscribers, Lucy and Audubon returned from England to the United States so that he could mount several expeditions to hunt more of the needed bird species. They now had many copies of the first volume of *Ornithological Biography* and the first bound, sixty-pound volume of *The Birds of America*. To say that Audubon's agenda was ambitious would be an understatement. He had in mind some two years of travel, hunting, and drawing along the eastern and Gulf coasts, then up the Arkansas River to the Rocky Mountains, and beyond to the Pacific, if possible. Victor, who was living and working in Louisville, traveled to Philadelphia to accompany his mother back to Louisville, where Lucy would live with her younger brother William and his family. To accompany and assist him, Audubon had brought from London a young taxidermist, Henry Ward. And he managed to persuade the Pennsylvania landscape painter George Lehman to travel with them as well to provide backgrounds and plants for the new bird drawings. Audubon's letter of September 20 to the engraver Havell reflects the dizzying speed at which he had been taking care of business since their arrival in New York seventeen days earlier. After a week with Lucy's sister and brother-in-law Berthoud, he explains, they moved on to Philadelphia to await Victor's arrival and prepare for the first leg of the trip south to Charleston and St. Augustine: "[A]s soon as he arrives we push off for Baltimore w[h]ere we will part and I will proceed for the Floridas as fast as Steam Boats or Coaches will allow" (*Letters* 1:136–37). From Baltimore, Audubon and Victor went to Washington to request permission for Audubon's men and equipment to travel on U.S. Revenue cutters whenever possible. On October 9 Audubon and his companions reached Richmond; Fayetteville, North Carolina, on October 13; and Charleston on the sixteenth. There he quickly met and befriended the Reverend John Bachman, who opened his home to Audubon and his companions, and they began the work of transforming living birds into preserved skins and engraveable images.

Audubon's success depended to a great extent on maintaining his celebrity and keeping the public informed about his movements and adventures while he traveled. Before leaving Philadelphia, Audubon and the editor of the fledgling *Monthly American Journal of Geology and Natural Science*, George Featherstonhaugh, made a mutually beneficial arrangement. The famous American naturalist would write occasional letters to the editor about his travels to publish in the journal, which would increase its general appeal and keep the public interested in where Audubon was and what he was doing (Berkeley and Berkeley 92–96). Lucy and their sons were equally aware of the need for good publicity. Writing from Wheeling on October 12, 1831, Lucy reminded her husband, using her pet name for him, that the world was watching him and that their future prosperity depended upon the success of this expedition: "I miss you, but on this journey I know much of your future fame and success depends for the World has its eyes upon you my Dear Laforest, but in the midst of fame remember the friend of 25 years standing and the prosperity of our beloved children" (Beinecke). Lucy and Audubon were staking everything on his ability to find, shoot, or otherwise procure and draw the needed birds and to do so as efficiently as possible. In his response to Lucy, Audubon assures her of his commitment to their project: "I will not forget the Friend of 25 Years standing I think—neither will I forget her precept for I am also aware that the World has an eye upon me" (*Letters* 1:145).

With these motives in place, Audubon, Ward, and Lehman—often accompanied by Bachman and a pair of enslaved men—went to work. On November 7 he reported to Lucy, "We have been here just three weeks and I have drawn 15 Birds which make 5 Drawings all of which are finished by Lehman with views Plants &c Henry has skinned and preserved 220 Specimens of 60 different species of Birds" (*Letters* 1:147). Following the very productive month in Charleston, based at the Bachman home, they moved south to St. Augustine, where the rapid pace of work continued:

"The transition of Idleness to hard Labour has operated upon me as if the electric fluid and I was very nearly *knocked up* last Sunday when I certainly drew faster than ever I have done in my Life" (1:155–56). And his "Young Men" were keeping up the pace, if in need of occasional prodding: "Lehman is as ever a most excellent industrious and Worthy Fellow—Henry is rather inclined to be Lazy but I pull him and have him out of his resting place as early as I do myself." Even though he was deeply concerned about the continued printing of the plates by Havell in London and would have liked to be there to assure the quality and be sure nothing slowed down production, he remained completely devoted to this expedition: "I cannot think of returning untill *my Journey* is compleated—and with me it is *Neck or Nothing!*" (1:157).

By January 4 they had collected and Ward had preserved 550 bird skins, and Audubon remarks, "I doubt if ever a man has undergone more fatigues than I now undergo and if I do not succeed I am sure it will not be for want of exertion" (*Letters* 1:171–72).

Farther south in Florida, Audubon found his progress slowing under more difficult conditions, but in his letters to Featherstonhaugh he stressed the extremity of his exertions and his determination to complete his work for the benefit of his readers and science. In a letter dated December 7, 1831, that was published in *Monthly American Journal of Geology and Natural Science*, he reports being warned by plantation owners not to hunt in the swamps in order to preserve his health: "[B]ut difficulties of this character must be disregarded by the American woodsman, while success, or the hope of it, is before him" ("Letter from Audubon" 360). He notes that his "physical constitution has always been good" and provides this context within which readers will consider the killing and skinning of birds: "I know that I am engaged in an arduous undertaking; but if I live to complete it, I will offer to my country a beautiful monument of the varied splendour of American nature, and of my devotion to American ornithology" ("Letter from Audubon" 363). In league with Featherstonhaugh's attempt to establish a new American journal of natural history, Audubon kills birds to advance

American ornithology. The gunning of the "American woodsman" therefore serves a nationalistic purpose. When he reports from a plantation in Florida being "anxious to kill some 25 brown Pelicans," he states his motive explicitly: "to enable me to make a new drawing of an adult male bird, and to preserve the dresses of the others." And twenty-five birds would not seem excessive when he reported being "in sight of several hundred pelicans" ("Letter from Audubon" [No. 2] 408). Since twenty-four brown pelican skins potentially could be sold for $100 or more, readers would understand that practical motive. In this letter, though, he is bragging, showing off. For his readers in the northeastern cities and England, Florida was a remote, exotic, tropical land they would never see themselves, and this great American naturalist and sportsman provided many of them vicarious pleasure on the stage Featherstonhaugh's journal provided. Even though he saw "several hundred pelicans," he still complains of the relative dearth of birds. One evening they decided to return to their host's home early "as the birds, generally speaking, appeared wild and few." His braggadocio reaches its height with this proffered definition: "[Y]ou must be aware that I call birds few when I shoot less than one hundred per day" ("Letter from Audubon" [No. 2] 408–9). While he is counting on his readers' envy and admiration, he is also reflecting the impatience he feels to advance his work. He is still imagining being in the Rocky Mountains in the summer of 1833, and he feels some frustration with every outing that is less productive than he hoped. As he explains in a letter to Lucy from John Bulow's home (the host mentioned above), dated January 4, 1832, "[T]he only thing of which I complain is the Scarcity of the Birds of which I am most in need of—I regret not having remained at New York and the Jersey Shore until the 1ˢᵗ of this month and then to have sailed for Charleston &ᶜ—it is too late and I must do my best" (*Letters* 1:169).

When Audubon wrote from Florida, he was also motivated to work diligently and efficiently by Lucy's letters from Louisville, where the culture of her brother's home and of Louisville generally was disappointing: "I do not see the few persons there are of any scientific or literary turn here, as the society of M^r & M^{rs} B, is quite of another sort, tho' W$^{\underline{m}}$ does when it comes in his way say all he can in its favour" (November 30, 1831, Beinecke). She also writes of being cold in William's home and of cracked fingers and feet. Only the fact that their sons lived there kept her from leaving: "[N]othing but our Children detain me here" (November 8, 1831, Beinecke). In fact, she writes that she might return to London with her sister and her husband before Audubon completes this expedition (Ford, *John James Audubon* 294). Even so, she encourages Audubon to persevere: "I think you are quite right to prosecute your journey *now* to the full extent *necessary*, rather than return again and I will do, or go any where to aid you I scarcely need say" (December 26, 1831, Beinecke). Nevertheless, the thought that Lucy might go to England without him alarmed Audubon, and he tries to reassure her: "Keep up thy Spirits I pray thee.—I am now on my last Journey after Birds in North America and I hope to be so successful as to be able to finish what I need for the Completion of our great enterprise." He urges on her a vision of their entering England together: "Our names and fame rising fast" (*Letters* 1:168–69).

But even after considering this array of motives, it remains true that Audubon enjoyed the sport of shooting birds. From aboard the U.S. schooner *Spark* on February 9, 1832, he complains to Lucy on this ascent of the St. John's River of their having been idle since December 25. Practically the only activity worth reporting from this time is the shooting of bald eagles, which presumably Henry Ward would skin and preserve: "We have had two frolics at Shooting White headed Eagles and killed 5 in 24 hours which is more than most Sportsmen can boast of" (*Letters* 1:181). He also enjoyed a fully anthropocentric sense of fun. In the second of the letters published in Featherstonhaugh's journal, dated December 31, 1832, Audubon narrates a dangerously cold night he and several others spent stranded by low tide on the mud of a marsh and their recovery the next morning by a warm fire and rising tide. Having survived so challenging an ordeal, "our spirits rose—and rose to such a pitch, that we in fun set fire to the whole marsh: crack, crack, crack! went the reeds, with a rapid blaze. We saw the marsh rabbits, &c. scampering from the fire by thousands, as we pulled our oars" ("Letter from Audubon" [No. 2] 410). He intends this, no doubt, as a playful act for his readers to enjoy, but it also reveals the utter liberty he granted himself in his relation to the natural environment.

His expedition to the coast of Labrador in 1833, on the other hand, brought out some of the most unequivocal statements he ever made about conservation. Following a voyage to the Keys on board the *Marion* in April 1832, Audubon retreated to Charleston because numerous disappointments and difficulties had caused him to abandon his hopes for an extended trip to the Rocky Mountains at that time. After reuniting with his family in Philadelphia, they traveled and then spent the winter in Boston, where they laid plans for the Labrador expedition. On April 20, 1833, Audubon wrote to Havell in London, "My Youngest Son and I are going a long & tedious journey this spring & summer.—I intend to Visit the whole coast of Labrador into Hudson's Bay and reach Quebec by returning over Land—No White Man has ever tramped the country I am about to visit" (*Letters* 1:213). In contrast to the relative scarcity of birds and humans in Florida, once his expedition ship, the *Ripley*, reached the gannet, auk, and puffin rookeries of coastal Labrador, he witnessed birds by the millions and scores of ships arriving for the purposes of cod fishing, seal hunting, and egg harvesting. His main purpose, of course, was to draw new birds, learn their life histories, and gather stories for episodes to be published in *Ornithological Biography*, but he and his traveling companions also killed birds to add to their immense collection of bird skins to sell and trade. Here, however, Audubon observed the killing of birds and egg harvesting on a scale so clearly unsustainable that he was moved to call not only for restraint but also for laws to compel the needed restraint.

For $350 per month, Audubon arranged the use of the schooner *Ripley*, which he had outfitted with a special drawing room and table. From Audubon's Labrador journal,[7] it appears that on the ship's first approach to the Gannet

7   All references herein to Audubon's Labrador journal are quoted from Lucy Audubon's biography of her husband, unless otherwise stated.

Rocks off the coast of Nova Scotia, their first business was to acquire birds by shooting them, which was done mostly by Audubon's son John, three other young men invited along, and several crew members. Nevertheless, this journal's dominant feature is the extreme contrast between the scale and purposes of the killing done by the Audubon party and that committed by the crews of the ships come to take eggs, birds, feathers, and seal furs back to various markets. Audubon's excitement builds in his account of their first approach to the Gannet Rocks off the coast of Labrador (June 15): "The air for a hundred yards above, and for a long distance around, was filled with gannets on the wing, which from our position made the air look as if it was filled with falling snowflakes, and caused a thick, foggy-like atmosphere all around the rock." But the following account of the first attempt to kill and collect birds from a boat manned by John and several others certainly renders their efforts as clumsy, ill-advised, and needlessly destructive: "The discharge of a gun had no effect on those which were not touched by the shot, for the noise of the birds stunned all those out of reach of the gun. But where the shot took effect the birds scrambled and flew off in such multitudes and such confusion that, whilst eight or ten were falling in the water dead or wounded, others shook down their eggs, which fell into the sea by hundreds in all directions." The party returned with only "some birds and some eggs" (311). Although Lucy's biography reports that the "journal gives a list of the names of one hundred and seventy-three skins of birds" (350), the number of birds killed on this trip by Audubon and his companions is much, much higher than 173. The following passage from Maria Audubon's edition suggests fairly clearly that much shooting for sport occurred. They have sailed to the ironically named "Paroket Island" to "procure the young of the *Mormon arcticus*," the common puffin:

> As we approached the breeding-place, the air was filled with these birds, and the water around absolutely covered with them, while on the rocks were thousands, like sentinels on the watch. I took a stand, loaded and shot twenty-seven times, and killed twenty-seven birds, singly and on

the wing, without missing a shot; as friend Bachman would say, "Pretty fair, Old Jostle!" The young men laughed, and said the birds were so thick no one could miss if he tried; however, none of them did so well. We had more than we wanted, but the young were all too small to draw with effect. (AJ 1:426)

Nevertheless, Audubon's Labrador journal repeatedly contrasts the killing done by his party with the nearly unimaginable scale of and the unworthy motives behind the depredations committed by the eggers and other market harvesters. Audubon records a report, for example, of an American fisherman whose crew would "kill thousands of guillemots in a day, pluck off their feathers, and throw their bodies into the sea" (340). With the possible exceptions of the passenger pigeons and golden plovers, Audubon had never seen destruction on the vast scale he witnessed and heard about on the Labrador expedition, and he understood the scale of it to be directly attributable to human greed. When he considered the declining populations of the northern fur animals, he was moved to blame and lament: "Thus the Fur Company may be called the exterminating medium of these wild and almost uninhabitable regions, which cupidity or the love of money alone would induce man to venture into. Where can I now go and find nature undisturbed?" (325). On a particularly illuminating evening, July 21, when Audubon and his companions were invited into the camp of several officers of the British Royal Navy—intelligent "gentlemen of education and refined manners"—the conversation Audubon records was not of sport and adventure, but rather "of the enormous destruction of everything which is going on here, except of the rocks." When someone said that evening that it was "rum which is destroying the poor Indians," Audubon made a causal connection between the greed of the market harvesters and the decline of the original human inhabitants: "I replied, I think not, they are disappearing here from insufficiency of food and physical comforts, and the loss of all hope, as he loses sight of all that was abundant before the white man came, intruded on his land, and his herds of wild animals, and deprived him of the furs with which he clothed himself." Rum, Audubon saw, was a convenient explanation for whites, for it implied a weakness inherent in natives and absolved whites of culpability. At this point in the journal entry, Audubon moves into a reverie of lamentation almost equal in eloquence and conservationist vision to his journal entry of December 12, 1826, when he called for a halt to the "ravages" of Anglo-Americans in the Ohio River Valley: "Nature herself is perishing. Labrador must shortly be depopulated, not only of her aboriginal men, but of every thing and animal which has life, and attracts the cupidity of men. When her fish, and game, and birds are gone, she will be left alone like an old worn-out field" (337). In the December 12, 1826, entry, America's original wilderness is losing its grandeur and becoming an emaciated, hobbling woman doomed to die unless humans restrain their behavior. Here in the remoteness of Labrador's coastal waters, the market harvesters are killing "Nature" for the short-term profits they make, and they will leave her behind, just as the Virginia farmers, Audubon once noted, were compelled to move westward to fertile ground after they had depleted the soil of their original farmlands (OB 1:31).

But it was the men who sailed to the bird rocks of Labrador to take the eggs of sea birds by the millions back to markets who inspired Audubon to make the most overtly conservationist statements of his entire life. By traveling to so remote and desolate a place, with its incalculable productivity of bird life—a place so productive largely because of its remoteness from humans—he witnessed human destructiveness on so vast a scale that he publicized it in hopes of reining it in. In his journal entry for June 23, he reports the gist of what was happening to the bird populations: "We heard to-day that a party of four men from Halifax, last spring, took in two months four hundred thousand eggs, which they sold in Halifax at twenty-five cents a dozen. Last year upwards of twenty sail of vessels were engaged in this business; and by this one may form some idea of the number of birds annually destroyed in this way, to say nothing of the millions of others disposed of by the numerous fleet of fishermen which yearly come to these regions, and lend their hand to swell the devastation" (315). Even though Audubon had witnessed the human harvesting of wild bird eggs elsewhere, and on a fairly large scale, he viewed that harvesting as sustainable. In his essay about the clapper rail, he reports that the gathering of this bird's eggs in coastal New Jersey "forms almost a regular occupation" and that he has "seen twenty or more persons gathering them by thousands during the season; in fact, it is not an uncommon occurrence for an egger to carry home a hundred dozens in a day; and when this havock is continued upwards of a month, you may imagine its extent." Yet his next gesture in this essay is not to call for restraint but to emphasize the great size of the nesting area as well as the population of the species: "The abundance of the birds themselves is almost beyond belief; but if you suppose a series of salt marshes twenty miles in length, and a mile in breadth, while at every eight or ten steps one or two birds may be met with, you may calculate their probable number" (OB 3:35). He also acknowledges that he himself has gathered "so many as seventy-two dozens in the course of a day" (OB 3:34). But in Labrador, the eggers take so many eggs that the birds can't recover: "The eggers destroy all the eggs that are sat upon, to force the birds to lay fresh eggs, and by robbing them regularly compel them to lay until nature is exhausted, and so but few young ones are raised. These wonderful nurseries must be finally destroyed, and in less than half a century, unless some kind government interposes to put a stop to all this shameful destruction" (315–16). If the four eggers mentioned above managed to sell all four hundred thousand of their eggs at Halifax at twenty-five cents a dozen, they would have raked in $8,333.25, enough to purchase eight complete sets of *The Birds of America* with change.

When Audubon published his episode entitled "The Eggers of Labrador," just fifty pages after the clapper rail essay in the third volume of *Ornithological Biography* (1835), he did not step back from an earlier idealistic and unrealistic position, as he did in "The Ohio" in the first volume in 1831. In "The Eggers of Labrador," he exceeds the earlier call for restraint in his journal by declaring a commitment (if heavily qualified) to activism in the cause of conservation. In the published account of the eggers, his figurative language makes the eggers into filthy, drunken thieves and murderers sneaking around in the hidden passages among the rocks. They are motivated only by money and never moved to pity. The episode concludes,

> The Eggers themselves will be the first to repent the entire disappearance of the myriads of birds that made the coast of Labrador their summer residence, and unless they follow the persecuted tribes to the northward, they must renounce their trade.
>
> Had not the British Government long since passed strict laws against these ruthless and worthless vagabonds, and laid a heavy penalty on all of them that might be caught in the act of landing their cargoes in Newfoundland or Nova Scotia, I might— (OB 3:85–86)

We don't know who decided to break this sentence off in a sort of rhetorical threatening posture, but it's the only such sentence in all of Audubon's published writings. The original manuscript version of this episode, however, does not end thus abruptly. The final sentence continues from "I might": "perhaps have been induced to have ere this humbly have prayed in behalf of the feathered tribe before the proper authorities in England for the extinction of the wasteful if not criminal barbarity of the Eggers of Labrador" (Lownes 103). Perhaps this closing line was not published because it is so feeble a declaration; still, for 1835 it is a rare voice. Thus, in 1835 Audubon added to his public persona an impulse toward conservation that was more direct and overt than had existed before in his writings.

By the mid-1830s, Audubon's drive and determination not only to complete the ongoing work (*The Birds of America* and *Ornithological Biography*) but also to develop additional projects are nothing short of staggering. He announced to his naturalist friend Richard Harlan of Philadelphia his intention to write an autobiography. He was certainly motivated by his lust for fame, but as he explains to Harlan he also would hope to convert his readers to a love of nature study: "Cannot, or will not, people of our days read with as much pleasure of the delightful Studies of Nature in our Woods, as they can of the frightful revolutions, bloody Battles, and the Carnage of Wars which overrun lands as the fearful Lava rushes down the Burning Mountains? Certainly Yes.—and for that reason alone I have determined to picture my Life on Paper as I have already done that of our birds!" (November 16, 1834, Beinecke). He would also hope to obviate the inaccuracies of future biographers of himself, who no doubt "would produce 10 chapters of untruths for my one of realities." And the autobiography could be quite large: "As to Numbers of Volumes, I cannot say but I know this well that had you read the Journals now around me you would be not a little puzzled to condense them even by the assistance of steam." Eight months later, we find him writing to Harlan about a "*Manual with plates* of the Birds of the United States and Northern America" that he intends to publish (July 13, 1835, Beinecke). "The plates will give the bill, leg & foot, one Tail feather and a primary, the ear of Birds of prey diurnal or nocturnal, Trachias and breast bones, with perhaps one egg of each species through out.—Thus you see that I do not intend to stand with arms a kimbo during my older days." No, indeed, he never intended to be idle. In the summer of 1837 he wrote to Joel R. Poinsett, the secretary of war, to request that the secretary write to the American Philosophical Society, where most of the journals of the Lewis and Clark expedition were housed, and to have those national treasures sent to him in England so that he could prepare a book treating the natural history findings of that expedition. John Vaughan, librarian of the society, thought it might be better for Audubon to study the manuscript journals just where they were when he returned to the United States (Poinsett-Vaughan correspondence, APS). He would need the records of Lewis and Clark to aid him in identifying new species on his western trip whenever he managed to find time for that. I offer all this as further indications of the fervor that drove the practicality with which Audubon worked throughout the 1830s.

The 435th and final plate of *The Birds of America* was printed on June 20, 1838, and the fifth and final volume of *Ornithological Biography* was published a year later. The Audubons returned to the United States in September 1839 and rented a house in New York City, and soon thereafter the family began work on the smaller, more affordable Royal Octavo edition of the *Birds* and the first plates of the grand Imperial Folio edition of *The Viviparous Quadru-peds of North America*. At the same time Audubon began extensive travels to obtain subscribers for all of the Audubon family publications. Throughout this time that was so tumultuous for his family, Audubon was planning what would become his final expedition, to the Upper Missouri River, on which there is no credible evidence of ethical restraint in the hunting of birds and mammals. Although several utterances of compunction about the killing were made, no one shot fewer animals because of ethical considerations. From all that I have read, however, Edward Harris provides the most insightful anatomy of the transition he and his companions had to experience when they entered the culture of hunting on the Upper Missouri. In a letter to his brother-in-law after the expedition was over he writes,

> I am almost ashamed to tell you that we left our Bulls, and fine fat ones they were, on the ground for the Wolves, carrying away nothing but the tongues, and novices as we were in this sort of murder, we were weak enough to feel more depressed than exalted at our triumph, and even ventured to utter some such treason as this, that Buffalo hunting was no better than the chasing a Wild Steer. But ere long our consciences became pretty well scar[r]ed and we had no more feeling at the death of a Buffalo Bull than at the demise of a Towhee Bunting, such you know is human nature all the world over. (EH 34–35)

They accepted their subordinate role as "novices" being trained by those already initiated and inured, and after once allowing their consciences to express their "treason," settled in and enjoyed—in Melville's phrase—"the fiery hunt."

To accomplish his great work, Audubon approached the need to kill like a liberty-loving American hunter-naturalist. From very early on, he had an innate sense of fair play and wished always to be worthy of his prey. He disparaged and avoided hunting methods that made the kill too easy for the hunter, for such diminished the honor or sense of accomplishment from a successful hunt. Because he had a Romantic's reverence for the American wilderness and natural beauty, he had early moments of unrealistic hopes that it might be preserved. Once he and Lucy, however, had committed to the family project of *The Birds of America*, putting their future at stake, the very practical need to generate an incoming flow of cash to cover the great costs of producing the double-elephant folio plates and the costs of his extensive travels to collect specimens led him to take advantage of opportunities when many birds might be taken in a short time. Preserved bird skins were as good as cash among nineteenth-century naturalists. He also just loved hunting, the satisfaction of a successful approach to one's prey and a good shot. Ultimately, though, he had an impulsive and competitive personality, which rendered his hunting behavior at times inconsistent with a strict conservation ethic.

But he was never just a hunter. He was the visionary creator of *The Birds of America* and *The Viviparous Quadrupeds of North America*, compelled to see these unprecedented works through to their completion. His "lived life" was a constant navigation through changing conditions toward the dream of his and his family's success.

In the next chapter I examine evidence of a complex "written" ethic by which he engaged the world publicly in *Ornithological Biography*.

THE WRITTEN ETHIC

My goal from the beginning of this project has been to understand Audubon's thoughts about how humans should live in relation to nature, particularly with regard to his beloved birds, without oversimplifying or avoiding complexities in the available evidence. Living when we do and having experienced the planet-altering environmental crises that have emerged since World War II, we naturally want this nineteenth-century icon of conservationist thought to have been a thoroughgoing conservationist. The evidence I gathered and studied about his "lived life," however, did not lead to such a conclusion. But Audubon had a "written life" as well, the public voice that directly addresses the readers of his essays about birds published in the five-volume *Ornithological Biography*. In the following pages, I explain my reading of this vast work as Audubon's subtle call for a humbler human posture in relation to the natural world and as the earliest argument for the ethical considerability of birds in the history of American conservationist thought.

From the beginning, however, it is important to understand that the voice, or persona, of *Ornithological Biography* is one that evolved over time, from the earliest versions of some essays in late 1826 to the last ones, written early in 1839. Because most who have read Audubon's essays have done so in the order in which they are presented in the much more accessible Royal Octavo *Birds of America*, where the essays accompany the images that appear in a completely different order than in the original *Birds of America*, they have not been able to see the larger patterns of development evident when one reads the essays in the order presented in the original five volumes of *Ornithological Biography*, published respectively in 1831, 1834, 1835, 1838, and 1839. While Audubon never was a systematic thinker, he was not nearly as erratic, inconsistent, hypocritical, or even schizophrenic as his persona can appear to be when his essays from 1831 and 1839 are read alongside one another.

Solicitude, his desire to serve and please the reader, is one of the main themes Audubon develops in the first volume of *Ornithological Biography*. In the opening paragraph of his first appearance on this stage, he is the humble servant of his "Kind Reader," who has developed his field notes into bird biographies "with no other wish than that of procuring one favourable thought from you" (OB 1:v). To further please the reader, he inserted a story of frontier life or travel narrative, referred to as "episodes," after every fifth bird biography. In the inaugural episode, "The Ohio," he explains his wish, "kind reader, to relieve the tedium which may be apt now and then to come upon you" as a result of reading endless essays about birds (1:29). The rapidity and precariousness of his rise to success since he arrived in Liverpool in July 1826 was still upon him; he knew it could vanish as quickly as it arose—hence the solicitousness. In his journal for 1826, he knows his world as one "wherein I may prosper but wherein it is the easiest thing to sink into compleat oblivion" (390). While he, MacGillivray, and Lucy composed and edited the bulk of the first volume of *Ornithological Biography's* 506 pages in just under four months in Edinburgh, he knew the first volume of his *Birds of America* (i.e., the first one hundred plates) was approaching its completion and would soon be before the public. In this period of heightening excitement and trepidation, he was especially mindful of how he appeared before his readers, whom he conceived of largely at this point as wealthy British patrons. As he presents one of the most dramatically beautiful birds in North America, the Carolina parakeet, in that essay's first paragraph he appeals to the reader to see more than vividly animated parrots in the plate; he wants his "kind" reader to see "that I spared not my labour" and that "I never do, so anxious am I to promote your pleasure" (OB 1:135).

He sounds like a courtier: "I have much pleasure in being able to give you an account of [the solitary flycatcher]" (1:147). His biography of the Stanley hawk opens with "allow me to submit to your consideration" (1:186). Taking almost excessive care not to be seen as presumptuous in the winter hawk essay, he writes, "I hope, kind reader, you will not lay presumption to my charge, when I tell you that I think myself somewhat qualified to decide in a matter of this kind, or say that I go too far, when I assert" that I can identify one from far away (1:364). This veteran field naturalist did not feel so humble or diffident as his rhetorical posturing suggests. Four and a half years earlier he had written to an Edinburgh supporter that, while European naturalists might be "no doubt far my superiors in point of education and literary acquirements, but not so in the actual courses of observations of Nature at her Best—in her Wilds!—as I positively have done." But then he cautioned himself to shape his public appearance carefully: "Yet, as I am but an Infant entering the Great World of Man, I wish to be submissive to its ways and not stubornly raise mountains betwen my connections with it and my own Interests" (*Journal of 1826*, 282). Thus the persona of volume 1 of *Ornithological Biography* deliberately reins in his self-confidence.

The second volume of *Ornithological Biography* appeared three years and nine months after volume 1, in December 1834. While a due solicitude remains apparent in the persona, especially in the introductory essay, the chief trait of the voice the reader hears now is a new, earned confidence. Since the first volume appeared, Audubon and Lucy had traveled to America, where he mounted his collecting and drawing expeditions to Florida and the Keys and then north to Labrador. He had been elected to membership in the American Philosophical Society. He was recognized everywhere he went, and his work was praised in newspapers and journals. He was undoubtedly one of the most experienced and accomplished field ornithologists in the world. It is not surprising that his voice would reflect a greater confidence than he expressed in the first volume.

Rather than the subservient courtier as in volume 1, now he casts himself as a favorite child of the reader, which also reflects his greater confidence of acceptance. In the introduction, he explains that when he once left his father in France, "he evinced his sorrow; when I returned he met me with an affectionate smile. If my recollection of your kind indulgence has not deceived me, I carried with me to the western world [that is, to the United States] your wish that I should return to you; and the desire of gratifying that wish, ever present with me as I wandered amidst the deep forests, or scaled the rugged rocks, in regions which I visited expressly for the purpose of studying nature and pleasing you, has again brought me into your presence" (OB 2:vi). Hereby he casts his readers in 1834 as the loving, grateful parents of the selfless, intrepid young naturalist who devotes his life to their edification in the workings of the natural world. In the rest of the introduction he recounts his amazing rise to success from unknown to honored naturalist, while dropping the names of dozens of friends, fellow naturalists, politicians, noblemen, even monarchs who have supported him in his work.

This new confidence is fully in place in volume 2's first bird biography, "The Raven." He opens with criticism for closet naturalists who become mere compilers of fables in their publications. And as he now resumes his work of delineating "the manners of the feathered denizens of" America, he will confine himself to his own original field observations, "which I have been able to gather in the course of a life chiefly spent in studying the birds of my native land, where I have had abundant opportunities of contemplating their manners, and of admiring the manifestations of the glorious perfections of their Omnipotent Creator" (2:1). Here, though, the confidence takes on a new dimension with more significant implications. He presents an image of himself traveling in all regions of America's diversified landscape—"amid the tall grass of the far-extended prairies of the West, in the solemn forests of the North, on the heights of the midland mountains, by the shores of the boundless ocean, and on the bosom of the vast lakes and magnificent rivers"—but he takes on the awesome role of intermediary between his reader and his god's creation when he claims that he has "sought to search out the things which have been hidden since the creation of this wondrous world, or seen only by the naked Indian, who has, for unknown ages, dwelt in the gorgeous but melancholy wilderness" (2:1). He is God's privileged servant bringing news from the creation to his readers—he and he alone. The second paragraph concludes with an even bolder claim: "And now, Kind Reader, let me resume my descriptions, and proceed towards the completion of a task which, with reverence would I say it, seems to have been imposed upon me by Him who called me into existence" (2:2). It would be difficult to conceive of a greater increase in confidence between volumes 1 and 2. In terms of a gunning ethic, he is now doing God's work when he kills a bird. This is an extreme statement, and it was omitted from the Octavo version of the essay in *Birds of America* (RO 4:79), presumably because someone felt that a reasonable reader might consider him—at least rhetorically—delusional here on this point. A few essays later, however, in "The White-Crowned Sparrow," when he reiterates his belief that God is pleased by his work, the rhetoric is more carefully qualified. In Labrador, when he heard the song of the white-crowned sparrow, it seemed to him "an invitation thus made to offer [my] humblest and most sincere thanks to that all-wondrous Being, who has caused [me] to be there no doubt for the purpose of becoming better acquainted with the operations of his mighty power." His focus here is on the quality of his original field observations of "the operations of his mighty power," the very thing that distinguishes him from all "closet naturalists." The result was that he "could not refrain from indulging in the thought that, notwithstanding the many difficulties attending my attempts—my mission I must call it—to study God's works in this wild region, I was highly favored" (2:88).

This confidence in God's approval of his observing and collecting practices empowers him later on in volume 2 to teach his readers moral and ethical lessons. In "The American Crow" he boldly admonishes American farmers for killing so many thousands of crows every year when they should be grateful to them for eating so many harmful organisms and much less presumptuous in their stance toward nature: "Must I add to this slaughter other thousands destroyed by the base artifice of laying poisoned grain along the fields to tempt these poor birds? Yes, I will tell you of all this too. The natural feelings of every one who admires the bounty of Nature in providing abundantly for the subsistence of all her creatures, prompt me to do so. Like yourself, I admire all her wonderful works, and respect her wise intentions, even when her laws are far beyond our limited comprehension" (2:318). As intermediary between the inner workings of nature and his reader, Audubon's persona in volume 2 of *Ornithological Biography* is authorized to lecture most American farmers—"Yes, I will tell you of all this too"—about the overweening arrogance with which they approach nature, which is beyond their comprehension. A few pages later, this prodigious collector of birds feels comfortable instructing his readers in a very fine distinction: it is wrong to take young birds when a parent is near. In "The Zenaida Dove," he narrates his attempt to capture by hand a female dove sitting on her nest, but she evaded his grasp: "She would then alight within a few yards of me, and watch my motions with so much sorrow, that her wings drooped, and her whole frame trembled as if suffering from intense cold. Who could stand such a scene of despair? I left the mother to her eggs or offspring." On another occasion, "however, I found two young birds of this species about half grown, which I carried off, and afterwards took to Charleston, in South Carolina, and presented to my worthy friend the Rev. John Bachman. When I robbed this nest, no parent bird was near" (2:356). This persona presents himself as one of an elite, privileged class of humans worthy of the moral responsibility to guarantee that no bird's life is taken in vain. When a soldier shoots a rare Key West pigeon for him, "How I gazed on its resplendent plumage!—how I marked the expression of its rich-coloured, large and timid eye, as the poor creature was gasping its last breath!—Ah, how I looked on this lovely bird! [...] Did ever an Egyptian pharmacopolist employ more care in embalming the most illustrious of the Pharaohs, than I did in trying to preserve from injury this most beautiful of the woodland cooers!" (2:383). This is a rather morbid comparison, but it shows his sense of ethical responsibility to ensure that the dead bird's beautiful body will not be damaged before he can draw it for posterity.

The confidence he exhibits in his relationship with the reader in volume 2 seems to reach a height late in the volume in his essay about the golden eagle. Here he is willing to gamble that he can emphasize the magnificence of the caged eagle he has purchased to draw, narrate at length his labored, bungling attempts to kill him without causing him pain or injuring his body or feathers, finally resort to a sure method of killing him, and still have his reader ready to empathize with him for his own sufferings that resulted from his long hours of labor and breathing of the fumes with which he hoped to asphyxiate the eagle: "I thrust a long pointed piece of steel through his heart, when my proud prisoner instantly fell dead, without even ruffling a feather." He then focuses on himself: "I sat up nearly the whole of another night to outline him, and worked so constantly at the drawing, that it nearly cost me my life. I was suddenly seized with a spasmodic affect, that much alarmed my family, and completely prostrated me for some days; but, thanks to my heavenly Preserver, and the immediate and unremitting attention of my most worthy friends Drs Parkman, Shattuck, and Warren, I was soon restored to health, and enabled to pursue my labours" (2:465–66). The persona's confidence in volume 2 is that of the favored son who knows that his doting parents will consider his needs paramount.

The authorial confidence of volume 2 swells occasionally a year later in volume 3 to overweening pride and arrogance, even cockiness. In "The Black Guillemot," where he could have reiterated his basic and by-now-common admonition to young naturalists that they rely always on their own direct observations of nature, never on someone else's claim, he instead belabors the erroneous reports of European naturalists that the black guillemot lays only one egg, rather than the three that he has regularly observed, and then sarcastically comments, "Thus, Reader, I might have been satisfied with the sayings of others, and repeated that the bird in question lays one egg; but instead of taking this easy way of settling the matter, I found it necessary to convince myself of the fact by my own observation. I had therefore to receive many knocks and bruises in scrambling over rugged crags and desolate headlands; whereas, with less incredulity, I might very easily have announced to you from my easy chair in Edinburgh, that the Black Guillemots of America lay only a single egg." It's the flippant sarcasm that makes the difference here. He concludes by cozying up to his reader: "It is the 'American Woodsman' that tells you so, anxious as he is that you should enjoy the pleasure of studying and admiring the beautiful works of Nature" (3:149). Thus he sets an elaborate stage to promote the "American Woodsman" as a brand, superior to all of Europe's closet naturalists. This is the most extreme moment of bragging I can recall in *Ornithological Biography*. Its manner here reflects poorly on his character. I'm surprised someone didn't talk him out of this for the Octavo edition of *Birds of America*, but it is repeated there (RO 7:273–74).

Other representative examples of this new glib conceit occur in his biographies of the white ibis and the whooping crane. A cypress swamp provides the vivid setting for dramatic action in "The White Ibis," where either Audubon or a companion wings an ibis that falls into the swamp. The flailing of its wings attracts three alligators, who pursue the panicked bird directly toward the gunners, who in the last possible moment save the bird's life: "It was the alligator's last chance. Springing forward as it were, he raised his body almost out of the water; his jaws nearly touched the terrified bird; when pulling three triggers at once, we lodged the contents of our guns in the throat of the monster." Audubon then represents the wounded, exhausted bird's next action as gratitude: "the Ibis, as if in gratitude, walked to our very feet, and there lying down, surrendered itself to us" (OB 3:178). That Audubon is willing to suggest to his readers that the bird he just shot could be grateful to him for saving his life might be funny if it were not so absurd, but it certainly reflects the easy self-assurance that the persona of *Ornithological Biography* has grown into in volume 3.

While "The Whooping Crane" presents the most comic moment in all of *Ornithological Biography*, its challenge to the reader was apparently deemed too extreme. The comedy arises from Audubon's victimization by a crane he has just wounded. After being wounded and then pursued by Audubon, the crane stands his ground and then chases the man, who left his gun behind: "The farther I removed, the more he advanced, until at length I fairly turned my back to him, and took to my heels, retreating with fully more speed than I had pursued. He followed, and I was glad to reach the river, into which I plunged up to the neck, calling out to my boatmen." While many birds are shown to evade Audubon's attempts to capture or kill them in *Ornithological Biography*, in no other essay does he become a victim of his intended prey, and the boatmen (surely many of his readers as well) are "highly delighted" by the sight of the great American naturalist up to his neck in the river while the crane is "up to his belly in the water, and only a few yards distant, now and then making thrusts at me with his bill." But then, despite the fact that the bird has proven himself quite admirable and accrued considerable pathos from the reader's point of view, the apparently unselfconscious overconfidence of the *Ornithological Biography* persona allows a surprisingly abrupt and offhand conclusion to the bird's life: "However, the battle was soon over, for, on landing, some of them struck the winged warrior on the neck with an oar, and we carried him on board" (3:207). While reviewing the bird biographies for their next publication in the Octavo *Birds*, someone saw this scene as presenting an unnecessarily extreme challenge to the reader and the entire scene was deleted.

An even more extreme instance of Audubon's apparent inability in the mid-1830s to see that the confidence he displayed in his relationship with the reader sometimes went too far occurs in the original manuscript for the biography of the American woodcock. While the opening paragraphs of both the manuscript essay and the version published in volume 3 of *Ornithological Biography* illustrate the courage and devotion of the mother woodcock while attempting to protect her young by mimicking a wounded bird to lure an attacker away, the published essay stops short of narrating the actions that Audubon once took to stop a group of boys from tormenting and killing a clutch of fledglings. The published version of the story stops just when the naturalist himself takes the stage to stop the boys: "The mother might have shared the same fate, had I not happened to issue from the thicket, and interpose in her behalf" (3:474). The essay then turns to an account of the habits of the species. The opening scene that Audubon originally wrote for publication, however, while personally one of my favorite moments in all of Audubon's writings, presents a total stranger emerging from a woods to kick and thrash several boys for torturing some birds. Because this manuscript of the essay has never been published, I quote it now at some length:

> Not the least disconcerted at my sudden sight, approach or talk, one of the youngsters caught the old bird, spoke loudly of his natural right to do as he might chuse with the birds, adding that *he* cared not "Two straws" for me. This rather raised my temper:—I remonstrated not, but at once and "sans façon," betook myself to kicking and thumping, and that pretty smartly too, the youngster who held the parent bird, a circumstance which so compleatly rendered him aware that any one being stronger than another could only act as a law which ought to be unwarantable at all times by beings of our Species, that tears rolled from his eyes and he let go the Woodcock! Sensible that I might change the notions of the Boys as regards humane and generous feelings towards every kind of creatures less powerful than themselves, I coaxed them after a few minutes to set down on the grass by the road side where I placed myself also. My feelings of anger were quite dispelled, I spoke to them of their own parents—of the sorrows these would experience were they to see some great Annimals either killing their children or carrying them away to be used as playthings. I told them how much greater would have been their joy if instead of spending the time at teazing the poor things, they had employed themselves in the gathering the raspberries so abundant around and taking these to their little sisters at home. At this moment the mother Woodcock which perhaps after a search for her lost young came back to the road as if to ask of us what had become of it. I pointed toward her and interpreted the meaning of her reappearance to them. The Boys hung their heads and remained silent. I had gained my point, and kissing every one of them walked away pretty confident that probably

never after this lesson would they torment neither old or young birds!
("American Woodcock")

I enjoy imagining Lucy reading this in their rooms in Edinburgh and won-
dering what had become of her husband's judgment. Did he really believe
that he had such power over his reader that he could boast of how "smartly,"
and in a fit of anger, he kicked and thumped these boys whom he did not
know, to the point of tears in the one rude boy? Audubon apparently was
confident enough to take this gamble, but someone talked him out of it.
The scene was not published until now.

Elsewhere in volume 3, however, there is evidence of a moderation of
such confidence, where I see a maturation in Audubon's evolving relation
to his reader. In "The Roseate Tern," for example, he opens with a scene
of his delight in a species of which he has just killed thirty-eight individuals: "I had never seen a bird of this species before, and as the unscathed
hundreds arose and danced as it were in the air, I thought them the Hum-
ming Birds of the sea, so light and graceful were their movements. Now
they flocked together and hovered over us, again with a sudden dash they
plunged towards us in anger; even their cries of wrath sounded musical,
and although I had carried destruction among them, I felt delighted." The
qualification is subtle, but he does acknowledge that his reader may find his
delight grating against the destruction. His next strategy, however, takes the
reader's perspective into consideration; he strives to assure the reader that
he will not take an excessive number of the birds and that he recognizes
that a certain ethical obligation arises from the killing: "[A]nd now I had
my cap filled to the brim with specimens. You may rest assured that I took
precious care of those which I had procured, but not another individual
was robbed of life on that excursion" (3:296–97). He knows the reader
wants to see him killing fewer, and here instead of assuming his authority
to kill whatever he needs, he reins in his behavior. In "The Pied-billed Dobchick," there is further evidence of a carefully shaped ethic in his persona
to appeal to readers. He compliments the grebes on their ability to hide

underwater, referring to them as "Cunning things!" who "have profited by [nature's] instructions" (3:359). His use of direct address here is extremely rare and shows a degree of deference unusual in these middle biographies: "Labour on, mind me not, I am a true friend and admirer of your race." The absence of any directly acknowledged killing in this biography is also rare and suggests that the author is rethinking his relationship with his reader.

What I am referring to as maturation in this relationship develops strongly in volume 4, which appeared in 1838, three years after volume 3. This emerging persona with a more mature confidence that I see in this later period does not need to puff himself up or overpower the reader with authoritativeness. The new relationship is almost egalitarian, one of partners, even companions.

In the introductory address of volume 4, Audubon surrounds himself with the names of many supporters in Philadelphia who helped him acquire the right to publish the new birds that the naturalist John Townsend had sent to Philadelphia from his western travels. The more egalitarian relationship with his reader is evident in his willingness to assume that his reader will stand with him in his denunciation of his former enemies in Philadelphia, those "*soi-disant* friends of science" there: "But, let me assure you, Reader, that seldom, if ever in my life, have I felt more disgusted with the conduct of any opponents of mine, than I was with the unfriendly boasters of their zeal for the advancement of ornithological science, who at the time existed in the fair city of Philadelphia" (4:xii). This is his first public reference to the opposition he encountered in Philadelphia, at first in 1824, although traces of it persisted but in weakened form on subsequent visits in the early 1830s. By 1838 he assumes that in his reader's mind he has prevailed.

This new development in Audubon's rhetorical stance before his reader is particularly evident and interesting in a continuous series of five bird biographies that begins with "Black-headed, or Laughing Gull" and concludes with "American Avoset" (4:118–74). This cluster of biographies can be seen as the rhetorical transformation of Audubon's reader into one whose ethic agrees with Audubon's continued killing of birds for the various purposes presented throughout.

The "Black-Headed, or Laughing Gull" is the essay in which I believe we see the literary apotheosis of Audubon the Ornithologist. His ornithology has become so good—combining amazingly extensive field observations with elegant inferences—that if anyone is still paying attention to the killing, he has no time for them. The reader, he now clearly assumes, is on his side. The following may be the most complex sentence in all of *Ornithological Biography*, but it is eloquent and displays a rhetoric precisely phrased to sing his victory over the closet naturalists he denounces in this essay for misidentifications they make from their comfortable rooms:

> Others, not at all aware that most Gulls, and the present species in particular, assume, in the season of pairing, and in a portion of the breeding time, beautiful rosy tints in certain parts of their plumage, which at other periods are pure white, have thought that differences of this sort, joined to those of the differently-sized white spots observable in particular specimens, and not corresponding with the like markings in other birds of the same size and form, more or less observable at different periods on the tips of the quills, were quite sufficient to prove that the young bird, and the breeding bird, and the barren bird, of one and the same species, differed specifically from the old bird, or the winter-plumage bird. (4:120)

While the sentence requires two or three readings for a just appreciation, it intricately weaves together, in a kind of counterpoint, the uninformed observations and erroneous conclusions of "Others" with the exact information they needed to avoid their mistakes, information that he, alone among all naturalists, had acquired by means of prolonged and necessarily repeated field observations. Although he thus places himself in a class by himself, he is careful to bring his reader along with him in the next, emphatically short sentence: "But, Reader, let us come to the point at once" (4:120). The two subsequent lengthy paragraphs thoroughly explicate the claims summarized in the sentence above and amid the thicket of details invite the reader in as fellow observer: "As to the white spots on the extremities of the primary quills of birds of this family, I would have you, Reader, never to consider them as affording essential characters. Nay, if you neglect them altogether, you will save yourself much trouble" (4:121). The persona of volume 4 finds ways to bring the reader over as his partner, here specifically as an advocate for original field observation and an ally against closet naturalists: "If all this be correct, as I assure you it must be, being the result of numberless observations made in the course of many years, in the very places of resort of our different Gulls, will you not agree with me, Reader, that the difficulty of distinguishing two very nearly allied species must be almost insuperable when one has nothing better than a few dried skins for objects of observation and comparison?" (4:121). At this point in the argument of this essay, the earlier "[b]ut, Reader, let us come to the point at once" becomes a masterful understatement. Audubon calculates his reader will be convinced that, clearly, no other naturalist can present such detailed and thorough knowledge and that he is the master of field ornithology. Furthermore, while his readers might say what they will about the killing, they will have to admit that he could not have learned all this without killing many and without the maturity of decades in the field.

His confidence that the reader is now on his side whatever the debate is about appears also later in the same bird biography. With wry irony he makes a joke while narrating his arrival on the revenue cutter *Marion* in the midst of many gulls: "The Gulls that laughed whilst our anchors were swiftly descending towards the marvellous productions of the deep, soon had occasion to be sorrowful enough" (4:125). Audubon's expectation here seems to be that his reader is in on the joke, and that this joke that he and the reader are sharing is at the gulls' expense. Together they enjoy a laugh at the laughing gull.

Through the next four biographies, the humor and jokes continue as Audubon solidifies the new partnership with the reader. The essay on the "Knot or Ash-Coloured Sandpiper" opens with the new partnership sustained: "I have been informed that several students of nature have visited its breeding places; but why they have given us no information on the subject, seeing that not only you and I, but many persons besides, would be glad to hear about it, is what we cannot account for" (4:130). This new allegiance manifests itself in a new, antic way in the next paragraph: "I myself am very conscious of my own remissness in this respect, and deeply regret the many opportunities of studying nature which have been in a manner lost to me, on account of a temporary supineness which has seized upon me, at the very moment when the objects of my pursuit were placed within my reach by that bountiful Being" (4:130). With his reader by his side, he comically chastises himself for having to nap from time to time just when God has brought him close enough to a bird to observe it. This is nearly the opposite of the early solicitation of the reader to allow him to serve him or her, vowing to work ceaselessly for the reader's pleasure; it's also quite different from the middle period of writing the bird biographies, when he regularly emphasized all the difficulties he confronted while making his field observations. It's as if he's satisfied now that he has proven his point, that his reader is now in league with him, with the result that he can satirically present himself as so relaxed that he might catch a nap while observing bird behavior.

The next stage of the humor develops in "Anhinga or Snake-Bird." Audubon comically sets up the reader to decline his invitation to slog through a swamp with him to procure anhingas. He also this time includes the reader among the anhinga's enemies, along with Native Americans and himself, the enthusiast: "There, trusting to the extraordinary keenness of their beautiful bright eyes in spying the marauding sons of the forest, or the not less dangerous enthusiast, who, probably like yourself, would venture through mud and slime up to his very neck, to get within rifle shot of a bird so remarkable in form and manners" (4:141). Far from sensing a need to apologize for or justify the killing, the American woodsman now even imagines handing the gun to the reader, so complete is their assumed alliance.

The reader becomes a veteran traveling companion in "Surf Duck": "Although several years have elapsed since I visited the sterile country of Labrador, I yet enjoy the remembrance of my rambles there; nay, Reader, many times have I wished that you and I were in it once more, especially in the winter season" (4:161). The conceit here is that Audubon and his reader have always traveled together. And in the next essay, "American Avoset," the reader has joined him in his room in Edinburgh where he writes this very essay: "Now, Reader, wait a few moments until I eat my humble breakfast" (4:169). Through these five biographies, Audubon's persona has handed the reader the gun, invited him to travel with him and assumed that he always had, and now the reader sits with Audubon in Edinburgh waiting for the next essay. I see this moment as a comic window into just how punchy Audubon, Lucy, and MacGillivray were becoming from working so hard to complete this vast project. They had amassed and published nearly two thousand pages of bird biographies and MacGillivray's anatomies. Three of the massive (i.e., sixty pounds each), awe-inspiring double elephant folio volumes of *The Birds of America* were in print and had been distributed in Europe and North America. The entire family had been heavily engaged in every stage of this work for six or seven years now. The last of the 435 plates would be printed on June 20, 1838. The end was in sight, and the reader as traveling companion and partner should make himself comfortable but give us a minute to finish breakfast.[8]

I also detect an interesting development in the risks Audubon is willing to take with his reader as he narrates various instances of his killing of birds. Parallel to the persona's increasing confidence through volumes 2, 3, and 4 arises a willingness to be bold and to startle, going even so far as to create conditions that make the reader complicit in the killing. The greater willingness to more boldly acknowledge the problematic ethic of some of his shooting appears tentatively in volume 2, but in volume 3 the author fully develops his rhetoric of reader complicity. Just as the more mature

8   Volume 5 contains no further developments in the persona's relation to the reader or in the persona's gunning ethic. This catchall volume contains brief treatments of species acquired from John Kirk Townsend's western collections, most of which Audubon did not observe in the wild; it also contains supplementary information about species already treated in volumes 1–4, frequently quoting accounts of distribution and behavior sent to Audubon by people he deemed credible. Species not included in *The Birds of America* are briefly mentioned. MacGillivray's anatomies comprise a much larger proportion of this final volume than they do of any of the previous volumes.

relationship with the reader emerges in volume 4, so too does the need for reader complicity diminish in that volume, which I'll explain later.

The outstanding example of the risky new boldness in volume 2 occurs in "The Mississippi Kite." There he presents his reader with the pitiable sight of a courageous mother bird trying to preserve her helpless fledgling from himself, the man with the gun firing from below their perch in a tree. He gambles, however, that he can maintain his reader's sympathy by showing himself as overcome by the flawed human desire to possess an elusive bird: "My feelings at that moment I cannot express. I wished I had not discovered the poor bird; for who could have witnessed, without emotion, so striking an example of that affection which none but a mother can feel; so daring an act, performed in the midst of smoke, in the presence of a dreaded and dangerous enemy. I followed, however, and brought both to the ground at one shot, so keen is the desire of possession!" He then placed the pair of birds "under a log," intending to retrieve them upon his return later that day; however, "On coming back, what was my mortification, when I found that some quadruped had devoured both! My punishment was merited" (2:109–10). He knows the reader sees what he's doing: developing pathos in the mother and young birds in order to exacerbate his sin, for which he is promptly punished and justice is done. This is an early example of a new result of the rising confidence of the voice of *Ornithological Biography*. With the first volume published, he is now experimenting, testing the reader's willingness to accept his proffered reasons for the killing.

In volume 3, the time for testing is over; he has apparently decided that shocking his reader with graphic accounts of killing runs so little risk that he can safely acknowledge the pleasure he derives from his gunning. Considerably more so than either of the previous volumes, volume 3 brings Audubon the unapologetic sportsman fully into view. The opening biography, "The Canada Goose," makes a grand gesture; it is one of the longest, if not the longest, of all the bird biographies and provides the most fully developed account of the life history of a bird ever published up to this time (3:1–19). Any reader would see that the field observations necessary to produce this account—of migrations, courtship, nesting, hatching, parenting, and the growth of the young—would be possible only because of Audubon's devotion to observing everything and traveling everywhere relevant to the species. In completely anthropomorphic terms, the first five and a half pages present a complex goose culture that emphasizes the affection between mates and of parents for their young. Yet, immediately after immersing and involving his reader in this fascinating world, he acknowledges his killing: "During my residence in the State of Kentucky, I never spent a winter without observing immense flocks of these birds, especially in the neighbourhood of Henderson, where I have killed many hundreds of them, as well as on the Falls of the Ohio at Louisville, and in the neighbouring country" (3:6). This is the highest number of personal kills that he specifies for any bird in all of *Ornithological Biography*. Later in the essay, however, he takes an even greater risk with equal confidence. Amid a jovial, jolly account of goose hunting "with one of the best sportsmen now living in the whole of the western world" (possibly his brother-in-law William Bakewell of Louisville), he allows himself an unrestrained expression of the great pleasure of such hunting for sportsmen: "Bang, bang, quoth his long gun, and the birds in dismay instantly start, and fly towards the spot where I am. When they approach I spring up on my feet, the geese shuffle, and instantaneously rise upright; I touch my triggers singly, and broken-winged and dead two birds come heavily to the ground at my feet. Oh that we had more guns!" No doubt many fellow American sportsmen would revel in such tales, but Audubon was aware that many of his readers might cringe at such indulgence in the pleasures of gunning. Yet the new degree of confidence evident in volume 3 permits him to conclude the scene with, not an apology, but a moment of feigned discretion: "In this manner we continue to shoot until the number of geese obtained would seem to you so very large that I shall not specify it" (3:16).[9]

---

9   A very similar moment occurs in "The Puffin": "I had two double-barrelled guns and two sailors to assist me; and I shot for one hour by my watch, always firing at a single bird on wing. How many Puffins I killed in that time I take the liberty of leaving you to guess" (OB 3:106–7).

A few pages later Audubon introduces a device for bringing his reader into complicity for all the killing. The biography of the clapper rail seems structured to first appall the reader with the suffering of wounded and dying birds, amid the apparent pleasures of the hunters, but then to admonish the reader for believing him- or herself free from all ethical implications. The essay asks the reader to consider the killing from the hunter's perspective. Because the daytime hunting of the clapper rail is "exceedingly pleasant," Audubon will "attempt to describe it." Following a description of the early morning approach to the marshes and the opening shots of the day, Audubon gives his reader images of the suffering birds: "It is a sorrowful sight, after all: see that poor thing gasping hard in the agonies of death, its legs quivering with convulsive twitches, its bright eyes fading into glazed obscurity. In a few hours, hundreds have ceased to breathe the breath of life; hundreds that erst revelled in the joys of careless existence, but which can never behold their beloved marshes again." Strategically, then, Audubon appears to agree with the reader's presumed repulsion at these images and the hunters' pleasures: "The cruel sportsman, covered with mud and mire, drenched to the skin by the splashing of the paddles, his face and hands besmeared with powder, stands amid the wreck which he has made, exultingly surveys his slaughtered heaps, and with joyous feelings returns home with a cargo of game more than enough for a family thrice as numerous as his own" (3:37). Recall that this is the sport he introduced as "exceedingly pleasant"; the disparity between that promise and the delivered scene of agony and cruelty is great. The persona seems to be deliberately puzzling the reader for a few moments. The essay's closing gesture, however, offers a reconciliation: "On speaking once to a friend of the cruelty of destroying so many of these birds, he answered me as follows:—'It gives variety to life; it is good exercise, and in all cases affords a capital dinner, besides the pleasure I feel when sending a mess of [Clapper Rails] to a friend such as you'" (3:38). Audubon takes the part of the reader by asking a hunter about the cruelty and thereby relieves himself from the need to win the case outright and allows the hunter's explanation to stand as the last word. Whether any given reader is persuaded is another question, but Audubon's strategy in this essay seems to be to implicate in the killing itself any reader who has ever benefited from the hunt in the form of killed game birds. With the last word in the essay's final phrase, "a friend such as *you*," the persona points a friendly finger of admonishment to the reader he has cultivated for several years now.

Establishing the reader's complicity in the kill becomes a recurring motif or strategy in volume 3 and reflects a greater confidence that his reader cannot condemn him for the killing with impunity. A subtle example occurs in "The Arctic Tern." On his first-ever approach to this species in the Magdalen Islands, the author felt the mounting excitement of an ornithologist: "Until that moment this Tern had not been familiar to me, and as I admired its easy and graceful motions, I felt agitated with a desire to possess it. Our guns were accordingly charged with mustard-seed shot, and one after another you might have seen the gentle birds come whirling down upon the waters" (3:366–67). His implication is that he assumes this burden of regret in order to produce the work that will please his reader: "Alas, poor things! how well do I remember the pain it gave me, to be thus obliged to pass and execute sentence upon them." If the next sentence alludes to the Reign of Terror that followed the French Revolution, throughout which period Audubon was a boy growing up in France, he may be suggesting that his reader holds a power over him to order the execution of birds comparable to that the Committee of Public Safety held over local judges charged with eliminating those deemed undesirable: "At that very moment I thought of those long-past times, when individuals of my own species were similarly treated; but I excused myself with the plea of necessity, as I recharged my double gun" (3:367). That would certainly be an outrageous analogy—comparing his reader to Robespierre—but the phrase "obliged to pass and execute sentence upon them" makes the possibility of this implication difficult to ignore.[10] In any case, his rhetoric is designed to show that the "necessity" of killing the terns comes from a power higher than himself.

Audubon develops his rhetoric of complicity quite overtly in "The Florida Cormorant." The author contrives to lay on his reader some of the blame for the "frightful havock" his and his companions' guns commit: "In a short time the bottom of our boat was covered with the slain [. . .]. You must try to excuse these murders, which in truth might not have been nearly so numerous, had I not thought of you quite as often while on the Florida Keys, with a burning sun over my head, and my body oozing at every pore, as I do now while peaceably scratching my paper with an iron-pen, in one of the comfortable and quite cool houses of the most beautiful of all the cities of old Scotland" (3:389). He's joking with the reader, of course, but he's also making sure that the reader takes his or her share of the blame for the deaths necessary to complete his work, as well as for his long periods of discomfort. Since Audubon was aware that many would object to much of the killing he narrates, in volume 3 he cordially manhandles the reader in order to avoid having all the guilt placed upon himself.

As I discussed earlier, in volume 4 of *Ornithological Biography* Audubon entered into a new relationship with his reader, one more equitable, resembling a partnership. I consider that a kind of maturation whereby the author presumes he has won over the reader and so no longer can benefit from startling, tricking, or bullying him or her. The need to establish the reader's complicity in the killing practically disappears in this volume of 1838. He seems so assured now of his reader's support and approval that other issues rise in importance. In "American White Pelican," for example, he invites the reader to view this species through the lens of deep geologic

10   The Committee of Public Safety, under the Law of Suspects, authorized local "surveillance committees" to detain and in many cases order the executions of "suspects," a term very broadly defined. During this period of "The Terror," anyone who failed to carry out an order of the Committee of Public Safety became a "suspect." See Andress, especially 210–43, who explains, "In the Paris of the *sans-culotte* Terror, there was to be literally no excuse for not being an ardent patriot" (212).

time (I believe for the first time in all of *Ornithological Biography*), thereby diminishing the importance of individual lives taken for science and art: "[M]ay this splendid bird wander free and unmolested to the most distant times, as it has already done from the misty ages of unknown antiquity" (4:88). With this perspective in the first paragraph, Audubon frames his discussion of the killing of numerous specimens in the overall history of their existence in North America as well as in the overall history of wildfowl breeding ranges and migratory patterns. If the reader is somewhat complicit in the shooting of several pelicans in this essay, so are all the other gunners mentioned: "Being anxious, when on my last expedition, to procure several specimens of these birds for the purpose of presenting you with an account of their anatomical structure, I requested all on board our vessel to shoot them on all occasions; but no birds having been procured, I was obliged to set out with a 'select party' for the purpose" (4:92). By foregrounding the scene of shooting that follows with his son as the lead gunner, Audubon suggests a much broader complicity: their guns "charged with buckshot, were in readiness, and my son was lying in the bow of the boat waiting for the signal. 'Fire!'—The report is instantly heard, the affrighted birds spread their wings and hurry away, leaving behind three of their companions floating on the water. Another shot from a different gun brought down a fourth from on wing; and as a few were scampering off wounded, we gave chase, and soon placed all our prizes in the after sheets" (4:93). This essay even further broadens the charge of complicity to include practically all Euro-Americans who have ever altered a pond or marsh used as breeding grounds by waterfowl. When he witnessed white pelicans breeding at the mouth of the Mississippi River and near Galveston in April and May 1832, he realized that the species bred at the same time north and south throughout the Mississippi River Valley as well as along the Missouri and other western rivers. This fact was new to him, together with his knowledge that numerous species of waterfowl had abandoned former breeding waters along the Atlantic seaboard, and it suggested to him that throughout "the misty ages of unknown antiquity" the many species of North America's waterfowl had nested and raised their young "in every section of the country that was found to be favourable for that purpose." The following conclusion that most Euro-Americans are to blame indicates that the author no longer needs to implicate his reader in his gunning ethic: "It seems to me that it is now on account of the difficulties they meet with, from the constantly increasing numbers of our hostile species, that these creatures are urged to proceed towards wild and uninhabited parts of the world, where they find that security from molestation necessary to enable them to rear their innocent progeny, but which is now denied them in countries once their own" (4:91). In volume 4 of *Ornithological Biography*, Audubon hopes to encourage his reader to assume a planetary perspective and a humbler human ethical stance toward wildlife and nature generally: "Ah, Reader, how little do we yet know of the wonderful combinations of Nature's arrangements, to render every individual of her creation comfortable and happy under all the circumstances in which they may be placed!" (4:90). The much-vaunted pioneers and frontiersmen do much more to destroy birds than do the American naturalist and all his readers.

This development in the author's relationship with his reader proceeds so far in volume 4 that he reverses the rhetoric of complicity found so prevalently in volume 3. In "Winter Wren" he gives an account of a truly rare find: the nest of the winter wren. (Audubon writes that he saw only two in his entire time afield.) After carefully observing two wrens for some time, he saw "a protuberance covered with moss and lichens" with a rounded aperture: "I put a finger into it, and felt the pecking of a bird's bill, while a querulous cry was emitted. In a word, I had, the first time in my life, found the nest of our Winter Wren." After he removes the eggs, however, he is moved to leave their nest and eggs: "The little bird called upon its mate, and their united clamour induced me to determine upon leaving their treasures with them; but just as I was about going off it struck me that I ought to take a description of the nest, as I might not again have such an opportunity. I hope, Reader, you will believe that when I resolved to sacrifice this nest, it was quite as much on your account as my own" (4:432). He leaves the nest behind—a "sacrifice"—for his reader's sake. The overpowering desire for possession that drove him in volume 2 in "The Mississippi Kite" to kill the mother bird and her young no longer holds sway. Audubon's persona claims now to have begun to listen to his newly partnered reader.

Perhaps the most important discussion taking place within the pages of *Ornithological Biography* concerns the human condition and how humans should live in relation to nature. As I have written earlier, Audubon was not a systematic thinker, yet through all of his writings a set of beliefs that resemble what academic philosophers refer to as "sentimental deism" seems to persist: the creator or author of all is good, and so is creation, or "nature" in the broadest sense of that word. Humanity, then, is the cause of evil. Humans are born good but acquire immorality or vice through selfishness and base motives that corrupt culture, which itself then becomes corrupting. All humans have an innate knowledge of the benevolent creator, and one's conscience is a sufficient guide to the moral life, which is the pursuit of the good and the simple, which are closely associated with nature. By contrast to animals, which are happy and fulfilled, humans create causes of worry and are miserable. This philosophy of nature and humanity was thoroughly codified in Rousseau's *Émile; or, Treatise on Education* (1762).[11]

Audubon showed an early propensity to view humanity in these terms on his descent of the Mississippi River in 1820. In his journal entry for December 10, 1820, he is reminded of his species' original innocence when he sees several Osages: "Whenever I meet *Indians* I feel the greatness of our Creator in all its Splendor, for there I see the Man Naked from his Hand and Yet free from Acquired Sorrow." Similarly, the condition of nature is essentially innocent. The numerous bird species persecuted by farmers who perceive them as pests are not actually evil or inimical. Even though plate

---

11 While Audubon likely would have rejected any one label to describe his philosophical beliefs, all his surviving writings suggest that he imbibed most of the ideas associated with Rousseau's influential natural religion. See Mossner for a brief overview of Rousseau's sentimental deism.

7 of *The Birds of America* boldly depicts a pair of purple grackles enjoying a decimated ear of corn on the stalk, like criminals caught in the act, in the bird biography for this species, he writes, "This is the tithe our Blackbirds take from our planters and farmers; but it was so appointed, and such is the will of the beneficent Creator" (OB 1:35). Even the most savage and sadistic of all his bird species, the bald eagle, is not guilty for its depredations: "The ferocious, overbearing, and tyrannical temper which is ever and anon displaying itself in his actions, is, nevertheless, best adapted to his state, and was wisely given him by the Creator to enable him to perform the office assigned to him" (1:160).

The only evil in the world is caused by humans. As he wrote to his son Victor on July 18, 1829, their species is motivated purely by selfishness: "[U]pon my word it is wond[erful] to live a long life and see the movements, thoughts, and different actions of our poor degenerated species.—good only through interest; forgetful, envious, or hatefull through the same medium" (*Letters* 1:92). In *Ornithological Biography* Audubon repeatedly characterizes his own species as the enemy of nature. As a representative example, the opening paragraph of the blue jay essay characterizes humankind as selfish, duplicitous, and malicious through overt comparison with the behavior of the jays: "[H]ow like beings of a much higher order, are these gay deceivers!" (OB 2:11). The red-headed woodpecker, of all species of birds, is the most "gay and frolicsome"; yet humankind is its "most dangerous enemy" (1:141). Much of the irony of *Ornithological Biography* is founded upon this disparity between the essential innocence of nature and humanity's apparent need to live in enmity with all that is not human.

There is no irony in nature, however, and it should be viewed, by default, as perfect. The opening paragraph of the cowbird biography elaborates this principle quite clearly: "The works of Nature are evidently perfect in all their parts. From the manifestations of consummate skill everywhere displayed, we must infer that the intellect which planned the grand scheme, is infinite in power; and even when we observe parts or objects which to us seem unnecessary, superfluous, or useless, it would be more consistent with the ideas which we ought to have of our own feeble apprehension, to consider them as still perfect, to have been formed for a purpose, and to execute their intended function, than to view them as abortive and futile attempts" (OB 1:493). He makes the same point in the American crow essay. After berating American humans for killing so many of this species, his reasoning approximates what today we call the "cautionary principle": "I admire all her [i.e., nature's] wonderful works, and respect her wise intentions, even when her laws are far beyond our limited comprehension" (2:318). His argument is that farmers ought not shoot crows because humans are ignorant of nature's purposes and laws and should assume that crows play an appointed role in the perfection of nature.

The ethic of *Ornithological Biography*, however, rests mainly upon its extension of ethical considerability to all birds. Even the word "biography" in the title suggests that something more than mere "natural history" is here attempted. A kind of premise for the ethical considerability of birds appears in a conversation Audubon had with an Edinburgh friend on February 12, 1827. Basil Hall challenged what he saw as Audubon's anthropomorphization of birds: "Captain Hall expressed some doubts as to my views respecting the affection and love of pigeons, as if I made it human, and raised the possessors quite above the brutes." But Audubon explained that he did not see so clear a divide between "brutes" and humans: "I presume the love of the mothers for their young is much the same as the love of woman for her offspring. There is but one kind of love; God is love, and all his creatures derive theirs from his; only it is modified by the different degrees of intelligence in different beings and creatures" (Buchanan 119). In February 1827, Audubon knew he must produce natural history essays for each of the bird species depicted in *The Birds of America*, and with this premise apparently at work in his mind, *Ornithological Biography* would grant birds a degree of consciousness and all the attendant emotions and motives of humans, but on a reduced scale, and the implication is that this is not merely metaphorical.

To begin with, birds are worthy of names that are not disrespectful and that do not reflect human ignorance of the actual meaning of their behavior. Audubon's chief objection here is that members of his own species regularly refer to some species of birds as stupid when they are easy to kill. About "Booby Gannet," Audubon complains, "I am unable to find a good reason for those who have chosen to call these birds *boobies*. Authors, it is true, generally represent them as extremely *stupid*; but to me the word is utterly inapplicable to any bird with which I am acquainted." People do not understand the real cause of the birds' apparent lack of fear of the most dangerous predator: "[I]t is only when birds of any species are unacquainted with man, that they manifest that kind of *ignorance* or *innocence* which he calls *stupidity*, and by which they suffer themselves to be imposed upon" (OB 3:67). Audubon makes the same point about numerous species, including the least bittern, the green heron, the woodcock. In the wood ibis essay he becomes nearly vitriolic about this misapprehension: "Talk to me of the stupidity of birds, of the dulness of the Wood Ibis! say it is fearless, easily approached, and easily shot. I listen, but it is merely through courtesy" (3:130). Viewing the wood ibis through his experience as a hunter of the bird often foiled by its wariness and vigilance, he represents this species as having the skills and intelligence necessary to survive in its habitat, traits wrongfully disparaged by others less knowing than he.

Perhaps the most emphasized trait that should create compassion for birds in humans is the bond of parent and young, the utter devotion, the powerful desire to protect and raise them. "The Mallard" presents one of the most thoroughly developed scenes of Audubon's ethical response to a display of avian parental affection. Once when walking with his well-trained dog, he came upon a female mallard "leading her young through the woods" on her way to the Ohio River; he ordered the dog to find and bring the young ones to him:

> On this the mother took to wing, and flew through the woods as if about to fall down at every yard or so. She passed and repassed over the dog, as if watching the success of his search; and as one after another the ducklings

were brought to me, and struggled in my bird-bag, the distressed parent came to the ground near me, rolled and tumbled about, and so affected me by her despair, that I ordered my dog to lie down, while, with a pleasure that can be felt only by those who are parents themselves, I restored to her the innocent brood, and walked off.

He then claims that a communication between species occurred: "As I turned round to observe her, I really thought I could perceive gratitude expressed in her eye; and a happier moment I never felt while rambling in search of knowledge through the woods" (OB 3:169). The American woodsman here becomes a seeker of knowledge, not game, who finds fulfillment in granting a duck's right to raise her young unmolested by him—this time, at least.

The volumes of *Ornithological Biography* report many other instances of avian parent behavior moving Audubon to mercy (as just a few examples, killdeer, ruby-throated hummingbird, great white egret, Canada grouse, Bartram sandpiper). Although Audubon provided an ethical basis for sparing parent birds in his conversation with Basil Hall mentioned previously, most sportsmen would go along with this leniency for parent birds for the purpose of game management, to assure that game birds reproduce and restore their populations for future hunts. In the 1830s, when Audubon was writing his bird biographies, there were scarcely any laws in the United States to protect birds or to limit sportsmen's behavior; these begin to appear in the 1850s (see Palmer). In the absence of such laws, though, Audubon's emotional admonitions to let the parent birds live to reproduce and raise their young serve a similar purpose—but from a moral rather than practical perspective. As he wrote about an incubating zenaida dove, "Who can approach a sitting dove, hear its notes of remonstrance, or feel the feeble strokes of its wings, without being sensible that he is committing a wrong act?" (OB 2:354).

Numerous testimonies of consciousness and intelligence in birds complement *Ornithological Biography*'s argument that humans should consider birds as entitled to ethical treatment. Much of animal behavior that humans tend to regard as produced by mere instinct, Audubon suggests, is actually attributable to a higher form of consciousness and intelligence. When he observed the custom of laughing gulls perching on the heads or beaks of brown pelicans in hopes of catching fish that escape from the pouches of the pelicans, Audubon perceived actual problem solving, a sign of consciousness: "To me such sights were always highly interesting, and I doubt if in the course of my endeavours to amuse you, I ever felt greater pleasure than I do at this moment, when, with my journal at my side, and the Gulls and Pelicans in my mind's eye as distinctly as I could wish, I ponder on the faculties which Nature has bestowed on animals which we merely consider as possessed of instinct. How little do we yet know of the operations of the Divine Power!" (OB 3:380). If it is his great pleasure to inform his reader of consciousness in the avian world, it is in the hope that the reader's capacity for wonder will be enhanced as well as his or her respect for animals. He describes himself as having been moved to mercy by a display of "sagacity" in a Canada goose near the shore of Labrador. With several men in his skiff, they were pursuing a goose that could not take to wing because he was in molt. As the men approached, the goose dived under the surface of the water and was not seen again until "the man at the rudder accidentally looked down over the stern and there saw the goose, its body immersed, the point of its bill alone above water, and its feet busily engaged in propelling it so as to keep pace with the movements of the boat. The sailor attempted to catch it while within a foot or two of him, but with the swiftness of thought it shifted from side to side, fore and aft, until delighted at having witnessed so much sagacity in a *goose*, I begged the party to suffer the poor bird to escape" (3:14).

The capacity for pleasure and fulfillment in birds is less than that in humans, but Audubon presents it as present even in the smallest birds. He writes of the tree sparrow preparing for its southern migration in the fall, "They had already tuned their pipes, which sounded in my ear as their affectionate farewell to a country, where these sweet little creatures had met with all of happiness that their nature could desire" (OB 2:512). Occasionally he pretends to give up on the possibility of convincing his reader that birds have a capacity for reasoning beyond mere instinct. When he observes that the razor-billed auks devise a variety of ways to prevent their eggs from being exposed to water in differing situations, Audubon dismisses his reluctant reader: "Call this instinct if you will:—I really do not much care; but you must permit me to admire the wonderful arrangements of that Nature from which they have received so much useful knowledge" (3:114). Nature, he implies here, has provided birds with "knowledge" that they use to respond to changing circumstances around them, evincing thereby some degree of consciousness. With the raven, he indulges in an extreme of anthropomorphization, but he presents this famously intelligent species as having not only consciousness but also memory and even religion. He describes a mated pair sailing side by side: "Would that I could describe to you, reader, the many musical inflections by means of which they hold converse during these amatory excursions! These sounds doubtless express their pure conjugal feelings, confirmed and rendered more intense by long years of happiness in each other's society. In this manner they may recall the pleasing remembrance of their youthful days, recount the events of their life, express the pleasure they have enjoyed, and perhaps conclude with humble prayer to the Author of their being for a continuation of it" (2:2–3). Unlike most humans, who despise the raven and destroy it, Audubon writes that he admires "the Raven, because I see much in him calculated to excite our wonder" (2:4). The degree of consciousness present in the birds of the world must sometimes be rendered metaphorically, but Audubon's creator has shared the gift of consciousness with more creatures than his humans, in part, at least, to enliven the natural world for the greater exercise of the human capacity for wonder, and Audubon sees the role of the naturalist-artist as that of mediator between his reader and the wonders of nature.

Audubon's purpose in *Ornithological Biography* is to teach his readers to read bird behavior against the backdrop of the wisdom or perfection of nature and not against the backdrop of human expectation. His purpose is to teach his readers to come down an ethical notch or two, to adopt a somewhat humbler stance in relation to nature. And this argument—as scattered, subtle, and muted as it may appear—stands in a dynamic relation to all the accounts of gunning and sometimes brutal killing that appear in his journals and that he published—or wrote and then suppressed. The journals report the realities of his world and of his struggle to make his way, whereas the published essays explore how things might be if he did not actually have to live in the real world. In the essays, by weaving throughout them his argument for a humbler human ethic, to a meaningful extent he can suspend the contingencies of the lived life long enough to imagine, and to encourage his readers to imagine, a redeemed existence—one that renders his violence innocent. And through the evolution of the relationship that he deliberately crafts with his readers through the five volumes, he attempts to make it difficult for readers to free themselves from complicity in the killing and to bring them into partnership, all mutually benefiting. Whether the strategy works or not is another question. But we are left with this: at a time when hardly anyone else was thinking in such terms, and not even Audubon himself consistently practiced what he preached, he was amassing a three-thousand-page monument to the ethical considerability of birds.

Epilogue

When, in the autumn of 1803, the eighteen-year-old Frenchman John James Audubon first arrived in the United States, Meriwether Lewis, William Clark, and the first recruits for Jefferson's Corps of Discovery from Kentucky and Illinois were still making their way in a keelboat to what would be the Corps' first winter encampment, on the American side of the Mississippi just north of the confluence with the Missouri River. Audubon's passion for drawing birds and his ability to "procure" them were already well established—as was the pleasure he derived from walking the woods, often with his father, with a fowling piece and bringing down a new, prized possession. He was also possessed of a driving determination to improve his ability to transfer the life of the birds he killed onto paper.[12]

The return to St. Louis of the members of the Corps of Discovery on September 23, 1806, marks a dramatic acceleration in the accumulation of new knowledge about North America as well as in the rate of changes to the landscape brought by settlers throughout the Ohio and the Lower Mississippi River Valleys. Audubon came onto this scene in 1808 with money and property, charged to build a financial basis for his and his family's future prosperity. But also looming in his mind—dimly at first but later ever more brightly—was a vision of fame and fortune won by means of his nature illustrations. By 1820, following the failures of his attempts in business, he was completely—if also desperately—committed to the risky, dangerous,

12 His essay *My Style of Drawing Birds* attests to this desire and the early improvements in his artistic ability.

and perhaps self-indulgent work he would have to perform to secure that future. By the time he reached Liverpool in 1826 in search of an engraver, he had lost his foreigner's perspective on the American wilds by having lived in them and traveled through them for more than twenty years. He had come to take for granted the natural abundance of wildlife in America and the freedom to pursue and take it. His success and burgeoning celebrity in Liverpool and Edinburgh overwhelmed him at times and made all the more tenacious his sense of identity as The American Woodsman. Thinking of himself as a native of the American forests like Daniel Boone and Natty Bumppo, he granted himself license to speak for his beloved Ohio River Valley by expressing in his semiprivate journal a visionary but unrealistic wish that the settlers' clearing of the forests could be stopped.

Among settlers and naturalists at this time, however, there was neither a mechanism (such as conservation regulations) nor a motive for restraint. And for the rest of his life, Audubon himself would rarely exercise restraint when it came to procuring the birds and mammals he needed to advance his work—or to preserve skins to barter, or to enhance his or others' natural history collections, or to eat and share with others. James Fenimore Cooper's 1823 historical romance *The Pioneers* calls for settlers to exercise some restraint (on sugar maples, passenger pigeons, fish), but it actively promotes the expansion of settlements. Beginning two decades later, calls for restraint would increase in volume, especially in the work of Caroline Kirkland (*Forest Life*, 1842), Susan Fenimore Cooper (*Rural Hours*, 1850), and Henry David Thoreau (*Walden*, 1854). And in 1864 George Perkins Marsh would publish the first formal study of how humankind around the world had altered and damaged the natural environment in ways that were no longer sustainable into the future (*Man and Nature*). I find it illuminating to view Audubon's career and the evolution of his ethic as occurring between the Lewis and Clark expedition, when Americans were eager to bring down the forests and change the land, and Marsh's declaration half a century later that they had already gone so far that informed, intelligent restraint must be imposed.

I see Audubon's Missouri River travels as occurring at the end of an age of relative innocence, in the last moments when it was still possible to be blind to the extent of human destructiveness in North America. His is the story of a people that spoiled everything they loved and touched, a country that did not know how to restrain itself, perhaps a species that does not know how to rise above primal, competitive urges long enough to leave an environment healthy and productive for future generations.

If the Audubon of the "Great Auk Speech" did not exist, the Audubon who supplied the voice and conscience of *Ornithological Biography* did, indeed, and does exist. The voice of his prose tells us two things. First, because he conceived of *The Birds of America* and *The Viviparous Quadrupeds of North America*—magisterial contributions to American ornithology, mammalogy, and art on an unprecedented, vast scale—he committed himself to a proportionate amount of work, on which he gambled all of his resources. Second, while he professed that he wished that he did not have to kill as much as he did, he advised the rest of us that there is plenty of reason to think of birds and other animals differently now—as deserving of much greater respect, even as having an inherent right to a fulfilled existence. He thought that if he could just get his work done before Americans did irreversible damage to wild animal populations, no one after him would need to repeat what he had done. He believed that his passion and his gifts made him exceptional.

This sounds like a double standard, and it was perceived as such in his own day. Audubon's good friend the Scottish naturalist William MacGillivray captured this aspect of Audubon's ethic in a lively account of a lovely May day of fowling near Edinburgh in 1839. The three speakers in this sketch are Audubon, MacGillivray, and his son. Just as MacGillivray's son raises his gun to shoot a lapwing, Audubon speaks up: "Don't shoot it [. . .]. It has a nest, and if you kill it you probably destroy five birds, or prevent four from being hatched. I hate to see birds shot when breeding." The young hunter's response is wryly incriminating: "By any person but yourself" (qtd. in Chalmers 182).

Audubon could not have realized it during his lifetime, but he did lay the foundation for a humbler, more sustainable conservation ethic by raising his countrymen's awareness of American bird and mammal species with his images and essays, by inspiring George Bird Grinnell to name his organization for bird species preservation after Audubon (as well as after Lucy), and by encouraging later generations to consider the lives of animals, especially birds, as complex narratives of survival in a wondrous but essentially dangerous world, where men and their guns figure among the gravest threats. Because of how early some of his conservationist statements were made, at a time before they might actually have effect, I tend to balance what can appear to us as the excesses of his gunning with the advocacy for the ethical considerability of birds at work in the monumental *Ornithological Biography*.

Audubon's reverend and often understandably exasperated coauthor John Bachman was right to see the 1843 expedition as a failure from a scientific point of view. While Audubon and his companions did bring back six new bird species, they did not find a single new species of quadruped,[13] which was one of the at-least-nominal purposes of the trip. When Audubon wrote to Bachman on November 12, 1843, he tried to put a smiling face on what he knew would disappoint his learned friend: "I have no less than 14 new species of birds, perhaps a few more, and I hope that will in a great measure defray my terribly heavy expenses. The variety of quadrupeds is small in the country we visited, and I fear that I have not more than 3 or 4 new ones" (qtd. in Audubon, *Audubon Reader*, 598). Writing over two years later, on March 6, 1846, Bachman's assessment was that, while Audubon brought back "in your Journal some Buffalo hunts that were first rate," in the final accounting, "your expenses were greater than the knowledge was worth" (348). Bachman had long suspected of the Audubon coterie "that all you think of is to get up the drawings," as he wrote to Victor on January 17, 1846 (342). Audubon was right about one thing, though, and Bachman

13 See Robert Peck's assessment of the expedition's scientific shortcomings, especially 84–92.

conceded this: "I have the best account of the habits of the buffalo, beaver, antelopes, bighorns, &c., that were ever written and a great deal of information of divers nature."[14]

The Missouri River expedition of 1843 was, in Michael Harwood's apt phrase, Audubon's "Last Hurrah." Supported by his four traveling companions and a rugged array of American Fur Company hunters, the great American woodsman was the impresario of what became an excessive, indulgent sporting trip on an epic scale that resulted in some of the most vivid and appealing accounts of hunting, adventure, and wild animal life on the distant western prairies that had been published. His experience on the prairies also enabled him to render his lithographic images of the large predators and game animals all the more bold and charismatic. This was purchased at great cost, however, and the income of the Audubon family publishing business could scarcely justify the expense and the patriarch's time away.

The two partial copies of the original Missouri River journals and the field notebook help us see Audubon much more clearly and accurately than we could when all we had was Maria R. Audubon's edition of those journals. Even though he was fifty-eight years old, he was not losing his enthusiasm for hunting, and he did not call for fewer bison to be killed—even if we wish he had. He was the same skilled and eager sportsman he had always been, if less capable of extreme exertions. If Maria succeeded in creating a narrative in which her artist-naturalist grandfather suddenly embraced the principles of conservation, she may also have distracted us—as well as herself—from his real accomplishments as a writer. Because the Audubon she created in *Audubon and His Journals* can seem capricious, inconsistent, and simply puzzling at times (and thus unlikely to produce worthwhile literature), and because *Ornithological Biography* is formidable in its size and for most people difficult to find in its original form, few people have been encouraged to study it closely as a whole. However, the hunter-naturalist we see in 1843 must be understood as more than an avid sportsman oblivious

14 Audubon to Bachman, November 12, 1843; quoted in Audubon, *Audubon Reader*, 598.

to any consequences of his and his companions' relentless gunning. I see him now as complex rather than hypocritical, containing within him the capacity to enjoy hunting as well as a vision of how members of his own species could improve their environmental ethic. The man Maria felt compelled to remake in the image of an 1890s conservationist had already through the 1830s articulated a truly advanced ethic, but she doesn't seem to have noticed.

PART V

Other Materials from the
1843 Expedition

The 1843 Diary of John Graham Bell

MARCH 13:

Left Philadelphia at 8 on Sunday morning & arrived in Baltimore at 3 oclock. Mr Audubon came to Baltimore on Monday.

MARCH 14:

Left Baltimore for Cumberland at 7, arrived at Cumberland at 6, by railroad took dinner at Harpers Ferry, at the junction of the Shanendowa & Potomac Rivers. Left Cumberland at ½ past 7 in stage, charged 30 dollars for extra baggage, 4 cents per pound. Traveled all night, snow.

MARCH 15:

Traveled all day & all night or untill 4 oclock in the mourning after a very tedious ride of 36 hours, 9 hours behind the usual time. The snow on the Alleghany Mountan was 2 feet deep.

MARCH 16:

Arrived at Wheeling this mourning at 4 oclock, left here at ½ past 1 in steamboat Evaline for Cincinnata down the Ohio River.

MARCH 17:

Saw General Harisons house & tomb at North Bend on the Ohio River. Saw great numbers of ducks & geese all along the river.

MARCH 18:

Arrived at Cincinnata at 3 oclock this mourning. Left here for Louisville in steamboat Pike. Arrived at Louisville at 10 oclock & remained on board all night. Went to the Galt House in the mourning, Sunday.

MARCH 20:

Spent the day in viewing the city & with one of our stage acquaintances, Mr Fellows, who was very polite to us indeed. He gave a party at his house & invited us all.

MARCH 21:

Accordion—$6.00

Sundries—.50

Spent the day in rambling about & shooting a few small birds. White crowned finches & shore larks very plenty.

MARCH 22:

Went out shooting, shot a few quails, woodcock &c.

MARCH 23:

Left Louisville at ½ past 12. Saw Porter the Kentucky Giant, about 3 miles below Louisville, where many of us walked to let the boat pass over the falls as the water was very low. Left in the steamboat Gallant.

MARCH 24:

Saw many ducks & geese & running very fast with the current.

MARCH 25:

Passed the mouth of the Ohio at 8 oclock this mourning. Saw hundreds of Canada geese and White fronted geese.

MARCH 27:

Shot at some ducks, &c.

MARCH 28:

Arrived at St Louis at 11 oclock. Put up at Glascow House, board 9 dollars per week each. It is one of the best regulated houses I have ever seen, & the best table.

MARCH 29:

Wrote to W J Bell. Received a letter from G Smith, N.Y. Amount of birds in shop in N.Y. $24.[**].

MARCH 30:

Wrote to G. Smith.

MARCH 31:

Went to see C. & D. Blauvelt (rain).

APRIL 1:

Preparing to go in the country on Monday. Saw Mr See of N.Y.

APRIL 3:

Went to try the repeating rifle.

APRIL 4:

Left St Louis for Edwardsville, at ½ past 8. Passed over the American Bottom, roads very muddy. Arrived ¼ of 6, distance 20 miles, fare 1.25 each.

APRIL 5:

Went out grouse shooting on horseback on the Prarie, found them very wild, shot 2. Saw about 40 grouse & 2 deer. Rabbits very plenty, shot 6, could have killed 20. Sprague fell in a crick & lost his gun.

APRIL 6:

Went to get Sprague's gun, distance 3 miles. Went on horseback with another young man, when about half way saw some ducks in a small lake, got off my horse to shoot them, the horse then got frightened for my gun & ran off. I held on the bridle, when he kicked me in my side & threw me 10 feet from him against a tree, but as I was so near him he did not hurt me seriously.

APRIL 7:

Made 2 grouse & 1 squirrel skin. Set some traps but caught nothing. Mr Harris shot one snow goose & 1 grouse & 5 lark finches.

APRIL 8:

Went to a lake to shoot geese about 6 miles on horseback. Shot 1 snow goose & 3 white fronted & 2 ducks. Shot them at 80 & 90 yards with buck shot with the large gun. Saw 4 deer & 3 turkies cross the road.

APRIL 10:

Made skins of 1 snow goose & 3 white fronted geese.

APRIL 11:

Went to lake. Shot some American snipe, Bartram's sandpipers & lark finches & a few ducks. Received a letter from W J Bell.

APRIL 12:

Made 15 lark finch skins. Went a short distance in the afternoon, shot 6 squirrels, 1 rabbit, saw a ground hog in a tree about 12 feet high.

APRIL 13:

Went after wild turkies, saw great number of tracks, shot 1 muskrat. Left Edwardsville for Bunker Hill at 2 oclock, arrived at 6, distance 18 miles. Heard the grouse booming between 9 & 10 oclock at night, *very bright moonlight.*

APRIL 14:

Went out early in the mourning, saw numbers of grouse, saw them strut & fight & make a very singular noise at least to me. It sounds something like *Boom Boom Boo,* the first 2 short, the last long.

APRIL 15:

Shot 1 fox squirrel & 1 blue wing teal. Made 1 grouse skin & shot some golden plovers, very plenty. Saw many sand hill cranes & thousands of ducks & geese.

APRIL 17:

Shot 3 grouse, 4 meadow larks, all very dark & very small, also 3 shore larks very small (breeding), 1 Henslow bunting. Rained all day & very muddy & unpleasant. Made 5 skins in the afternoon (1 doubtful savanah sparrow).

APRIL 18:

Went after squirrels, saw 2 fox squirrels, started 2 turkies. Shot at 1 with no. 8 shot small load, second barrel snaped. [Blank space in MS] started from under an old log as I steped on it.

APRIL 19:

Shot 16 grouse from a wagon & 1 skunk on the prairie.

APRIL 20:

Made skins of 14 grouse. Mr Harris, Mr Squires & myself taken sick about 10 oclock. I then threw up what I had eaten for breakfast & then felt as well as usual.

APRIL 21:

Made 1 squirrel skin & 1 Henslow bunting. Packed up my birds & started for Alton, distance 18 or 20 miles, at ½ past 8, arrived at 1 oclock. Left Alton at 3 in steamboat for St Louis, arrived at 5. Put up at Glascow House, made 4 goose skins at night.

APRIL 22:

Made 3 goose skins & unpacked all the grouse &c and laid them all out to dry. Spent the remainder of the day in purchasing articles for our journey, pair slippers &c—$1.00. Wrote to W J Bell.

APRIL 24:

Whole day spent in purchasing articles for our trip. Made 3 gofer skins. Saw first yellow warbler. Wrote to G. Smith. Received 2 letters from G. Smith, 1 from W J Bell.

APRIL 25:

Started at ½ past 11 oclock amid shouting, firing & waving of handkerchiefs & friends bidding each other adieu, many no doubt forever. Current very strong & boat rather slow. Saw first tanager & Baltimore oriole & yellow billed thrush. Very warm & fine, arrived St Charles ¼ 4.

APRIL 26:

Rain & rather cool, current so strong as to turn the boat ashore, being impossible to steer her. Shot one wild goose, saw where a house & many acres of land had fell in the river, arrived at [**]th Point ¼ of 4.

APRIL 27:

Under way at 6 oclock, quite cold. Passed Washington at ½ past 6. A man brought a pelican alive. Passed the Gasconade River at ½ past 5. Saw first redstart.

APRIL 28:

A most beautiful mourning. Passed Jefferson City at 7 oclock a.m. Saw a marmot & a few squirrels, also a blue wing warbler & several others. Saw a very small blue heron unnown to me.

APRIL 29:

Arrived at Boonville at ½ past 9 a.m. Sent a letter home. Rained nearly all day, water still rising, rained 3 inches during the night. Passed Glascow at ½ past 6 p.m.

MAY 1:

Stoped at 9 oclock on account of current & head wind, went on shore, killed 28 rabbits, 2 ground hogs, 2 partridges, 2 squirrels & several small birds. I killed 14 rabbits myself. Wounded 2 turkies, started a fine buck, made skins of 4 rabbits, 2 marmots, 2 squirrels. Saw some paroquets.

MAY 2:

Rainey & rather unpleasant, boat very very well, arrived at Independence at 1 oclock, remained there 20 minutes. Shot 2 paroquets while taking in wood, had a little difference with Squire. Passed the Kansas River at 6 p.m. Sent a letter home.

MAY 3:

Arrived at Fort Leavenworth at 6 oclock a.m. Remained 2 hours & shot 1 paroquet. Saw 1 turkey & a number of Cicapoo Indians. Got aground at 5 oclock, got off at ½ past 7, got on again and remained there all night.

MAY 4:

Got off at 9 oclock & went on shore to cut wood. Shot 1 squirrel & 1 new bird & several other small ones & 17 paroquets. Remained there untill 4 oclock on account of head wind. Landed the Iowa Indians & young men. Saw about 30 species of our common birds. Very warm & rained in the mourning.

MAY 5:

Arrived at Black Snake Hills at ¼ of 1. Landed 3 squaws & 2 chiefs of the Iowa Indians. Very high wind with a little rain in the afternoon. Killed 1 new bird, 1 squirrel & several Lincolns finches, white crowned sparrow &c. Made 14 skins.

MAY 6:

Shot 2 squirrels, 1 new vireo & white eye & warbling vireo & another of the new finches. Kentucky warbler. Fox Indian country. Saw about 100 Indians of the Ioways & Foxes. Very windy & cold. Made 3 skins.

MAY 8:

On Sunday Mr Harris killed another of the new finches. Yesterday & today has been very pleasant indeed. I killed 1 black squirrel, sent a letter home, saw 9 of the Ottowe Indians following the boat & making signs for us to stop.

MAY 9:

Passed the Great Platt River at 12 oclock. Passed the Otoe Village, about 20 lodges. Saw several Potawatamies Indians. Shot 2 red breasted grossbeaks, arrived at Bel View at 4 oclock, saw great numbers of the Otoes, spoke with the chief. Arrived at the bluffs at the Garrison.

MAY 10:

Stoped by an express sent to examine the boat. Detained 5 hours, was introduced to several officers &c. Killed 4 yellow head tropioles & several small birds. Saw the first deer. Passed the original Council Bluff, abandoned on account of the river leaving it 1 mile.

MAY 11:

Stoped twice to cut wood. Rainy all day, saw a wolf on a sand bar. Saw some large trees fall in river, a few shells at a small lake, 6 species, shot nothing.

MAY 12:

Saw 4 black tail deer. Passed Woods Bluffs, saw elk tracks, also wolf, deer & turkey. Rained in the mourning, but very pleasant in the afternoon.

MAY 13:

Saw several Omaha Indians. Passed the grave of Black Bird, a celebrated chief of the Omaha Tribe, buried on horseback, the horse alive. Passed Lieutenant Floyd's grave, one of Lewis & Clarks party, called Floyds Bluffs. Saw several turkies, killed 1 & 2 lark buntings. Stopped at the Big Sue River at night, rainy & warm.

MAY 15:

Saw a bear cross the river on Sunday {mourn}. Stoped at 1 oclock on account of wind &c. Rained until ½ past 4 p.m. Went out and shot 5 blue herons & 1 raven, muddy & rushes. Great number of elk tracks. Stoped again at ¼ of 8. Got 1 bat.

MAY 16:

Started at 3 oclock. Saw great numbers of cliff swallows nests, arrived at the Vermilion Post at ½ past 12. Burned a hole in the boiler & stoped. A few hours after, went out, saw 7 deer & 1 wolf, broke 1 deers leg & also the wolfs. Saw buffaloes floating down the river. (very pleasant)

MAY 17:

Went out at 4 a.m., shot 1 doe & 3 fawns & 2 chestnut colared lark finch & a female of Harrises finch, 1 lark bunting. Saw 9 buffaloes, 1 deer & 1 antelope floating down the river. Beautiful day. Shot at 1 deer & 1 turkey with the repeating rifle & missed them.

MAY 18:

We are still detained mending the boiler. Water falling very fast. Sent a letter to W. J. Bell by a macanaw boat. Shot another of the new verioes. Saw great numbers of black head gulls. Clear & fair.

MAY 19:

Started at 3 this mourning. Passed the mouth of the Vermilion River at 7 a.m., also some Indian graves below it. Saw a deer cross the river, shot a Baltimore oriel & a few other small birds while cutting wood. Stoped at 8 oclock having run only 35 miles. Clear & fine.

MAY 20:

Passed the Jacques River at 5 this mourning. Saw 3 deer, 3 buffaloes & a wolf before 12 oclock.

SUNDAY, MAY 21:

Saw 50 or 60 buffaloes at 6 oclock this mourning. Stopped at Fort Mitchel at 10 a.m., which we found abandoned, and tore down some old building for wood at the mouth of the Running Water. I was the only one that went to {Gevaues} grave. He was killed by the Sues at this fort. Shot 1 doe, saw 3 elks & 3 antelopes & 1 wolf.

MAY 22:

Saw 30 buffaloes before 7 a.m. & 2 wild cats. Saw 14 or 15 Indians on shore who fired 4 balls. 2 struck the cabin door & 1 went through a mans pantaloons & 1 in front of the boat. Saw 6 antelopes, sent out 7 men to shoot buffaloes, after we stoped at night at the Great Cedar Island.

MAY 23, THERMOMETER 92:

The party returned this mourning having killed 4 buffaloes. Saw at least 200 along the river, also several antelopes. Got off at 3 oclock and got on again soon after. Remained on all night. Warmest day we have had.

MAY 24, THERMOMETER 78:

Went out this mourning & killed 1 red shafted woodpecker, 2 Arkansas flycatchers, 1 blue grossbeak, 1 small verio, & 4 meadow larks. Saw Says flycatcher, 4 antelopes, 1 hare, plenty of buffaloes & Lasuely finch. Windy & pleasant, prarie very broken.

MAY 25, THERMOMETER 50:

Rainy & cold. Saw great numbers of buffaloes & a few antelopes, & wolves, & but nothing of importance, being very unpleasant.

MAY 26, THERMOMETER 46:

Went out & shot 3 Tennessee warblers & 1 arctic finch. Saw 2 black-shouldered hawks, & a buffalo swim the river. Sent 2 men to Fort Pier, distance 70 miles. Got aground several times during the day. Started with Mr Audubon, Mr Harris, Mr Sprague & 3 other men across the Great Bend. I shot at 4 black tail deer. One of the men killed one, which we cooked & ate. Part went to sleep at ½ past 9. Saw a village of prairie dogs.

MAY 27, PLEASANT / TEMPERATURE 70:

Rained in the night a little, also this mourning & cleared off. Went after buffalo with 2 men about 7 miles. Shot at a large bull with 2 balls with a double gun & 1 rifle. They all struck him but did not kill him. Got on board again at 3 p.m. Saw at least 5000 buffaloes on a beautiful level prarie & a few wolves & deer.

SUNDAY, MAY 28:

Arrived at Fort George at 3 p.m. Saw about 100 Indians & a few lodges encamped there. Was introduced to Major Hamilton.

MAY 29, PLEASANT / THERMOMETER 92:

Saw the house in which one of the principals killed 2 of his men, of the opposition fur company. We are detained here at Fort George on account of low water. Shot some black head grossbeaks, lark finches, Arkansaw flycatchers, and Tennessee warblers. Went to the fort, saw 7 Indian lodges & saw another prarie dog village & 1 burrowing owl. Was introduced to Major Hamilton & other officers at the fort.

MAY 30, PLEASANT / THERMOMETER 88:

Some of the men brought us a buffalo calf about 2 months old. Some of the men came to meet the boat from Fort Pier, being still detained at Fort

George. Got off at ½ past 3 p.m. Was introduced to Mr Chardon & Mr Peacot, of the Fur Co.

MAY 31, PLEASANT / 84:

Started, & soon broke something & stopped, started, & soon got on a bar, about 3 miles below Fort Pier. Arrived at Fort Pier at 4 oclock, went to the fort & Mr Peacot presented me 2 pair mockasins. Wrote to W J Bell & G Smith.

JUNE 1, PLEASANT / 88:

Left Fort Pier at ½ past 1. This afternoon killed 1 cow bunting (large) & some black head grossbeaks & some Arkansaw flycatchers. Saw several wild geese & 1 buffaloe calf & 3 young wolves & 1 old one. Run about 30 miles, left half the men & cargo at the fort. Boat runs much faster.

JUNE 2, PLEASANT / 82:

Started at 2 this morning. Passed the Shienne River at 8 a.m. Shot at a wolf crossing the river, saw 13 others on shore. Saw some geese with young ones & some gadwall ducks. Killed 1 buffalo from the boat & stopped & got him. Shot some lark finches &c.

JUNE 3, PLEASANT / 72:

Shot a large wolf from a sand bar, when we stopped to cut wood. Saw 10 others all within a few 100 yards. Saw 7 elk together, also 3 young wolves, but few buffalo. Saw several coots & wild geese, teal & 1 magpie. Run about 55 miles, rained in the night.

JUNE 5, RAINEY / 60:

Sunday cool & rainey, thermometer 66. Shot 1 Henslow bunting (different). Cool & rainey. Passed Cannon Ball River. Fired several shots at 4 buffaloes crossing the river, wounded them all, but killed none. Passed Beaver River, saw several wild geese & passed the old Ricaree Village.

JUNE 6, RAINEY & COLD / 47 X 52:

Very cold & plenty of white frost this mourning. Saw 4 antelopes & a few wolves. Rained nearly all day. We have a Ricaree Chief by the name of the

Four Bears going up to the Yellow Stone with us, also one by the name of the Iron Bear of the Grovons.

JUNE 7, RAIN & COLD / 52:

Arrived at Fort Clark, old Mandan Village at 7 a.m., now occupied by the Minaterees or Grovons who left their own village after the Mandans was nearly all dead with the small pox. They are now 3 miles from this place about 100 lodges & 1200 souls. Captain made them presents.

JUNE 8, CLEAR & COLD / 37 X 67:

Left Fort Clark at ½ past 3. Past the Mandan Village at ½ past 4, about 150 persons & 15 lodges. Arrived at the Manatarees or Grovons or Groseventres at 7 a.m. Remained 1 hour. Had a smoke with the principal men. Captain made them some presents &c. Saw an elk with her young one. Shot 1 young hare. Passed many curious clay hills. Run 65 or 70 miles.

JUNE 9, 40 X 75:

Started at ¼ of 3, passed the Little Missouri River at 12 oclock, saw 3 elk swim across it, killed 1 cow buffalo in the river, while cutting wood. Mr Harris shot at 2 elk. I shot 5 times at 2 buffaloes & killed 1 young bull & 1 calf. One of the men shot a hare. Saw several wolves.

JUNE 10, PLEASANT / 61 X 82 X 65:

Killed 2 buffalo bulls before breakfast. Saw 35 elk, 5 big horns, & antelopes & several wolves. Shot 2 wolves, caught 1 young elk by the neck with a string in the river but could not hall him up. The string broke & I lost him. Stopped & made some Assinneboins Indians some presents.

JUNE 11:

Killed a buffalo bull from the boat and shot at some wolves. Packed up the birds, saw 2 big horns.

JUNE 12, PLEASANT / 53 X 65:

Found a pair of elk horns, with the skull. Saw several wild geese with their young ones. Saw 22 big horns. Passed Fort Mortimer at 6. Passed the mouth of the Yellow Stone at ½ past 6 p.m. Arrived at Fort Union at 7 oclock amid roaring of cannon from fort and boat. Passage 48½ days.

JUNE 13, PLEASANT:

Missouri River, Fort Union: We arrived here in the steamboat Omega, Captain Soare, in forty eight days and a half, making the trip in 20 days less than any other boat that ever arrived at this place, and in that time we lost 7 or 8 days. The fort is situated on a beautiful bluff about 5 miles above the mouth of the Yellow Stone River with plenty of timber on the opposite shore, which is about half a mile wide at this place, the timber on this side haveing been all cut off to build the fort and boats &c. The fort is 220 feet square with stone bastions & pickets about 20 feet high made of square logs set about 4 feet in the ground and as close together as possible with all their buildings inside. They have 3 buffaloes, one year old perfectly tame, and some young calves of this year, and a young fawn quite tame. Spent nearly the whole day in moving our things from the boat to the fort and looking round, Wrote to W. J. Bell. There was a party of Assiniboins [of about 25 lodges] arrived on the opposite side of the river this afternoon, and such howling of dogs I never heard before. Their dogs are so much like wolves that were I to meet one on the prarie alone I would shoot him for one. They often tame the wolves and cross them with the dogs.

JUNE 14, RAINEY:

Went down the river about 3 miles where the steamer stopped to take in wood to see the Captain off & bid him good by. It rained a little a great part of the day so that we did not go out to shoot. In the afternoon we saw some wolves passing near the fort, when Mr Culbertson mounted his horse, and with his gun started in pursuit of him, when we all ran to the gallery of the fort to witness the result, but the wolf taking to the hills we soon lost sight of them, but in about 20 minutes he returned at full speed with the wolf on his horse. We then saw another and 3 men on horses started after him, but he also took to the hills and soon escaped.

In the afternoon the Indians came over and pitched their tents near the fort. In the evening Mr Culbertson gave a ball and we were all invited to witness the scene. They commenced the music with claronet, violin & drum, and each one choosing a squaw commenced dancing something like our common country dance. This they continued for several hours, but I soon got tired of looking on and went to bed, not to sleep, for that was out of the question, for the noise as they were all in the next room

JUNE 15, PLEASANT:

Went out shooting & shot one crow and a few small birds. Saw 1 sharp tail grouse. Mr Chouto & Mr Murray arrived with two macana boats from the Crow Indians station with robes. One of the squaws had a young badger alive, which we purchased of her for 2 strings of white beads, and intend to take him to New York alive if possible. He is very tame and playfull.

JUNE 16, COOL & PLEASANT:

Went over the river after antelopes, saw 15 but found them exceedingly shy and dificult to approach, being in the middle of the day when they generally keep on the top of the hills where they can see all around them for some distance. Also saw 2 magpies. One of the hunters that went with me caught a young fawn, which we killed and made a drawing of it. In the afternoon we saw 1 Arkansa flycatcher, 1 lazuly finch, & a few other common birds. The Indians all left early this morning.

JUNE 17, WARM:

Shot 1 red shafted woodpecker, 3 lazuly finches, 1 striped squirrel, & a wolf from the top of the fort in the evening. They come round the fort as soon as the gates are closed to pick up the bones &c that are thrown out during the day. Sometimes 4 and 5 are within rifle shot at the same time. I shoot them with the large gun loaded with 20 buck shot, which generally makes sure work of it, putting from 3 to 8 shot in them.

JUNE 18:

Two hunters went out and killed two antelopes & 2 deer. I then had some dificulty with Mr Audubon about working on Sunday.

JUNE 19:

Shot a new tit lark, in the afternoon & found the nest & eggs of the lark bunting. Shot a wolf from the top of the pickets in the evening. Shot a gold wing woodpecker with red cheeks, which I think is a new species, which I cannot describe haveing only one specimen. Mr Audubon & Sprague made a drawing of the head of the male antelope. I put the skins in pickle.

JUNE 20:

The tit lark rises to a considerable hight in the air & sings, something like the European sky lark, for an hour in succession. Shot a new grakle and two chestnut collared lark buntings & several clay coloured buntings & some arctic finches &c. I and Mr Harris heard some small birds singing on the prarie but found it impossible to find them, as we walked from one hill to the other for a long time and finally gave up, for when we got on one hill they appeared to be on the one we had just left.

JUNE 21:

Went out over the river with Mr Kenzie. Shot a new rabbit with a rifle, and cut him in two. I however brought his head and legs with me. Also shot a turkey buzzard flying with a rifle and found a very singular deer horn. Sprague shot an arctic blue bird. I shot a lazuly finch & a wolf from the fort in the evening. Saw the little tit lark singing in the air which gave us so much trouble yesterday in looking for him on the prarie. They rise to a considerable height in the air, and with an undulating motion, they keep mooving about generally in circles, but not at all regular, as they sometimes go round one way and sometimes the other, and when they sing, which is about every ten seconds, they moove very slowly with a quick flapping of the wings, something like a hawk when he is watching his prey & remaining stationary in the air. In this way they will remain in the air for an hour or more at a time, singing to their mates, who are sitting on the prarie beneath them.

JUNE 22, PLEASANT:

Shot 3 female arctic blue birds, 1 lazuly finch, bay shouldered bunting, cat bird, red eyed flycatcher, Maryland yellow throat warbler, black bill cuckoo, house wren, chipping sparrow, clay coloured bunting, ferruginous thrush, Wilsons thrush, cow bunting, & arctic finch.

Heard that the opposition fort at the mouth of the Yellow Stone River was falling in by the current washing away the bank in front, which is very rapid at this time and the water rising very fast. Mr Chardon & Mr Harvy started in a keel boat for the Blackfeet station this mourning which is about 600 miles from this place, which will take them about 35 days as they pull the boat up by hand, like a canal boat with a long rope.

JUNE 23, CLOUDY:

Shot a marsh hawk *very dark on the breast*, 1 lazuly finch, 1 Says flycatcher, 1 Arkansa flycatcher, 1 male arctic blue bird, 3 of the new tit larks, 1 pigeon, 1 chestnut collared lark bunting. Provost killed another doe. Saw the round cactus in flower.

JUNE 24, BEAUTIFUL / 70:

Went out with Provost very early this morning to shoot deer by imitating the cry of the fawn in distress when the does will instantly run to rescue them and are easily shot down by the hunter. I also saw the wolves hastening to the spot thinking to get something to eat, but often looses his life by so doing as the deer and the wolves will come within a few feet of the hunters if they do not smell them. The wolves will also come when they hear the report of the gun and eat the game of the hunters if they leave it for a short time where they can get at it. They tell me they have killed 3 buffaloes and by the time they could skin & cut up 2 of them, the third would be eaten up by the wolves.

Mr Harris shot 1 sharp tail grouse & Sprague found the nest of the new lark. I went with Mr Audubon in a wagon down to the other fort, called Fort Mortimore. On our way killed 4 of the new larks and 1 white bellied swallow, and found the nest & eggs of the arctic blue bird with 6 eggs, also the nest & eggs of the meadow lark in the mourning, 3 eggs. They told us at the fort the water had risen 3 feet the day before but had now fallen 14 inches. They had removed part of the pickets but not the house as the bank had not fallen in any in 24 hours. Before the bank fell in the fort was about 60 yards from the river, now it is about 20 feet.

This has been one of the warmest days we have had. The thermometer stood at 70 early in the morning.

JUNE 25:

Spent nearly the whole morning in writing. Mr Harris went over the river but has not yet returned. Squires went out with the hunters of the fort after buffalo on Friday, but have not yet returned, which they generally do the next day after they start.

The hunters returned this afternoon haveing killed 3 buffalo, 3 antelopes & 1 hare. They went about 60 miles.

JUNE 26, BEAUTIFUL / 76 X 82:

Went out & killed 1 black head grossbeak. Provost killed a doe by imitating the young fawn which we measured, weighed &c and saved the skin. It weighed 240 lbs. In the afternoon I shot 1 prarie lark bunting, 1 prarie lark finch, 3 lazuly finches, 2 shore larks, & 1 young red shafted woodpecker.

JUNE 27, PLEASANT / 64:

Went out on horseback with Mr Owen about 30 miles to look for buffalo but found none. When we started we shot a wolf on horseback. We was off our horses taking a drink in a ravine when we first saw him on a hill a short distance from us, when we instantly mounted and gave chase to him and overtook him in a very few minutes, when he turned short round and let me pass him some 40 or 50 yards. I then turned round and fired at him & missed him, when he passed Mr McKenzie who fired at him and wounded him in the fore foot. When I ran up to him and fired 2 pistol balls in his back he then laid down and I dismounted. I pulled him about and soon found that he was not dead. I then gave him a kick and up he jumped and ran off, but not very fast. I then ran on foot to his side and shot him through the lungs with a pistol & killed him on the spot.

We then started 2 sharp tail grouse & saw a great many chestnut collared lark finches & shore larks & the new tit larks and about 20 magpies.
We then started a long tail deer about 200 yards before us & instantly took
after him over a beautiful level prarie. We ran about half or ¾ of a mile and
in that time we gained at least 140 or 50 yards on him, when we came to a
deep ravine and was obliged to give up the chase, which I was very sorry
for as we would have killed him in a few minutes. We went on along the
Muddy River & killed 2 gadwall ducks and saw great numbers of pigeons.
On our return we tried to shoot some birds while at full speed & I tried to
load my gun while running, which I could do better than I anticipated. We
got home at 4 oclock haveing rode at least 60 miles. Mr Audubon presented
me with an Indian shirt.

JUNE 28, PLEASANT / 78:

The hunters killed one elk, one doe, & one big horn. I went down to the
other fort in a wagon & shot at a wolf in returning with a rifle and just
touched him. Mr Culbertson lost his flat boat last night, supposed it to be
untied by some of the men of the opposition fort. I shot another wolf from
the top of the fort in the evening.

JUNE 29, VERY PLEASANT / 62:

Finished the macanaw boat 60 feet long, which they commenced 6 or 7
days ago and have commenced loading it for the Crow station. Mr Kipp &
Mr Chouto started at 11 this mourning up the Yellow Stone for the Crow
Indians. Mr Harris killed a doe and a grouse & Canada woodpecker & red
shafted woodpecker. Provost also killed a doe. I got the young of the gold
wing & red shafted woodpeckers. Each of the young ones were like the old
ones, either red or yellow. Felt very sore and stiff from riding on Tuesday.

JUNE 30, 60 X 62:

The express from the opposition fort came and told us the steamboat Omega
passed the Mandan Village in two days after she left here. Shot 2 black head
grossbeaks, 2 meadow larks, and one red shaft woodpecker. Rained all the

afternoon. This afternoon the opposition fort fell in the river by the bank falling in, at least 60 feet.

JULY 1:

Went out on horseback with Mr Harris, shot 1 grouse, 2 arctic blue birds, 1 meadow lark, 1 pigeon and one Arkansaw flycatcher and found the nest and eggs in the afternoon. Shot 5 very remarkable gold and red shaft woodpeckers, some with red shafts and black cheeks, others with red shafts and red cheeks and others with gold wings and black cheeks, and others varying in the shade of the shaft very much and one very similar to our common gold wing. Mr Culbertson gave me an Indian coat & shot pouch & powder horn.

JULY 2:

Heard Mr Collins was very sick. Mr Audubon and Mr Culbertson is going to see him. Shot a wolf in a hole a few yards from the fort where he had got in but was not able to get out. Mr Culbertson told me he has killed 4 in the same hole in one morning.

JULY 3, VERY PLEASANT / 60 X 90 X 67:

Went over the river after deer with Mr Audubon, Mr Harris, Provost, Owen, Squires. Mr Audubon shot at one 3 times and missed it. The rest saw none. In the afternoon the two hunters returned with one big horn and a black tail deer, both females. In the evening we had some fireworks in the fort, such as rockets, wheels, roman candles, serpents, &c which they brought from St Louis for the fourth, but as we expected to go out hunting and not return the next day, they celebrated the third instead of the 4th.

JULY 4:

Went out after deer with Mr Harris. We started a very large buck out of a small bunch of bushes in the prarie. I was on one side and Mr Harris on the other, when he came out on my side and I fired 12 buck shot at him at 40 yards and wounded him very severely and fired a small load of no. 6 shot at his head. He then ran about a quarter of a mile and laid down in the open prarie in the worm wood, while we expected to find him in the bushes, and by mere accident I saw his horns above the worm wood, which were still in the velvet and yet very large. As soon as I saw him I went back to tell Mr Harris I had found him, but before I had gone 20 yards he got up and ran along a deep ravine about 75 yards, when he crossed it, and I at the same time ran towards him and fired both barrels at him but did not stop him as he was too far off. I then went to the spot where he lay and on examination found that he had been bleeding at the mouth by the froth &c on the ground. We then went on after him a short distance and Mr Harris started him from a small bush some distance from the others and also from me, when he fired both barrels at him and missed him, as he did not see him when he first started so that he was rather too far when he fired. We then went on after him again in the direction we last saw him but was obliged to give him up as lost.

We then went home a mile and a half to dinner, and told them what had occured and they all told us they knew he was dead and that they could soon find him, if we would go with them to where we shot him, so after dinner we started with Owen on horseback and soon found the bones of my fine buck completely eaten up by the wolves. We found them by the buzzards flying round over the wolves, while they were picking the bones, in quite a different direction from what we saw him runing when we last saw him, which is nearly always against the wind. I cut off part of his horn and we started back.

On our way we chased a wolf and was within 20 yards of him when he got in a deep ravine and hid himself. We then took after a fox, and he ran up hill and left us behind. I then shot a very large snake.

JULY 5, RAINEY:

Cold and rained nearly all day. Made and set some rabbit traps & snares. Made a whistle to imitate the cry of the fawn in distress to call the does to us. In this manner we have killed several females, who supposing a wolf or something else to have hold of the fawn, rushes to the spot and is often killed.

JULY 6, PLEASANT:

Went out with Owen and Mr Harris after deer. I saw one doe and her fawn but too far off to shoot. We then saw a wolf. I fired at him and wounded him, and then we took after him, but as we had some steep hills to run up, he soon got off. We went about 5 miles up the river, tied our horses and went on foot in a large point where we saw many elk & deer tracks. Owen saw 2 deer too far off to shoot. On our return we saw a fox about 200 yards from us on a level bottom covered with wormwood. We then went on slowly towards him untill he started when we done the same. We ran up to him within 40 yards when the bottom becomeing quite soft, he began to leave us, when Owen, being on the best horse, began to put on the whip and gained on him a little, and as he was now near the woods he fired at him with a rifle and missed and gave up, but if we had been on smart horses, we would have overtake him easily and no doubt have killed him.

In the afternoon I and Mr Harris went over the river and shot at a doe at 50 yards and wounded her. We then went on and gave a call when a doe came rushing past us, when we fired and killed her, and now we had killed her and the next thing was to get her to the boat, which was a mile and a half off. We cut a pole and tied her legs together and passed the pole between them and carried her a short distance when Mr Harris said he could not carry it. We then laid it down and I put my handkerchief over it and went on a short distance to look for more while Mr Harris went after a man that was burning coal near the boat. He then returned with him and we carried her to the boat. It is one of the finest skins we have got.

This afternoon Provost and Squires went up the river 3 miles after deer and elk.

JULY 7, VERY WARM / 90:

Went out with Mr Audubon and Mr Harris about four miles after rabbits but saw none. Shot a raven, found the service berry just beginning to ripen which grow somewhat like our huckle berry and are very good tasted but hard to digest. When green they are quite red, and when ripe a reddish blue. The gooseberry and currants are also in great abundance, but there will be but few buffalo berries this year on account of the late frosts in the spring. Provost & Squires went up the river 3 miles after deer & elk.

JULY 8, VERY WARM / 86:

This morning I saw a wolf a short distance from the fort and went out and fired [at] him with buck shot and wounded him. He then started off, but very slow, and I took after him and found I was gaining on him, but soon got tired and stopped and fired at him with small shot at 20 or 25 yards and gave up the chase when he ran a short distance and laid down. In the afternoon I and Mr Harris went a short distance after deer, saw none. Provost and Squires returned haveing killed one doe and saw 5 others. The weather has been very warm for 2 or 3 days, thermometer above 90.

In the evening we killed a very large wolf. Yesterday there was a war party of Assiniboins here. Haveing taken one scalp they were singing and dancing nearly the whole night. They all had their faces painted black, except their cheif who was red, and I think the most frightful looking objects I ever saw.

JULY 9:

I have been writing nearly all the morning and Mr Audubon is making a drawing of the wolfs head killed last night. It is not as warm today as it was yesterday. I caught a little cold last night and feel a little stiff and sore today. The Indians are all gone and things are all quiet.

JULY 10, 60 X 90:

Caught a cold on Saturday night and felt very stiff and sore next morning. I was quite unwell this mourning, had quite a fever last night, and a headache all day and no appetite, took some pills in the evening.

JULY 11, 66 X X 74:

Still have the headache and a pressure of blood at my hart, got Mr Harris to bleed me, and felt better, took some salts.

JULY 12:

Felt a little better but very weak, took some quinine. Mr Harris killed a wolf in the evening. Took some more pills.

JULY 13:

Felt rather better, but still had a little headache, took some salts in the mourning and some quinine at 12 oclock. Went over the river in the afternoon, and in crossing we saw a wolf lying under the bank where he could get up, when we rowed up to him and I killed him with my pistol, shooting him between the eyes and then through the lungs. I then shot a male gadwall duck and Mr Harris shot a crow and we returned after eating as many service berries as we wanted, which we found very fine and in great abundance. Shot a wolf in the evening.

JULY 14:

Felt much better this mourning. In the afternoon some of the men and squaws dressed themselves and their horses in full Indian dresses and rode out on the prarie to have a race. The two squaws had a race, they riding straddle on the horse like a man. In a few moments after they were out they saw a wolf, and they all took after him and soon killed him. They then came back and started again and killed 2 more. Mr Harris killed a young hare. I got a new mouse.

JULY 15:

Started with Mr Audubon, Mr Culbertson, Mr Harris in a wagon in tandam, Squires, Owen, and Provost on horseback and three men with the mule in the cart in which was placed our skiff and a tent and our baggage. Started after breakfast on our expedition to the Yellow Stone we have been talking of for some time after elk, big horn, beaver &c. We were provided with nothing in the way of eatables but about a dozen hard biscuits, such as are used by sailors. Our intention was to hunt on the north side a while and then take the skiff and drop down to a good place among the hills on the other side for big horns and then to the mouth of Charbonau River where there is said to be a beaver lodge & dam. We crossed the Missouri at the fort in the scow which we left well fastened to the bank to be ready on our return. We then followed the level prarie on the bank of the river for about 20 miles from the fort. We reached this spot early in the afternoon and encamping in the open prarie, directly on the bank of the river. While the rest were arranging the camp and pitching the tent, I, Mr Harris & Owen took our guns and went to hunt but saw nothing. On our return we found Mr Audubon had caught 4 small catfish, and with a small piece of biscuit we made out to make a light supper, not having eaten since breakfast. The day had been excessively hot, but with a good breeze, which failed at night and we all spread our beds outside the tent, and were soon comfortably sleeping beneath our mosquito nets. About 12 oclock we were aroused by the approach of a thunder gust and by the time we could tear up our fixings and bundle ourselves and other things into the tent, it came on to rain very hard with a strong wind and incessant thunder and lightning. I rolled myself in my blanket and spent the rest of the night as comfortable as circumstances would admit.

JULY 16:

We were stirring at an early hour this mourning, and while the rest were spreading out the beds and blankets &c to dry, I took my gun to try to kill a deer for our breakfast. I went about a mile and found the mosquitos so thick that I was obliged to return, where I found them taking a cup of coffee without anything to eat. I tried to shoot a small bird for fish bait when we saw a wolf and Owen was started on one of the hunting horses to run it down and shoot it but soon found he had an ugly customer, for we have discovered that there is a very great difference in the running of wolves. It was not until he had chased him several miles and fired 7 or 8 shots at him that he could claim him as his own. As soon as we found he had killed him, an old horse was sent to bring home the game. While we were watching the chase, Mr Culbertson discovered a buffalo at the distance of 3 or 4 miles. The Blackfoot man was soon saddled and he started off in pursuit.

He stopped on his way for Owen to join him as he returned from the wolf hunt and the two went on slowly to where we had lost sight of the buffalo at the foot of the hills. They proceeded slowly to give Owens horse time to rest after the hard chase she had had, and we continued to watch them with great anxiety. It was nearly an hour before I saw Mr Culbertson in chace of the buffalo who passed into a ravine from which I could see that they did not emerge, until I saw Mr Culbertson standing on the edge of the hill. I then felt sure that the bull was killed, although the distance was so great that we did not hear the report of any guns nor see the flash. We immediately ordered the cart to go for the meat and Mr Harris and I had a horse put to the wagon and drove to the spot. We met Mr Culbertson returning who told us that it was a remarkably fine bull, and that he thought that we should not get a better specimen for skinning. On reaching the spot we found a fine large bull of very good colour, haveing shed all his winter coat but the summer hair haveing only commenced growing over the hinder parts and the mane and long hair of the legs and throat much shorter than others we had seen killed on the river. The head too was somewhat damaged by cutting out the tongue which Mr Culbertson carried to the camp. Under these circumstances we determined to carry back the skin of the bull and had taken his measurements and commenced opperations when a messenger arrived from Mr Audubon with orders to bring in the skin. At this stage of the proceedings Owen discovered another bull making his way slowly across the prarie directly towards us. I was the only one of the party who had balls for his gun, and I would gladly have claimed the privilege of running him, but fearing that I might make out badly on my first trial with my large gun, which is too heavy to run with, and supposing the meat could be carried to the fort as it was much wanted, I handed the gun and balls to Owen and Mr Harris and I placed ourselves on an eminence to view the chase. Owen approached the bull who had continued to advance and was now about a quarter of a mile distant. The bull did not see him or did not heed him and they advanced directly towards each other until they were 70 or 80 yards apart, when the buffalo started at a good run and Owens mare, which had already had two hard runs this mourning had great dificulty in preserving her distance. Owen soon perceived this and applied the whip pretty freely. He was soon within shooting distance and fired a shot which sensibly checked the progress of the animal and enabled him quickly to come along side of him when he discharged the second barrel into his lungs passing through the shoulder blade which brought him to a stand. Mr Harris and I then ran to within speaking distance, called to Owen not to shoot again. The bull did not seem to be much exhausted but he was so much stiffened by the shot in the shoulder that he could not turn round quickly, taking advantage of which we approached him. As we came near he would work himself slowly around to face us and then make a pitch at us. We then stepped on one side and continued discharging our six barrel pistols at him with little more effect than increasing his fury at every shot. His appearance was now to inspire terror, had we not felt satisfied of our own ability to avoid him. Mr Harris came however very near being overtaken by him. Through his imprudence he placed himself directly in front of him and as he advanced, he fired at his head and then ran directly ahead of him not supposing that he was able to overtake him, but in looking round he found Mr Bull within three feet of him, prepared to give him a taste of his horns. The next moment he was off the track, and the poor beast being unable to turn quick to avenge the insult, I then took the gun and shot him directly behind the shoulder blade, when the blood flew out at the same instant 2 or 3 feet. He then bled most profusely at the mouth and nose. He then stood tottering for a few minutes and fell easily upon his knees and rolled over and was soon dead. He was a very old animal in poor case and even if he had been fat, with one cart we could not have taken all the meat to the fort. When I saw this I was sorry I had not run him myself, feeling satisfied I could have easily killed him running him alone, without the excitement of competition. We left the bull for the wolves and birds of prey and returned to the skinning of the first which was much the best skin of the two.

I went to work skinning when the men arrived with the cart we soon got off the skin and as much meat as we wanted and started for the camp. When I got there I found nothing to eat, and we soon concluded to pack up and return to the fort with the skin, where we arrived safe about 7 oclock with the skin and meat &c. We all stopped on the opposite side of the river on our return and ate as many service berries as we wanted, which I am very fond of.

JULY 17, PLEASANT / 66 X 91:

Spent nearly the whole day in cleaning the buffalo skin and put it in pickle.

JULY 18, PLEASANT 68:

An express arrived this mourning from Fort Pier, two men on horses, and two men were sent from here immediately on mules to overtake Mr Kip who started from here in a boat for the Crows. Mr Audubon killed a rock wren this afternoon. Mr Harris killed 2 Says flycatchers

JULY 19:

Went with Audubon and Mr Harris on horseback after rock wrens. I shot three and 2 Says flycatchers and a remarkably light coloured horned owl, Richardsons great white horned owl. Then there came up a thunder shower and we hastened home, where we arrived just in time to save ourselves from getting a good drenching.

JULY 20, 90:

Made skins of the birds killed yesterday and spent the remainder in preparing for our long talked of buffalo hunt which is to take place tomorrow. We have been choosing our horses and Mr Culbertson has given me his favourite pied mare. He bought her from the Blackfeet Indians.

JULY 21, 99:

We were all up early this morning so as to get off as soon as possible. We crossed the Missouri in the scow and started with Mr Culbertson, Mr Audubon, Mr Harris in a wagon with two horses, one in front of the other.

The rest followed in two carts, two mules each leading our running horses. On our way we saw several antelopes and stopped to shoot them but was unable to approach them near enough to shoot guns. We went on very slowly for about 6 or 8 miles when we discovered four buffalo bulls. We immediately drove towards them, and when within a short distance of them behind a hill, we stopped and took our guns, balls &c and mounted our horses. There being four of us and four bulls when we all agreed that each one should take his bull and kill him if he could, we then rode round the hill opposite the bulls and then wrode up to them in Indian file to within 200 yards of them when they started and we all started at the same time and was soon up to them, when two of them took off to the right and Mr Harris and Squires took after them. The other took off to the left and I and Mr Culbertson being on that side, we took after them. They ran round and round a small hill like circus horses four times, and much faster than I had any idea of, and so close together that we could not part them. I then cut across and turned them, which gave Mr Culbertson a fine shot, when he fired and shot him in the hip. He then told me to shoot, which I did and struck him in the hip and the other in the thigh as he was running from me. Mr Culbertson then shot his in the lungs and left him. He then came to me and fired at mine and struck him about 3 inches below where I first shot him. I then told him to leave him, that I could easily kill him myself and which was contrary to our agreement at the start. He then returned to his bull, and I haveing reloaded both barrels, I ran to the side of my bull and fired a ball in his lungs which brought him on his knees. He then got up again and I wrode up to him, when he made a rush at me when I touched the horse with my heel and was out of his way in a second or two when he passed behind me and fell down a small bank. He then got up and staggered about when I fired again and he fell dead on the spot. I then got off my horse and went to him and found I had no knife to cut out the tongue. I galloped back to Mr Audubon and Mr Culbertson who was skinning the one Mr Culbertson had killed, when I learned Mr Squires had been thrown from his horse by the bull rushing at him which they are very apt to do at this season if you ride close to them. Mr Harris ran his a short distance and killed him at the second shot and then caught Squires horse and then they both went in pursuit of Squires bull, but Squires soon gave up the chace and returned, but Mr Harris went on and followed him about 4 miles when the bull stopped and made a rush at him when his horse jumped on one side and threw him, and the bull passed on. Mr Harris then got up and snapped both barrels at him and left him. The reason the gun snapped was the cocks got full of dirt in the fall, but Mr Harris was not hurt. He then returned. We then loaded the cart with the meat and started it for the fort, between 3 and 4 oclock.

Mr Audubon was seated on a hill where he had a full view of all our movements and two of the bulls were killed within a few hundred yards of him, and no doubt he will give a glowing description of the whole scene. We then went on several miles farther where we intended to camp, when I discovered 8 more buffaloes a short distance from us. We were all saddled and mounted ready for action in a few minutes and started after them and approached them as before, within 200 yards on the open prarie, when they began to move off and we all started. We again let Mr Culbertson take the lead, he being on the right and I next to him. As we came near to them one large bull took to the right and Mr Culbertson after him. All the rest went to the left. I then took the lead and selected a large bull on the right and fired two balls in his shoulder, which checked his speed considerable. Mr Harris also shot another large bull in the same place, and when his came up with mine, it was almost impossible to tell them apart but I kept my eye on mine and reloaded and gave him another shot in the hump, and what was quite remarkable, Mr Harris shot his in the same place. I then fired and my horse jumped down a small bank at the same time I missed him. I then reloaded and by that time, my bull was very weak and scarcely able to run, when I ran up to him and shot him through the lungs, when the blood spouted out of the wound and his mouth and nostrils most profusely. I was then very anxious to see how Mr Harris and Squires was making out, when I fired another ball near the same place and he fell dead.

I then went to Mr Harris and found his bull had charged upon him and his gun had snapped at him as he was standing, he haveing previously haveing fired another ball in him. We stood and watched him some ten or fifteen minutes when he laid down to rise no more. He then commenced to cut out the tongue when I returned to mine and commenced the same opperation. After we had the tongues, we cut off the tails, which was all we took off these fine large fat animals and returned to camp very much regretting what we had done when they told us they could not carry any of the flesh home. Squires at the same time followed his for more than a mile beyond us and returned without killing any as did Mr Culbertson also. We then got something to eat and spread our beds and mosquito bars and tied our horses and laid down to sleep. It grew quite cold towards morning, but I slept very comfortable.

JULY 22, IN THE SUN 108:

We was up very early and got something to eat and started off after antelopes. We shot at them several times and wounded them but did not kill any. I saw several buffaloes, one was rolling in the dirt like a horse, quite near me and I could easily have crept up to him and shot him but I thought of the day before and left him. I then returned to camp and on my way found the nest and eggs of the chestnut collared lark finch and the shore lark. We then packed up and started for the fort at half past ten where we arrived at 3 oclock. Stopped on the opposite side of the river and ate as many service berries as we wanted, which I am very fond of. In the evening there came 20 Assineboin Indians to the fort. One of them gave Mr Culbertson a whip and Mr Culbertson presented it to me.

JULY 23:

This morning we saw a large party of Indians on the top of a hill a short distance from the fort. They then sent 3 of their number to the fort after whiskey, and told us they were Crees and 70 in number. After breakfast they all came to the fort. They had only two horses.

In the afternoon we discovered 4 buffaloe bulls on a hill a short distance from the fort and as they were going out after buffalo the next day, Mr Culbertson was ready in a moment and wished me and Mr Harris to go with him and kill them as they were very much in want of fresh meat. I told him I would go with him and see the sport, but he insisted upon my taking my gun, which I did, with three balls in my pocket and two in my gun, so away we started, with all eyes in the fort upon us and some betting we would not kill any, others that we might kill one or two of them. Mr Audubon bet we would kill all four in less than one hour from the time of leaving the fort. We went on following the ravines &c untill we came very near them when we went up the hill close to them, and as soon as we got on the top, they saw us and started, and we done the same. They took down a ravine and Mr Culbertson followed them. I soon saw them turn up another and I and Mr Harris cut across and met them. We then all fired into them and wounded two of them. One large one then turned to the right and I followed him. The others then turned to the left and Mr Culbertson and Mr Harris followed them. Mr Culbertson soon killed one of the wounded ones and they then ran the other two about a mile and each killed one of the others.

I had a young colt three years old and very small and lazy, and haveing no whip, he would run, trot or walk just as he pleased, and haveing the very worst kind of ground to run in, up hill and down constantly, I soon found that the bull ran as fast as I did, and I began to think I should not come up to him. At last I saw him follow a ravine round a hill and I cut across and met him and fired at him and missed him by a few inches as he was going down a hill. I then ran up to him as he was going up a small hill and gave him a shot in the side, which killed him dead on the spot. In examining the hole in him I found it the size of a twenty five cent piece. My gun carries a ball 16 to the lb., and when we took off the skin I found a hole in the flesh nearly the size of my arm. I then went home to send another cart after the other bull, but they said they had meat enough. Mr Audubon had the flesh taken off the bones so as to make a skeleton, and left the back bone on the prairie.

JULY 24, HEAVY SHOWER / 90:

Went to see the bull we left yesterday and found the wolves had not touched him. I then went home and we concluded to go weigh him which we did and found he weighed 1777 lb. The head weighed 118 lb. We then had a heavy shower. In the afternoon we went after the skeleton and found the wolves had spoiled it. I preserved the tails of all the bulls I have killed.

JULY 25, VERY PLEASANT / X 82:

Provost and the other hunter returned haveing killed 1 doe and 2 female antelopes. I packed up all the birds we had on hand. Owen and the other person that went with him after the Crow boat returned and reported have-ing seen plenty of buffalo cows about 30 miles up the Yellow Stone River, and we all made up our minds to go after them tomorrow.

JULY 26, VERY PLEASANT:

Started at six this morning in the wagon with two carts, leading our runners as before. On our way we saw great numbers of antelopes, shot at them sev-eral times. Saw several bulls but passed them by, went about 25 or 30 miles when we saw a bull alone, and proposed to Squires to go and kill him as he had not succeeded in the two other races. He was ready in a few minutes. Mr Culbertson, Mr Harris and Owen went to see the sport, Mr Harris taking no gun. Squires ran by his side and fired but to no effect when Owen fired at his head to stop him to give Squires a better chance and Mr Culbertson gave him a shot in the hind parts and checked him considerable, when Squires gave him another shot and then they all fired into him, and Squires comeing up to him again the bull made a rush at him and frightened him very much and he came very near falling off and being killed by the bull. He let his gun fall and clung to the mares neck, he haveing the same spotted mare I had in the first two races. They then ran him a mile haveing run him half a mile or more while I saw him and fired 18 balls in him before he fell. Owen then saw another bull and went after him and gave him 5 shots and killed him. We then took the tongue of the first and left him and went to the other which was much better, and cut out the best pieces for our supper &c.

At the commencement of the first race I was in the wagon with Mr Audubon on a hill where we saw nearly the whole race, and when they had killed him I drove on that way, the carts following. We then went on again for several miles when we came upon 7 bulls and one cow. We then all mounted our horses and started to kill the cow for our supper, instead of eating the bull meat we had, as the cows are much the best. I had a young colt 3 years old and very small, but I soon found that by applying the whip a little that the rest did not leave me behind as I expected. Owen was soon by the side of the cow, but his gun snapped several times when Mr Culbertson gave her a shot in the belly. I then put on the whip and was soon at her side and gave her two balls in the lungs when she stopped, and was bleeding at the mouth very much. Mr Harris then shot her as she was standing and she fell dead. We then commenced skinning her, and while we were at work some of the bulls came within 70 yards of us and that several times, and I as often mounted my horse to shoot him, in case he made a rush at us. I am told they have often driven men away from the cows after she was dead at this season of the year, and they roar tremendously when there is a large band and fight very much.

Mr Harris started after Mr Audubon and the carts and in returning to the cow started a small red fox, called here the swift fox, and said to outrun almost anything in this country. As he started very near Mr Harris, and he being on a very fair runner, he started after him and overtook him in a few moments and snapped both barrels at him. I then started after him, but Mr Harris gave up the chace before I could get in sight of him. We then started to find water and to camp which we soon found and made a large fire, boiled some coffee and roasted some meat and roasted a sharp tail grouse that I shot with a rifle this morning and made my supper of it. I then went and eat some service berries. The men pitched the tent &c and after securing our horses we went to sleep. We were then about 45 or 50 miles from the fort. During the day I killed 3 small finches, or buntings very similar to the Henslows bunting, the note very much like the marsh wren, and while cutting up the cow we had killed we saw a large band of buffaloes, but it was too late to go after them.

JULY 27, VERY PLEASANT / X 78 X 75:

Had some hump ribs and marrow bones for breakfast and found them very fine indeed. I then skined the birds I had killed yesterday and shot a titmouse and saw the tracks of a grizly bear about 50 yards from our camp, heard the antelopes blowing or snorting the first thing in the morning. After breakfast we packed up and started after the cows we saw yesterday and after riding several miles we saw 8 bulls and 4 cows. Mr Culbertson, Mr Harris, Owen and I started after them. Squires concluded not to run. We soon found they were traveling very fast so that we ran at least three miles before we came up to them. We then all started and soon left the bulls behind as the cows run much faster than the bulls. Owen rushed in among them and scattered them and shot one when the one he shot and another came near me and I took after the one leaving Owens behind. I ran about ¾ of a mile before I could get close enough to fire, on account of my horse being afraid of the cow. I whipped him up to her several times, but the instant I let go the reins to shoot he would turn from her. At last I got within 5 or 6 yards of her and she turned a little towards me to follow 2 bulls that was before me, and I fired both barrels at her, and I saw at once that I had mortally wounded her. I then reloaded still following her. She soon came to a deep ravine and had just strength enough to crawl up the other side when she fell in a small bunch of bushes, which they will always do if they can find any near, as if to hide and die. 2 of the others done the same thing, run for a ravine & fell in the bushes. After I shot her she ran about 150 yards, and when I came to her she was not quite dead and being anxious to return to the rest, I fired another ball in her lungs and killed her in a moment. I then tied my horse, took my knife and cut out her tongue and went to the others, which I found a few hundred yards from me. Mr Harris, who was behind me, not knowing that Owen had shot one and left her and took after another, ran after the wounded cow, but soon found that she was shot and left her and by that time all four of the cows were dead. Consequently Mr Harris did not kill any as Owen killed 2, Mr Culbertson killed 1 and I the other. The one Mr Culbertson killed was a very fine one and I skined her and saved the skin and the others they cut up and loaded the two carts. I cut off the tail of my cow and saved it, also the two bulls. I have also the tails of my 3 bulls.

The cows are generally shot within 2 or 3 yards and often so near that they can touch them with the gun, but the bulls at this season are very dangerous to approach too near as they will turn and rush on you, but if you keep 10 or 12 yards from them, there is very little danger, but they are much worse when wounded. We then put all the meat in the carts and spread the skin on top, and as we had considerable baggage, and the carts full of meat, we put some of it in the wagon, and Owen, Squires, and I rode home on horseback. We started at two oclock from where we killed the cows and rode until sundown, when we camped at some water, about nine miles from the fort. The wind was from the north and very cold, and we had not a particle of wood to make a fire, but we had a substitute at hand, and in abundance, which was the dried buffalo dung and soon had a first rate fire, and some water boiling before the carts reached the camp. On our way back we killed a hare which I skined and we roasted. As soon as the carts arrived some of the men cut off pieces of liver and threw it in the fire to roast which it soon did and they took it out and commenced eating it as though it had been cooked in the very best manner and that is not half as disgusting to me as to see them eat it raw. I have seen them cut off pieces as large as my hand as soon as it was taken out of the animal and eat it and the blood running down the sides of their mouths, and I have seen them take what they call the many folds and wash it off a little and then dip it in the blood and eat it, and in the same way they cut open the skull and marrow bones and eat it raw while it is still warm from the animal. This I have seen at every buffaloe that I have seen skined and cut up, which they do in a remarkably short time. I think I saw one young bull cut up in 15 or 16 minutes by two men with knives only, leaving only the head behind. They turn the animal on his belly and split him on the back and skin him without paying any regard to the skin as they often cut holes in it to put the fingers in to pull by. They then cut off the fore legs commencing by the shoulder blade, then the head and hind legs &c, but it does not look like our city beef when cut up. The Indians all skin them in this way, and then split them on the belly, as you see all the robes are sewed together on the back. It is also much more convenient in handling while preparing the skin.

Now to return to the camp. After supper, we secured our horses and was glad to go in our tent to sleep as it was very cold, and no moschetoes, we slept very sound.

JULY 28, COOL / 56 X 65 X :

Early this morning we got something to eat, mounted our horses and started off at a pretty fast gait, sometimes as fast as our horses could run. We soon got over the nine miles and stopped on the opposite side of the river and got as many berries as we wanted before the wagon came up to us. We then crossed the river and all got safe back again at 10 oclock. I then cleaned the cow skin &c.

JULY 29, COOL / 58 X 63 X :

Went over the river at ½ past 3 after rabbits with Mr Harris and two others, but saw none. After breakfast I and Mr Harris went after rock wrens, shot 3, and one female grouse and one young one and 2 logger head shrikes, one raven and one young striped squirrel and 2 spotted grossbeaks, 1 red headed woodpecker. They had a dance in the evening.

JULY 30, X 87 X :

Several Indians arrived last night and said they saw some buffalo cows near the fort. Mr Culbertson, Mr Harris, Owen and I started after them this morning but saw nothing of them. The brother of one of the Indians died at the fort a day or two since and they put him in a box and put it up in a tree. The brother went there and cut his legs in 100rds of places. Several of us went over the river and found a horse.

Went with 8 others in a point to shoot a rabbit but saw none. Provost and another man started after antelopes

JULY 31:

Several Assineboin Indians arrived at the fort and told us they saw some buffaloes a few miles from the fort. Mr Culbertson, Mr Harris, Owen and myself started after them and went several miles but could see nothing of them and returned. One of the Indians that arrived last night when he went away he left his brother very sick and he died day before yesterday and they put him in a box and put him up in a tree according to the Indian custom, as I have seen them in the trees in many places when out hunting. His brother learning he was dead went under the tree and cut his legs with a knife in 50 or 60 places, but not very deep, and returned to the fort with his legs all covered with blood.

In the afternoon some of the men went over the river after rabbits and found a horse. We then went after him and brought him home.

AUGUST 1, SHOWERY / 75 X 78 X 76:

Went up in the Point after rabbits, but saw none, then went after prairie turnips or pomme blanche, seeds &c. Provost and another man returned with two fine male antelopes. Owen told us that when he was up the Yellow Stone about 70 miles he saw some of the large cock of the planes, and I concluded to go with him in the morning after them. He said it was two days jouney.

AUGUST 2, PLEASANT / 61 X :

Started with Owen on horseback with a few biscuit and some pepper and salt, Owen carrying the repeating rifle and I a double barrel gun and our blankets, robes &c. We crossed the river and started at half past seven this morning. The first things we saw worth naming were some large and small yellow leg snipe and several young blue wing teal quite small, in a small lake or pond supplied by a very large spring, but the water not very good. Saw 3 bulls near where we camped on our first trip up the Yellow Stone, afterwards saw several single ones. I then shot two common cerlews in the afternoon and started two fine male elk, quite near us. They ran a few yards and stopped and Owen fired at one of them, and they started off over the open prairie. We then saw some small bands of buffalo, and started a flock of young sharp tail grouse, about two thirds grown. I got off my horse and killed one. We then went on half a mile and came to Pellows campment where we found plenty of good water, and it being now half past 5 and no water for 12 or 15 miles farther, we concluded to camp here for the night, as we have plenty of wood. Owen then commenced making a fire and I took my gun and went back where we started the grouse and drove up three and killed two and marked the other down and went and killed it, so that I had three in as many minutes. I then returned to camp, and we roasted our birds, and I think I never tasted better birds. They feed upon grasshoppers, and they are here by millions. We then made our supper, and made our bed, then tied our horses &c and laid down, but could not get to sleep very soon for the bulls bellowing. They kept roaring all night long near us. At 12 oclock I was awoke by my horse partly treading on my foot. She had the rope round her legs and came very near falling on us, but I was up and out of the way in a moment. We then loosened her and went to sleep again.

AUGUST 3, PLEASANT:

At half past three I was awoke again by the horses snorting and raising up. I saw a large bull within a few yards of us, going to drink. I then awoke Owen and we took our guns and crept up to him and Owen fired and wounded him. He then ran back and as he passed me I gave him a shot. He went on about 50 yards and stopped. We then gave him two more shots and he fell and was dead in a few minutes. Owen then went back to the camp for our knives and I took a seat on the bull, and looking round I saw a bull at a short distance comeing towards me. I then sat still and he came within forty five yards of me before he saw me. He then stopped and I fired and he fell at the report of the gun. I then went to him and he got up and fell again in a few seconds, and was dead in a few moments. He being the best of the two, we cut out some of the best pieces and returned to camp and made up the fire, and in looking round I saw a grizly bear on the prairie, making for the bushes just above us. We saddled one of the horses and Owen started in pursuit of him with my gun, and I followed him on foot with the rifle, but he got in the bushes before Owen overtook him, and made his escape. We then returned and roasted our meat and got our breakfast. There were hundreds of buffaloe near us this morning.

We then saddled our horses and started at half past six. We then saw 10 elk on a sand bar, and went in some thick willows to get a shot at them, but they got the wind of us and started off. In returning I heard a grunting in the bushes near me, and thinking it was a bear and the next moment I saw it in the bushes near me and I fired, when out rushed a bull towards me. I had mortally wounded him. He ran a few yards and stopped. Owen then gave him a shot and he fell dead. He was a remarkably fine one, but we left him as we had done the other two for the wolves. We then started again at 8 oclock. We went on a few miles and saw an elk. Owen took the rifle and crept up to her and fired and broke her back. I then went to him with the horses, and we then killed her as she was struggling very much, and making a mournfull noise. We then skined her and tied the skin around a large tree as high as we could reach and left it to drye until we returned. We then started again at ten oclock and saw hundreds of buffaloe all along our route.

When we arrived at our camping place, we discovered that we had lost our meat that we roasted this morning. Owen proposed to go and kill a bull that was then crossing a branch near us, and we took our guns and crept to the bank near where he would come up, and as soon as he was on the bank Owen fired. I then gave him two shots and still he did not fall, but stood still. I then gave him three pistol shots, with the same effects. I then went to him and attempted to pull him down by taking hold of his tail. Owen then stuck him in the side with his knife. I then stuck him in the other side and still he stood still. Owen then cut a large hole in his side and run his hand in and cut him in the lungs. He then soon fell, but attempted to get up three or four times, and I then cut off his tail, and it was several minutes before he was dead. We then commenced taking off part of the skin, so as to cut out some ribs, when we saw a band of about fifty comeing towards us. We then crept to the bank to see them run in the water, and I was quite surprised to see with what ease they dashed through the water and mud and how quick they got up the bank, which was seven or eight feet high. Some came up within ten feet of me, and on both sides. Some of them were so much frightened that they fell back in the mud and ran along the water a short distance and then came up again all covered with mud and dust &c. Owen was for killing a cow and had I not prevented him would have done so, as we could have killed three or four had we been so disposed. We then made a fire and roasted some ribs and ate our dinner.

We then let our horses rest and eat and started again at half past three, saw a sharp tail grouse with 12 or 15 young ones quite small. Saw three elk and wounded one of them, then saw five more, also a hare, but could not get a shot at him and saw hundreds of buffalo. I counted 185 in a band near us, and another 250. We then arrived at our camping place at six oclock, saw hundreds of buffaloes a few hundred yards from us. We then unsaddled our horses and let them go, and went along a ravine untill we crept within 8 or 10 feet of the path they were traveling and laid ourselves down to see them. We remained there for some 7 or 8 minutes, when Owen raised his head a little above the wormwood, when a large bull saw him and came within 5 or 6 feet of him to see what it was, but as we remained perfectly quiet, he passed on. This occured two or three times. Once they were alarmed and ran up the bank, but as they saw nothing more, they passed on again in the path. Presently Owen raised his gun which was the signal and we both fired at a fine fat cow. She then ran with the rest about 75 or 100 yards and stopped, and several bulls collected around her. They remained there a few minutes and left her and she in attempting to follow them fell down. I then crept up to her and gave her a load of buck shot in the side. She then got up and fell down a bank, rolling over and over some six or seven times before reaching the bottom. She had one leg broke and was quite dead when we got down to her. We then cut out some ribs and marrow bones and started for our camp and saw a fine doe a short distance from us. Owen took the rifle and attempted to get a shot at her but failed.

It is now half past seven and our ribs are roasting finely as we have plenty of wood and a good supply of water. I am very sorry Mr Harris could not come with us as he would have enjoyed it very much. Owen is a very good man to camp out with and a very good hunter. There is a rattle snake rattling away near our camp at this moment, the first I have heard. We also saw a bear tearing at something on a sand bar in the river at 2 oclock today. We then made our beds, tied our horses and laid down, but I could not sleep very well on account of the bulls bellowing all night long and a few wolves howling near our camp.

AUGUST 4, PLEASANT:

We were up at half past three this morning and in a few minutes had a good fire. I then commenced packing up my things, and while doing so I saw the horses looking at something and raising up I saw an elk passing quite near me. I stood still untill she passed behind a bush, then took my gun and ran towards her. I fired a ball at her. It went in her side and lodged in the shoulder blade on the opposite side. I then crept up to her within forty yards and gave her a load of buck shot, and she walked off. I then ran back to the camp and got the rifle and returned and soon found her and gave her another shot in the hind quarter as she was walking from me. She then went a few yards and laid down. I then went near her and shot her through the head just below the eye, when she got up and walked 50 yards more and again laid down. I then fired a ball in the back of her head and killed her instantly. I then took the measurements and skined her for Mr Audubon as it was a very large and fine female. While we were taking off the skin two black tail deer came near us and Owen wounded one of them with buck shot. There also came a bull within 20 yards of us. We then carried the skin to our camp and got our breakfast at half past 6.

We then hung the skin in a thick bunch of bushes after cleaning it &c. We then went after our horses but could not find them for some time. We then packed up and started at nine oclock, saw hundreds of buffaloes all along the bluffs and arrived at the place where Owen saw the grouse at 20 minutes of eleven. We then saw a large buck and got off my horse to get a shot at him and just as I got within 30 yards of him he saw Owen and ran off just as I was raising my gun to shoot. We then went to the bank of a branch of the river to camp, when I saw 5 elk laying on a sand bar about 40 yards from the bank. We then let our horses go and started after them and concluded both to fire at the same one, which we did. Owen put a ball in her shoulder and I put 15 buck shot in her side. They then all got up and walked to the bank, and Owen gave her another shot in the neck. The four then ran off and left her and Owen gave her another shot in the shoulder. She then walked off and laid down in a few minutes but as we came near her she got up again and walked in some thick bushes. I went on one side and Owen the other when she started out and ran from me. I fired and broke her thigh with buck shot and in falling and jumping she broke a fore leg. We then went to her and as she was strugling very much we cut her throat to let her bleed more freely so as to keep the skin as free from blood as possible. We then took off the skin and returned to camp. It was rather remarkable that while skinning all three of the elks, the bulls came within 25 yards of us.

We then started to look for the cock of the plains, saw hundreds of buffalo all over the same ground. We then started the buffalo and they started a fawn, and as it passed me about 50 yards off I fired and it fell. I then went that way and it got up and ran in some thick bushes and briers so that we could not find it. We then went on for some time without seeing a sign of a grouse, when it commenced raining. Still we kept on for some time, but as it was very wet and unpleasant, we returned to camp and concluded to return as we could do nothing without a dog, and even then it would be very uncertain if we saw any. We then started and hunted round on our horses for some time but saw none. We then gave it up and started for home, on a brisk trot.

We then saw another elk cross the branch of the river, near where we shot the last one, also a great many buffalo as we passed along, and four elk among them, also several antelopes on the bottoms and a few wolves. We then came to the last cow that we had killed and saw a wolf tearing away at her. Owen jumped off and took my gun and shot him, when he run and yelled for about 40 yards and fell. We then went to him and I gave him another load of buck shot, and left him. We then got the skin that we hung up this morning and went on. We then came to a very narrow passage between the bank and bushes and there I met a bull. He stopped within 25 yards of me, and I hallooed at him, but he stood still, seeming inclined to dispute the path with me. I then raised my gun and fired a ball at his head, but apparently without doing much damage, as he turned and ran ahead of me for some distance. We then went on untill we came to where we camped at 12 oclock the day before, saw 2 sharp tail grouse, and as there were plenty of buffaloes near our camp, we concluded to kill a cow, which we did within three hundred yards of our camp. There was a bull and a calf with her and we were obliged to wound him severely before he would leave her. He then walked off about 150 yards and stood still. We then took out some ribs, and I milked a little in my cup to taste it and found it very sweet and rich. The calf then went to the bull where we left them and no doubt but the bull died there as he had three balls in him. We then returned to camp at 8 oclock and made a fire and roasted our ribs and made our beds, then ate our supper and laid down, and slept very soundly. I forgot to say that it cleared off fine, which pleased me very much as I expected to spend a dismal night in the rain on the open prairie.

AUGUST 5, PLEASANT:

We were up and on our way at half past three this morning, saw hundreds of cattle all around us. At six oclock we came to an eagles nest with one young one in it. We at first attempted to drive him out with sticks but in vain. Although he was nearly full grown, Owen went up to him and drove him out. He then flew about three hundred yards and fell to the ground. I then ran to him and caught him. We then tied his bill and claws and Owen carried him home alive. We were then near where we killed the first elk, but our minds were so occupied with our eagle, that we entirely forgot it untill we were several miles past it and concluded to go on and leave it.

We then saw an elk male among some buffaloes, and as we approached them they ran off, and the elk came running to us. He came within gun shot and I fired at him and wounded him in the shoulder. He then ran a few yards and stopped. I then reloaded and went after him and crept for several hundred yards without getting a shot at him as he walked off as fast as I could creep and nearly within gun shot all the time. He then came to some very thick willows and went in, and as soon as he was out of sight, I ran to the spot and followed him. I could hear him within ten yards of me, cracking the sticks, as he had very large horns in velvet, and they very seldom go in the bushes while they are in that state, but keep on the open prairie, but the females keep in the willows. I then followed him some 15 or 20 yards, when I came suddenly upon a very large porcupine and, Mr Audubon being very anxious to procure one, as we have only one small one, I hesitated for a moment what to do, but soon made up my mind that a bird in the hand is worth two in the bush, for it was uncertain if I got a shot at the elk if I could kill him or not in such thick bushes, so I took out my gun and struck the porcupine over the nose. He then turned his head between his fore legs, and raised all his quills. I then twisted the screw of my rod in his back and took out my small knife and stuck him in the side. He then ran off and pulled out the screw, and in running after him I broke my rod. I then overtook him and got the screw in his back a second time. I then stuck him in several times and killed him. I then tied a string to his foot and draged him to where I left Owen. I measured him and took off the skin and turned it inside out and tied it behind the saddle and started.

We then saw six sharp tail grouse and several antelopes, afterwards nine more grouse, then ten wild geese, and fired at them from my horse. They were feeding on the open prairie. We then came to our first camp where we killed the two bulls, and drove 15 wolves from them, and I forgot to say that as we came to our camp where we killed the first cow, we saw some wolves on the hills around her and one tearing away at her, up to his ears in blood. I then told Owen to take my gun and go and kill him. He went and gave him a load of buck shot, when he ran about forty yards and yelled out something like a dog, which is the first I have ever heard make any noise when shot, and on enquiring of old hunters, they tell me they never heard one make a noise when wounded. I then went to him and he raised up his head. I then gave him a kick, but he was too far gone to resent it. I then shot him through the lungs and killed him instantly.

We then stopped at half past eleven to eat our breakfast and let our horses rest and eat, hundreds of buffalo within a few hundred yards of us. We then started again at half past one being anxious to reach the fort before night, so we trotted our horses nearly two thirds of the time, saw several antelopes, and a very large band of buffalo, and as we came opposite to them they got the wind of us, and started to pass in front of us. We then rode on until we parted them and passed through. After about 300 had passed us, the others then passed behind us, in a line, and they kept comeing and comeing from the River and running to the hills. They formed a complete line, one, two, and sometimes three deep, for more than two miles long with scarcely an opening between them. They are like sheep, when one starts the others will all follow. To me it was a most beautiful sight, to see them passing within 150 yards of me running as fast as they could, with their tongues hanging out, in the distance as they went up the hill, which was quite gradual in its ascent. They would collect in large bodies and move slowly to the right and left. They looked like as many soldiers marching along. We stood still until they were nearly all past, and as they passed us we would halloo at them to make them run faster.

We then hurried on towards the fort, saw some magpies. There then came up a violent thunder shower with hail, rain, wind, &c. Sometimes our horses would turn round with their tail towards the rain, and it was with great difficulty that we could get them to moove. I had on my India rubber cloak and kept perfectly dry, except my feet. It lasted about an hour and a half and cleared off. Saw two rainbows. We were about 12 miles from the fort when it commenced raining. Owen and his eagle got wet to the skin. We then heard a small fox barking and saw a doe just before we arrived at the river. We then crossed and arrived safe at the fort at seven oclock much pleased with my trip. We went 75 or 80 miles up the Yellow Stone River. The next day being Sunday.

AUGUST 6:

While I was away up the Yellow Stone, some Cree Indians and Assineboins arrived at the fort and 1 of the Assineboins shot one of the Crees horses from under him, and some Indians fired into the men belonging to the other fort. The ball passed through {1}.

I wrote in my journal and filled the porcupine skin. I was not stiff or sore from riding, and felt as well as I ever did.

AUGUST 7, VERY WARM:

This morning I started with Provost and Lafleur after big horns up the Missouri River about 20 miles. First I saw a bear, then a black tail deer (male) and Lafleur fired at him. Then saw a white tailed deer feeding, also some geese and a swan on a sand bar in the river. We then arrived at our camp ground at one oclock on the bank of the river, and looking across the river saw a doe come and drink. We then took something to eat, and started up among the hills to look for big horns. Saw bear tracks along the river and hundreds of cliff swallows nests and young birds, also a duck hawk feeding her young on the top of a very high hill. We traveled over the hills untill half past six without seeing any big horns, and started for our camp. On our way saw 3 prairie wolves. We then took something to eat, spread our blankets &c and laid down to sleep.

AUGUST 8, VERY WARM:

We were up early this morning after a good nights rest. The first thing we saw was a bull making his way towards the hills about a quarter of a mile from us. We then got our breakfast and started for home at half past four. Saw four fine bucks, afterwards saw four more or the same ones half a mile from where we first saw them, but could not get a shot at them. We then saw two young bucks and killed them both. We then took the skins and hind quarters, as they were very fat, and started again at half past six, then started a black tail deer quite near us, and in the hurry I fired a ball instead of buck shot, or no doubt I would have killed her. We then saw two more bucks, afterwards a doe and fawn, then saw some beaver tracks, and some young green wing teal quite small.

I also saw a vein of coal on fire. In returning along a path at the foot of a hill, I found it suddenly blocked up by the bank falling down from above, and in looking up saw smoke issuing from a vein of coal which was about 75 feet from the bottom and 40 from the top. The coal or some dark substance was about 2 feet thick and burning very slowly in several places for about 50 yards in length emitting a whiteish smoke, like sulphur, and I have no doubt there is more or less sulphur mixed with it. It must have been burning for a long time, as it had passed along the hill for some distance, turning the clay red, or brick colour. In some places I saw red banks 20 feet high. Where it was burning, it was red above, varying from one to three feet in thickness, and in some places it would undermine the bank above, so that several hundred tons would fall down at the same time. In some places I found the red clay hard, in others quite soft and scaly. I then shot 2 magpies, and lost one. We then got back to the fort at 12 oclock.

AUGUST 9, PLEASANT:

Made skins of several small birds &c. Gave Mr Audubon a short account of the habbits of the elk as I saw them while up the Yellow Stone. Mr Harris took cold while in bathing and is not very well. Provost, Lafleur, and Squires went after big horns this morning. Several Assineboin cheifs arrived at the fort this morning and was badly disappointed at not being able to obtain any whiskey, or *fire water* as they call it. They told them that their Great Father would not allow them to drink any more. They then said their Great Father was a fool because he had made them tell lies, for when they left home they told all their people they were coming to the fort to drink &c, and that now their people would laugh at them.

AUGUST 10, SHOWERY / X 78 X77:

Went out with Mr Audubon after rabbits, but saw none, killed a striped squirrel. Exchanged a pair of boots and two pair pants for an Indian saddle.

Rained this morning, went out in the afternoon, shot three gold & red shafted woodpeckers, very curious ones, one young black headed grossbeak, one young arctic ground finch, one young lark bunting, one Arkansaw flycatcher and one male sharp tail grouse.

AUGUST 11, X 72 X :

Made skins of the birds I shot yesterday, then made some balls to go after buffaloes tomorrow, and packed up some horns &c for Mr Audubon, as we expect to leave here on Wednesday next, 16th. Squires returned and left Provost & Lafleur, haveing seen only 2 big horns.

AUGUST 12, RAINEY & COLD / 61 X :

Started this morning at eight oclock, with Mr Harris and Owen with two carts to run buffalo, went about six miles and saw a band of 100. We then stopped and took our guns, balls &c and started after them. We always approach them on the lee side so as not to let them get the wind, or smell of us, and then advance in a line, one close behind the other untill they see us and start and run, then we rush up to them and shoot into a cow. Then they generally scatter in several directions and we take after the largest band as they do not run as fast as single ones, males. We wound one, then we take after her and give her another shot, but when we see the blood comeing out of the nose or mouth, we know they will not run far so we leave them & go after another and so on, then return and look for them we had shot. Sometimes they are dead and some we find standing still. We then go and shoot them again, then take off the skin, but they never save the skin but cut it to pieces. In this race we each killed a cow, and shot down several more but could not find them when we returned.

As we were cutting up the second cow, Provost and Lafleur passed us at a short distance and I started to overtake them and return home with them as I had a young small Indian pony three years old. Mr Harris and Owen concluded to go on a few miles farther and kill some more as we had only three, and wanted six, so I left them and soon overtook Provost & Lafleur, but on my way to them my horse fell and threw me over his head. I was up in a second, but there lay my horse. I then whipped him up and started again in less than half a minute. On our way we saw a large grizzly bear a few hundred yards from us. He then got the wind of us and started for the bushes and I after him, but as I came near him my horse was afraid of him and would not go near enough for me to shoot. He then raised upon his hind legs and looked at me when Lafleur came up and was taking aim at him as he stood, with his rifle, when he started off again and was in the bushes in a moment. We then came on and stopped and got as many berries as we wanted and got back to the fort at 3 oclock. Mr Harris, Owen and the two carts arrived at 3 oclock having killed 4 more cows.

AUGUST 13:

Wrote a little in my journals, made a skin of a small squirrel. It was quite cool this morning.

AUGUST 14, VERY PLEASANT / 65 X :

Moncrevier and Pike started this morning for the Mandans by land with six horses. Exchanged some needles and a vest for some macasins. Got a cabbage and 2 dollars worth out of the store, also an old pair of leggings. Spent nearly the whole day in packing up. Got a few cactus.

AUGUST 15, 69 X :

Spent the whole day in packing up &c. Exchanged a thin coat for a garnished pair of leather pants, and a handkercheif for a cabbree skin dressed, and a scarf & a knife for an elk skin dressed, and a breastpin for a pair of mocasins. Loaded the boat this afternoon &c.

AUGUST 16, COOL & PLEASANT / 65 X :

Packed up the remainder of our things & bid good bye to Fort Union at 12 oclock, & just as we were all in the boat, a party of Assiniboins arrived to trade. Stopped at 2 for a pair of elk horns, shot 5 teal, saw 1 bear and several geese. Camped at 5 haveing run about 30 miles. Saw many grizley bear tracks.

AUGUST 17, PLEASANT:

Saw 4 big horns, antelopes, pigeons. Stopped, went after big horns, saw geese, bear tracks & 5 deer & 30 swans & shot at them. Shot 1 elk & 2 bulls & several ducks. Mr Audubon shot at 2 bulls. I then fired and killed them. Saw 2 elk come to the river to drink, also some deer. Camped at 5 oclock. I shot 1 of Townsends small striped squirrel. Found some mint that tasted like annis seed.

AUGUST 18:

Started at 4, & stopped at 6 to take breakfast. I went out & snapped at a very large male elk within 30 yards. Saw thousands of geese, some buffaloes, wolves & 20 elk on a sand bar. We then landed. I then shot one and followed it ½ a mile by the blood. Saw bear tracks, started with Mr Audubon and Mr Culbertson across a bend and shot at an elk in some thick willows. Saw 4 deer & many bear tracks. Walked about 7 miles.

AUGUST 19:

Started at 4. Saw several bands of buffaloes & some wolves after. Saw several very large bands of buffaloes, killed 1 cow at 10 oclock & a doe on an island, where we stopped to kill elk. Saw some but killed none. Heard an elk whistle, wounded a fawn. Mr Culbertson and Mr Harris killed a bull & a cow. Camped at 8 oclock.

AUGUST 20:

Started at 4 & stopped at 6 on account of wind. Saw hundreds of buffalo & some elk, went after cabbon, broke the hind leg of 1 and lost it. Wounded a bull, returned to the boat at 9 oclock. Shot 2 bulls. Sprague shot a bull & made a drawing of him. Saw a fox. Mr Culbertson killed 2 cows. The men killed several bulls. Shot some of Townsends squirrels &c.

AUGUST 21:

Started at 4, stopped at 8 on account of the wind at the Snake River. Saw bear tracks & hundreds of young ducks (mallards & dusky). Killed 2 bulls. Mr Harris killed 1 grouse. Saw more bear tracks and heard an whistle or call, too. Caught great numbers of cat fish, remained wind bound the whole day.

AUGUST 22:

Started at ½ past 3 & stopped again at 8 wind bound. Took breakfast & all started after elk. In a large point, saw a great number of deer. Mr Culbertson killed a large buck, & I killed a large elk, very fat, on the bank of the Little Missouri & wounded another. Could have shot several deer. Dropped down a few miles to a prairie dog village and set some traps & toward evening killed a grizley bear.

AUGUST 23:

Remained wind bound below the mouth of the Little Missouri, started after elk, saw 3 females & 2 deer & heard a grizley bear in the bushes near me. Shot 2 prairie dogs. Started at 2, saw a grizley bear coming to a dead bull in the river. Camped at ½ past 5.

AUGUST 24:

Started at ½ past 3 and stopped at 10 on account of wind. Went out hunting but found nothing. Saw 2 wolves & found some petrified wood, impressions of leaves &c in red sand stone. Remained wind bound all day. Heard more wolves howling at the same time than we ever had before.

AUGUST 25:

Started at 5 this morning, fog & white frost, no wind. Saw several wolves & swans, & killed a bull while crossing the river. Arrived at the Grosventre Village at ½ past 12 on the Knife River. Stopped at the little Mandan Village at ½ past 2, arrived at Fort Clark, old Mandan Village, at ½ past 3, bought 2 robes, 1 lasso &c. Sprague was inclined to stay at the fort.

Arrived at the Mandan Village at ½ past 2. Many of the Grovents were there to meet us & among them the Four Bears, chief of the Grovents, the same that went up to Fort Union with us on the steamboat. Many of them swam out to us & came on board before we landed. We then gave them some tobacco, & left them. The chief remained on board, and went down to the

Ricaree Village with us. He was very much pleased to see us & recognized us at once. He had his horse shot under him and a lock of his hair taken off with a ball in a fight with the Rees a short time since. He is very much of a gentleman for an Indian. Several of the Mandans had light coloured eyes & hair & skin. Went through the Village with Mr Audubon and Mr Harris. They gave us some pemican, boiled corn, dried pumpkins &c. This was the first green corn we had seen this season.

AUGUST 26:

Started at ½ past 3, stopped to breakfast at 8. A canoe & 2 men passed us. Stopped at 10, wind bound, got some yellow buffalo berries, saw some swans. Started again at 3, camped again at 6.

AUGUST 27, WARM:

Started at 3, saw a band of buffalo cross the river, landed & went after them, but saw none. Stopped (wind) and went after buffalo for meat. I found a single cow and killed her. Saw beaver tracks, shot at a bull in the river. Started at 3, saw a doe & beaver signs & some small terns.

AUGUST 28, COOL:

The wind blew very hard last night & is still blowing. Remained camped all day on account of the wind, went out after bears, saw plenty of track &c, very fresh, but no bears.

AUGUST 29, PLEASANT:

The wind continues to blow very hard & we are still in the same place. At 6 oclock we dropped down the river about 2 miles & camped on the opposite side of the river so as to have a better place to hunt. We found plenty of mosquitoes &c.

AUGUST 30:

The wind is still blowing so that we cannot travel. Went out, killed 1 cow, 1 calf & wounded a deer. Saw a buff breasted sandpiper. Started at ½ past 5, saved the calf skin, shot 1 gold wing woodpecker with black over the eye, & otherwise different.

AUGUST 31, WARM:

Started very early, no wind & very warm. Stopped at Cannon Ball River at 12 oclock & dined, saw hundreds of stones as round as cannon balls, saw some bulls, swans, geese, ducks, pigeons, wolves, & small snipe. Traveled about 60 miles, camped at the mouth of Beaver River, saw a large prairie on fire.

SEPTEMBER 1, RAINEY:

Rained last night & very warm. Started at 7 this morning & made a very good run about 60 miles. Rained a little during the day, & again in the evening. Went after a band of buffaloes on a sand bar, killed 1 cow & lost her, saw 2 elk & 1 deer.

SEPTEMBER 2, COOL & WINDY:

Started at 4 this morning, stopped at 8 on account of wind. Found several shells, petrifactions &c. Caught some fish, killed 4 young black ducks. Wind blew very hard all day.

SEPTEMBER 3:

Started very early, passed the old Ricaree Village at 7 oclock. Went on shore after a bull, afterwards went after a band of cows but got none. We then camped at the mouth of Moroe River, plenty of pigeons & small snipe.

SEPTEMBER 4, VERY FINE:

Started at ½ past 4 and stopped at 10, wind blowing very hard. Went out & killed a fine buck & 5 pigeons, found several trees that were cut down by the beavers, & parts carried to the river to eat. Found a beaver lodge & broke it up, & saw many elk tracks, set 2 beaver traps, caught none. Remained camped all day.

SEPTEMBER 5, PLEASANT:

Started at 5, after getting our traps & some stumps of the trees cut down by the beaver, also some of the chips &c. Saw the beaver leave the lodge. Went about 20 miles & stopped, wind bound. Killed 1 male elk, saw many others. Remained all day, found some ripe thorn apples or haws, found them very good tasted.

SEPTEMBER 6, WINDY:

Went out early after elk, saw none, saw many pigeons & 2 wolves. Shot 4 pigeons & 1 grouse. Wind blew all day. Caught some fish, saw some blue birds, gold wing woodpecker, brown thrush, cat bird, arctic finch, black & white creeper, Maryland yellow throat, yellow crowned thrush &c.

SEPTEMBER 7, WINDY & COLD:

Had a tremendous thunderstorm & high wind last night. All hands called on shore for fear of the boat sinking. Shot a new whippoorwill, and a common titmouse. Started, and went to an island a few miles below to shoot elk. River rose 8 inches in about 2 hours. Saw several young turtle doves, quite small.

SEPTEMBER 8, COLD:

Started very early this morning. River rose 5 feet in 24 hours. Very little wind. Heard the note of the new whippoorwill and a new owl last night. Passed the Big Cheyenne at ½ past 9, arrived at Fort Pierre at 5 oclock, took tea at the fort & remained until 8 in the evening. Rained a little all day. Saw Mr Laidlow, Mr Buoyce, Mr Murray, Major Drips, Indian agent at this place.

SEPTEMBER 9, WINDY:

Remained at the fort all day. Rained a little & the wind continued to blow all day. In many places the rose buds or berries are so plenty that it gives the prairie quite a reddish appearance for several hundred yards around. The buffalo berries are getting quite good & nearly ripe, also the grapes.

SEPTEMBER 10:

Remained at the fort all day. Saw several yellow headed tropioles, got 3 pair mockasins of Moncrevie in exchange.

SEPTEMBER 11, WINDY & RAINEY:

Went out & shot 7 yellow headed tropiales &c, saw a prairie dog village 2½ or 3 miles long & ½ a mile wide. Wind blew very hard all day. Mr Audubon had the boat unloaded, for fear she would sink. It was done in a few minutes & in the {worst} confusion, & wholy unnessary. Had a elk skin dressed. Attended a ball in the evening, had 7 pair mockasins made and got 1 buffalo calf skin.

SEPTEMBER 12, BEAUTIFUL:

Catlins book very much condemned by the people in this country. Exchanged some knives &c for some mockersins & other things. The water has fallen several feet. Saw great numbers of turtle doves, golden plover, red backed sandpipers, semipalmated sandpiper, spotted {br****}.

SEPTEMBER 13, RAINEY:

It cleared off in the evening & is a most beautiful morning. Made skins of 7 yellow headed tropiales. Exchanged some more knives &c for 2 robes, 1 bow & arrows &c. Was very much annoyed by the grass hoppers flying in my face while out hunting going against the wind.

SEPTEMBER 14, PLEASANT:

Started from Fort Pierre at ½ past 3. Mr Laidlow & party started at 11 for Fort Union with 2 wagons & Mr Murray, Mr Brewyer & old Battieste. Mr Culbertson & Mr Kellog go to Fort John on the Platte River in a few days. Stopped at the farm 10 miles below & got some corn, potatoes & a pig, then went a few miles below & camped.

SEPTEMBER 15, VERY PLEASANT:

Started & soon stopped again on account of the fog, arrived at Fort George at 9 oclock, remained ½ an hour. Mr Ellingsworth exchanged a few things with Mr Harris. Stopped at 2, wind bound, got some very fine wild plumbs, wild cherries &c. Saw no service berries for several days past.

SEPTEMBER 16, BEAUTIFUL, WARM:

Started at 5. Passed Mr Abbotts farm at ½ past 6, heard some parrots, passed Medacine River at 7, arrived at our old camp at the Great Bend at 11. Stopped to kill a black tail deer, saw none. Killed 1 old & 5 young rattlesnakes. Sprague made a sketch of our camp. Found some fossiles, took a piece of bark off of the tree under which we camped.

SEPTEMBER 17:

Left the old camp at the Great Bend at 5 this morning & stopped at 7 on account of the wind. Went out & killed 3 sharp tail grouse & a few other birds, then started again at 3 oclock. Caught 1 field or meadow mouse. Went round the Great Bend & camped.

SEPTEMBER 18, WINDY & CLEAR:

Started at 5 & stopped at ½ past 8, wind bound, appearance of the hills & country quite changed, dark clay & hard stone commenced. Saw fish hawks & gulls. Started again at 2, saw some common tit larks, camped at sun set & went out & killed some small birds for the fox & badger.

SEPTEMBER 19, WARM:

Started at 6 & stopped at 8. Saw great numbers of snipe. Dug up a tree of yellow buffalo berries, male & female. Started again & stopped again at 3 wind bound about a mile below White River, & remained camped the remainder of the day.

SEPTEMBER 20, WINDY & WARM:

Wind still blowing too hard to travel, went out & killed 1 red shafted wood-pecker [****] the [****] & 1 downy, & 1 buzzard. Found some very large peas. Started at ½ past 12, saw thousands of ducks & shot a few, camped at 5.

SEPTEMBER 21, CLOUDY & COOL:

Started at ½ past 5, stopped to examine the hill that had been burning for 3 or 4 years, picked up some specimens of clay &c. Saw 11 bulls, went out & killed 1 & wounded another. Started at 12, skinned 1 gold wing woodpecker female, found some curious seeds, camped at the Great Cedar Island at 5, cut some ores & canes & went out after buffaloes & deer. Rained in the afternoon.

SEPTEMBER 22, RAINEY:

Rained all night. Started at ½ past 7, stopped again at 8 on account of rain, saw large flocks of willets, started again at ½ past 3. Rained all day. Camped at 4 oclock. Saw a beautiful rain bow at 6. Found great quantities of fine wild plumbs. Saw many gar fish shells on the bank of a small stream.

SATURDAY, SEPTEMBER 23:

Started at 6, very cloudy, no rain or wind. Saw several wolves, ravens, robins, eagles, geese, ducks &c. Passed the tower at 2 & at ½ past 3 took in 3 men that had come out to hunt from the steamboat. Met the steamboat at Ponchas Island where we all camped together. Went on board & took tea with Mr Cutting &c.

Had a very fine run, cloudy & rainey but little wind. Stoped & took in 3 men that had went up the river to hunt belonging to the steamboat that was on a sand bar 8 miles below. We then stopped at 5 oclock at Ponchas Island to hunt. It is the same island on which I killed a doe & saw nine elk. I and Provost went out a short distance. The grass & weeds being very high & wet make it very unpleasant. I soon found that the men from the steamboat had been hunting on the island. I saw one doe, but not near enough to shoot. When I returned to our camp, we saw the steamboat coming up the river. They came up and camped near us. We then were invited by Mr Cutting to come on board & take tea, which some of us accepted. Mr Cutting then came on board of our boat & remained some time. Young Mr Livingston came on board & handed us some letters to carry to New York. Mr Cutting brought Mr Audubon a letter from his son Victor, which informs us of all the folks being quite well. Mr Cutting then left us.

In passing down the river we saw in many places trunks of cotton wood trees 2 feet in diameter sticking out of the bank some 8 & 10 feet below the surface & trees of the same diameter growing over them, & then underneath perfectly sound, & when used as timber &c it decays very soon.

SEPTEMBER 24:

Left Ponchas Island at 5 this morning. The steamboat left at the same time. Cloudy, cool & windy at ½ past 6. Saw about 100 white pelicans on a bar, fired at them but killed none. Passed the Poncas River at 7 & the Fort Mitchel at 8 at the mouth of the Running Water. Stopped at ½ past 9 (wind). Went out, shot 5 wood ducks, 1 raven. Wind blew all day & we remained camped on the same place making cedar oars.

SEPTEMBER 25, RAINEY:

Started at ½ past 9, rained all the morning. Passed the Bon Homme or Good Mans Island at 2. Rained a little. We are now in the country for wild turkies. Deer & elk are very plenty on the islands, where they resort for the wild peas which grow in great abundance on these islands. Camped at ½ past 4. Mr Culbertson gave us some papers &c which was very acceptable here.

SEPTEMBER 26, CLOUDY & COOL:

Started at 5 this morning, heard the note of the towhee bunting ({first}). Passed Jacques River at ½ past 10, saw thousands of geese & found a few shells at the mouth. Killed a white pelican & wounded another. Points larger & more timber. Stopped at 6, killed 1 goose at 7 as they came over our camp. They are on their way south. Killed 2 ducks.

SEPTEMBER 27, CLOUDY:

Started at 6, shot some ducks & geese, saw rackoon tracks, passed the Vermilion River at 7, met a keel boat at 10. Micho, a half breed, left it & returned with us. Arrived at the Vermilion Fort at ½ past 11. Mr Pascal let us have a beef & some potatoes, corn, pumpkins &c, corn 4 dollars a bushel.

SEPTEMBER 28, BEAUTIFUL:

Left Fort Vermilion at 8 & stopped at an island at ½ past 12 to hunt elk, saw only 3 females & 1 male, killed 1 turkey. This is the finest day we have had for some time. Saw some very curious places where the banks had slid down along the river. Some had slid down many feet and left the surface undisturbed, some with large trees &c.

SEPTEMBER 29, RAINEY & WINDY:

This morning we left our camp & crossed the river to be more secure from the wind &c & camped at the mouth of the Soo River, where we saw the shells of the gar fish along the bank. Remained camped all day, rained & windy all day.

SEPTEMBER 30, WINDY:

Left our camp this morning at 8. Rained all night. Saw a fox squirrel at 9 oclock & a great number of geese, ducks. Landed to get a pair of elk horns at 10. Stoped at 2, wind very high. Camped about 3 miles above the Sue River.

Saw the first squirrel & blue jay, common titmouse, towhee bunting, & gold wing woodpecker, gulls &c. Stopped 3 miles above the mouth of the Sue River, where we found some curious beans, 2 kinds on the same vine, some large ones on the ground & others up in the trees. Also found thousands of the Jerusalem artichokes, which are here so thick that we could scarcely walk through them.

OCTOBER 1:

Started at 7 this morning. It rained a little but no wind. Passed the Sue River at 9. {heard} a pileated woodpecker. Stopped at ½ past 9 and killed 3 turkies. Went a few miles & stopped opposite Floyds Grave for high wind at 10. Went a few miles & stopped again, started again at ½ past 4, camped at the Omaha crick

OCTOBER 2, COOL & WINDY:

Started at ¼ of 6 & stoped at 8 on account of wind, started again at 3, found some curious seed pods, with the ends joined together. The bank of the river here is formed of yellow & white sand stone. Passed Black Birds grave at ½ past 4. Shot some turkies where we camped near Woods Bluffs at ½ past 5. Saw 3 deer & a beautiful moonlight night.

OCTOBER 3, CLEAR & COOL:

Started at ½ past 5, saw 6 deer & hundreds of geese. The geese are beginning to migrate. Saw a few prairie wolves, passed the Soldiers River at ½ past 1. We amused ourselves by shooting duck & geese & pelicans. Camped at the mouth of the old channel leading to old Council Bluffs. Thousands of ducks & geese on the old bed of the river. Went within 2 or 3 hundred yards of the old site, shot some duck & geese.

Passed Woods Bluffs, below which we found much more timber, some oak, black walnut, ash, elm &c. Saw very few wolves, some gulls, snipe & 1 phalarope, but hundreds of geese & ducks. This was 1 of the finest days we have had for some time. The trees here begin to look quite yellow.

OCTOBER 4, CLOUDY & WINDY:

Started at ½ past 5, passed Bowyers Creek at ½ past 8, stopped at 10, wind bound. Here we found some very large elm trees, walnut, hickory, beach, maple & oak & hazlenuts. Shot some wood ducks, saw 2 woodcocks. Remained all day wind bound.

OCTOBER 5, CLEAR & COOL:

Started at ½ past 5, water falling very fast. Stoped on account of wind at 8, started to walk to Fort Croghan, 6 miles, where we arrived & took dinner. The boat started at 2 & arrived at the fort at 4. Collected several seeds &c. The officers also gave me some mountain [****]. Beautiful moonlight night.

We were stopped by the wind about 6 miles above Fort Croghan, when Mr Harris, Squires & myself started at ½ past 8 to walk to the fort. On the way I shot 1 of the new finches. We arrived at the fort at ½ past 10. I found Captain Berguin, Lieutenant Noble, Lieutenant Carlton, Lieutenant McCrate & Lieutenant Madison all well & glad to see us. They told us one of the men had a ball pass through both hands the day before we arrived. I walked around & collected some seeds they had in the garden &c, also some berries &c. Lieutenant Carlton gave me some mountain [****]. We found all the officers very polite & very gentlemanly. We took dinner & tea with them & spent the time very agreeable, & as they are to abandon this post, they are to leave with us tomorrow, part by land & part by water, to Fort Leavenworth, where they will remain.

OCTOBER 6, CLEAR & COOL:

Took breakfast at the fort, then started at 8. The lime stone commences here, also the honey bees. Stoped at Mr Sarpees at 9, the other boat at 10. Saw a great number of the Pawnee Indians & Major Miller, Indian agent. Passed the Platt at ½ past 12, shot 1 white pelican. We here crossed the line & are once more in the U. S. Camped at 5, saw a number of turkies. The soldiers camped with us.

We took breakfast with the officers & then left them at 8. Lieutenant Carlton is to join us with 18 men in the boat, at Belview, a few miles below, where we arrived at 9, where we saw great numbers of Pawnee Indians, among them the principal or greatest cheif of the 4 bands, and several inferior cheifs. The 1 cheifs name is Sa cha re re cari so, the witty cheif. They are the Pawnee Soup, Grand Pawnee, Republican Pawnee, Tapaye Pawnee. They had come in to receive their annuities, but as they had stolen some 400 horses from different tribes, their annuities were withheld untill they returned the horses, which they did, and started with all their baggage while we were there, for their village on the Platt over 100 miles from Belview. We also understood that the Sues, 1200 in number, stole 1100 horses & killed 80 Pawnees, & the interpreter of the Pawnees told us he killed 3 of the Sues with a common horse pistol. We then bid good by to Mr Sarpee & Belview at ½ past 10. Passed the mouth of the Platt River at ½ past 12, where we saw a few Otoe lodges. The soil here is very rich, the grass & weeds very high, some 7 & 8 feet in the bottoms, 4 & 5 on the uplands & in many places in the points the trees had a very singular appearance being covered with ivy (now red) all over the body while the tops & branches remained green, and in almost all the points it is almost impossible to hunt for the pea vines, grape vines, rushes, weeds &c. The trees here are quite yellow, there haveing been 2 white frosts. The hills are also much more covered with small trees & grass than above & the country more level. Saw the prairie on fire.

OCTOBER 7, CLEAR & COOL:

Started very early this morning, saw the first parrots & the sycamore trees & a few gulls & iron wood, bass wood &c. Shot some wood ducks, saw 1000^{nds}. The banks are now red clay, red sand stone & lime stone. Thousands of geese going south. Camped at the mouth of the Nishnebottana River at Mr Bowmans farm.

We are now in the state of Missouri on our left & the Otoe country on our right. We saw the first parrots this morning, plenty of black walnut, hickory, elm, oak, sycamore, iron wood, bass wood &c. Saw thousands of Canada geese, white fronted geese & ducks on their way south as this is one of the coldest days we have had.

OCTOBER 8:

This is the second night we have camped in the state of Missouri, & we now see a great number of log cabins all along the shore where there is a good place. This is on the left, on the right is still the Indian country. The people here raise plenty of corn & potatoes for their own use & gather great quantities of honey which they find in the woods. Saw great numbers of barn swallows, ducks, geese, parrots, pelicans & a few swans nearly every day for the last few weeks past. Started with Mr Harris, Sprague at ½ past 12 to walk across a bend 12 miles round & 4 across to Mr Roubedous at the Blacksnake Hills. We shot 6 squirrels, 3 parrots & 2 quails. Arrived at the Blacksnake Hills at 3 & started again at ½ past 3, found some poppaws, stoped at ¼ of 5. Shot 1 grey squirrel, which is smaller than ours & more red & as they are nearly all of the same colour, I think they may prove a new species. Also saw some red bellied woodpeckers & a distillery in the woods where they made whiskey out of corn & saw the first apples.

OCTOBER 9, BEAUTIFUL:

Started at 6, stoped at the Nemaha agency & shot 1 grey squirrel. Saw first ground squirrel, plenty of log cabins. Saw barn swallows & parrots in abundance. Started with Mr Harris & Sprague to walk across the bend at the Black Snake Hills. Shot 6 squirrels, 3 parrots, 2 quails & found some pawpaws. Camped at ¼ of 5, shot 1 squirrel & saw great numbers of squirrels & some turkies.

OCTOBER 10, BEAUTIFUL:

Started at 6, saw great numbers of parrots & 2 cardinal grossbeaks. Found some quails drowned in the river. Heard a new locust, saw hundreds of parrots, arrived at Fort Leavenworth at 4 oclock, was introduced to Major Wharton & several of the officers at this post. Saw the exercises &c & the music in the evening.

Arrived at 4 oclock at Cantone Leavenworth in company with Lieutenant Carleton & 18 dragoons from the Bluffs in a macanaw boat. We all went up to Major Whartons house where we remained for some time. We found the Major very much of a gentleman & done everything in his power to make us comfortable. He presented us some peaches, & some musk melons, the first we had seen. Canton Leavenworth is situated on a beautiful high bluff, with a row of buildings for the soldiers on the east, majors & the other officers houses on the north & east & the stables on the south, forming a hollow square to parade in with the magazine in the center & plenty of forest trees, for shade. At sun down the gun was fired & the music commenced & the soldiers went through their exercises which done them great credit, haveing been well drilled. There were 4 companies of infantry stationed here besides the dragoons. Captain Berguin had not arrived with the dragoons, by land, who left the Bluffs at the same time we did. The steam boats run to the cantonement every few days.

OCTOBER 11, CLOUDY & PLEASANT:

Left Canton Leavenworth at ½ past 6, found abundance of pawpaws, passed the Little Platte at 11, saw some Delaware & Shawnee Indians. Passed the Cansas at ½ past 12, arrived at Mrs Chouteau at 1 & Liberty at ½ past 4. Camped at ½ past 5. A steam boat passed us at 8. Saw plenty of quails.

Left Leavenworth at ½ past 6. We stopped at several places along the river & found them suffering severely from billious fever & fever & ague all along the river but not so much back from the river some 20 miles.

OCTOBER 12, CLEAR & COOL:

Started at 6, passed several small settlements. Very sickly, fever & ague. Passed snag boat at 9, old Fort Osage at ½ past 9. Arrived at Lexington at ¼ of 3, bought some eggs &c. Passed steamboat Tobacco Plant at ½ past 4. Camped at 5 oclock, shot 2 quails & 1 parrot, saw some ruffed grouse & common blue snow birds.

Passed a snag boat & saw her take up some snags, also another steamboat going up to Leavenworth. Saw the first common snow birds.

OCTOBER 13, COLD & CLEAR:

Started at 6, foggy & plenty of frost. Passed several towns, stopped at Greensville at ½ past 2, saw about 100 pelicans, fired 14 shots at them without killing any, wounded several. Saw a number of titmice & yellow rump warblers. Saw immense flock of geese on the bars, some white fronted. Bought good beef at 2½ cents, butter 10 cents. Camped at 6.

We had a severe frost this morning & a little fog on the river. Saw vast numbers of geese on the sand bars feeding on the small young cotton wood trees, which are 3 or 4 inches high. They are very shy & dificult to approach in such situations as the sand bars afford no concealment to the sportsman. We also saw several common gulls & cormorants & a great many sand hill cranes, common crows, blue jays, wood ducks, meadow larks, golden plover, spotted breast sandpipers, parrots, quails, purple grakle, gold wing woodpeckers, a few red head woodpeckers, towhee buntings, red bellied woodpeckers & downy woodpeckers, ruffed grouse, eagles, red tail hawks, turky buzzard, & a few other small birds, titmice, yellow rump warbler.

OCTOBER 14, COOL:

Started at 6 & stoped at 11 on account of the wind, found some very large locust & coffee trees, & bur oak. Killed 7 parrots, 2 Hutchins geese. Saw some turkies, white throat sparrows, Carolina wren. Got some apples &c. Remained camped all day. Steamboat Omega passed us at 12 oclock at night.

Saw a Carolina wren, white throat sparrow, crested tit, king fisher. At the towns here they have house boats at the ferries. The river here is quite free of snags. Mornings cool, & in the middle of the day warm.

OCTOBER 15:

Started at 6. Passed the Charaton River at ¼ past 7, arrived at Glascow at ½ past 7, passed steamboat Admiral at ¼ past 8. Shot at some pelicans & geese, passed Arrow Rock at 11 & Franklin & Boonville at ½ past 1, & Rushfort at ½ past 3, camped at ½ past 5.

OCTOBER 16, CLEAR & COOL:

Started at 6, passed some small villages, arrived at Jefferson City at ½ past 11. Passed the Osage River at 2, saw 4 deer on a sand bar. Passed the steamboat Satan at ¼ past 3, camped at ¼ past 5. Saw a corn field over ½ a mile long & at least 1000 geese in it. Saw a rabbit.

Passed several small towns, stopped at Jefferson City, met some steamboats, saw 4 deer &c. The hills here are something like the Palisades on the north river, formed of lime stone, but not so high or so regular, but the trees are very much the same. The river is remarkably free of snaggs this year & the water rather higher than usual at this season. Saw a ground squirrel & some grey ones, saw the first butter nut trees & some peccans.

OCTOBER 17, COOL & FOGGY:

Started at 6, saw 2 deer, stopped at {Armond} town at 11, eggs 6 cents a dozen. Passed several small towns, made some pipes & repacked my trunks &c. Saw a ground squirrel & some grey ones, blue birds, nuthatch, crested tit, cardinal, siskin, purple grakle. Camped a few miles below Washington.

OCTOBER 18, BEAUTIFUL:

Started ¼ of six, arrived at the Tavern in the Rock at ¼ past 9, arrived at St Charles at 3 oclock & camped 6 miles below. Went out & in a short time shot 6 squirrels, saw great numbers of them. Sold my pistol.

Was one of the most beautiful mornings we have had for some time.

OCTOBER 19, FOGGY & WARM:

This morning we started very early but was stopped by the fog. 3 steamboats passed us, arrived at the mouth of the Missouri at 11 oclock. The water looked very curious to us, being clear after seeing muddy water so long, and the water does not mix for several miles, but each keeps on its own side. Arrived at St Louis at three & went up to the Glascow House and engaged rooms, then had the boat unloaded & had the things stored at Mr Berthouds, & after tea took a warm bath, met several acquaintances who appeared to be very glad to see me. Went to look for C W {Blanvelt}. Went to the post office &c but found no letter.

OCTOBER 20, VERY PLEASANT:

Had some new [****] made for the fawn, fox & badger &c. Packed up the elk horns &c. Bought 1 robe & 1 grizly bear skin for 6.00, sundries 25 cents. Mr Audubon paid 10.50 cents for me & received 16.00 from me.

OCTOBER 21, VERY WARM:

Finished packing up & sent all our things on board, then found C. W. {Bla**te}.

OCTOBER 22, BEAUTIFUL:

Left St Louis at ½ past 1 in steamboat Nautilus for Cincinati, Captain Martin.

OCTOBER 23, PLEASANT:

Stopped several hours on account of the fog, arrived at Cape Girandeau at ½ past 8, saw a black squirrel, arrived at the mouth of the Ohio at 1.

OCTOBER 24, RAIN:

Arrived at Shawnee Town at ½ past 7, 220 from the mouth of the Ohio. Passed the mouth of the Wabash at 9, stopped at Henderson at ½ past 3, stopped at Evansville at 5, got acquainted with a Mr Thomas, who promised to write to me concerning buffaloes.

OCTOBER 25, CLOUDY:

Ran all night, making about 95 miles. Passed Salt River at ¼ of 5 in the afternoon. Arrived at the canal a few miles below Louisville at ¼ past 8.

OCTOBER 26:

Passed in the canall (3½ miles long) at 6 this morning, arrived at Louisville at 8, saw Mr Bakewell. Left Louisville at 10, arrived at Madison at ½ past 4. Toward night it commenced snowing & raining.

OCTOBER 27:

This morning we had about ½ an inch of snow, arrived at Cincinati at ¼ of 8. Left Cincinati at 12 oclock in steamboat Ohio Mail, Captain Bailey. The trees had a very beautiful & singular effect, produced by the snow on the leaves &c. Found Major Michel on board. I was introduced to Mr Wickliff—Buffaloe.

OCTOBER 28:

Stopped at Portsmoth at ½ past 7, stopped at Hanging Rock at 1 to take in some pig iron. Found some buck eye nuts &c. Started again at ½ past 6, took in 85 tons, had a band of music on board.

OCTOBER 29:

Stopped at Point Pleasant, mouth of the Big Kanawa at ¼ past 7, saw no more cotton wood.

OCTOBER 30, PLEASANT:

Had a severe frost last night, passed many small towns, arrived at Wheeling at 10 minutes of 6. Major Mitchel & Mr Wickliffs family left us here. Young Wickliff is going to Europe as minister to—

OCTOBER 31, COLD:

Stopped at Beaver Point at ½ past 8, had ice over half an inch thick. Stopped at Economy at ½ past 10, arrived at Pittsburgh at ½ past 3, at the mouth of the Alleghany & Monongahala. Shipped our things, engaged our pasages & took tea at the [U****] Hotell.

NOVEMBER 1:

Arrived at Freeport at ¼ of 7, then crossed the Alleghany River, then followed the Kiskiminetus River, arrived at Saltsburgh ¼ past 2. Passed in the Conemaw Tunnel at ½ past 4—130 yards.

NOVEMBER 2:

Arrived at Johnstown at ½ past 4, ascended 4 enclined planes, 1 wire rope, the last at 12 oclock. Arrived at the Summit Mansion House at ½ past 12. Descend 5 enclined planes, arrived at Holidays Burgh at ¼ past 2, started by canal at 3, passed along the Juniata River, 4.

NOVEMBER 3, COLD & FINE:

Traveled all night & this morning along the Juniata River, passed some small villages & some iron works, mountains very high, in some places 1500 feet. Followed the Juniata River all day, saw many fine horses, & mules very cheap.

NOVEMBER 4:

Arrived at Harrisburg at 20 minutes past 6 a.m. & left in the cars at 7 a.m. Stopped at Middletown at ¼ of 8, at Elizabethtown at ¼ of 9. Passed through the tunnel at 9, saw many superb farms &c. Stopped at Mount Joy at 10 minutes past 9 and Lancaster at 10, arrived in Philadelphia at 3.

NOVEMBER 5:

Left Philadelphia at 20 minutes of 8 a.m. by rail road, arrived in New York at ¼ of 1. [****] went over to G Smiths, with Sprague.

NOVEMBER 6:

Sent Mr Audubon's baggage home. Met W J Bell at the boat, gave Sprague 2 small tools worth 1.25, then went home. W J B {ln}. Piermont Cartage 55 cents.

NOVEMBER 7:

Went to see D H, A. Smith, A Mabri, & E [****] &c.

NOVEMBER 8:

Returned to New York.

The 1843 Diary of Isaac Sprague

Boston to N York	
N. York to Phil.	96
Phil. to Balt.	98
Balt to Cumb^d	187
Cumberland to Wheeling	139
Wheeling to Cincinnati	360
Cincinnati to Louisville	131
Louisville to M. Ohio	394
M. Ohio to St. Louis	175

FEBRUARY 16, 1843:

Left Hingham this morning for Boston; thence in the afternoon—

FEBRUARY 17:

Arrived at New York about 10 o clock this morning.

MARCH 10:

New York: New York is a very large & fine city and contains many splended buildings, but as a whole, its appearance is not superior to Boston. It covers a greater extent of ground and is much Dirtier, the streets at this time being filled with accumulations of filth, mixed with snow & ice—They have also here a custom which Does not improve the cleanliness of the place, that of keeping their hogs in the streets in front of the houses, instead of in pens in the rear as in Massachusetts.

MARCH 11:

Left New York at 5 o clock in the afternoon—& arrived at Philadelphia at half past 10 same evening—Having crossed the State of New Jersey—The land along the road is low and level—We are obliged to remain here over Sunday—

MARCH 12:

Philadelphia is the finest city I have seen. The buildings are mostly brick and many of the houses have a basement—and the steps to the Front of white marble—which are kept washed very clean. This together with the great regularity of the streets gives the city a very neat appearance—

Here we were joined by Mr Harris—our party now consists of five persons viz.

John J Audubon
Edward Harris
Isaac Sprague
John G. Bell
Lewis M. Squiers

MARCH 13:

Left Philadelphia this morning and arrived at Baltimore in the afternoon. We were detained a short time to day in consequence of the engine's coming in contact with a cow, that happened on the track. The animal was instantly killed, but no other damage done.

Baltimore is called the city of Monuments, and they have several fine ones. We went to the top of the Washington Monument, which is built of white Marble, about 180 feet high and commands a view of the city and surrounding country.

MARCH 14:

Left Baltimore this morning at 7 o clock, and arrived at Cumberland about 6 in the evening—having traveled thus far by railroad. At this place we took coaches and proceeded at once to cross the Allegheny Mountains, continuing to travel all night without stoping except occasionally to change horses—

Coal is very abundant along the road, and at some of the houses where we stopped they have the largest grates I ever saw.

MARCH 15:

This morning about 10 o clock we reached the highest point called Laurel Hill, and commenced our descent into the great valley of the Mississippi. The first descent is quite steep for 5 miles and when covered with snow & ice rather dangerous, winding as it does along the brow of the hill, with a wall of rocks several hundred feet high on one side, and a precipice of great depth on the other.

This is the hardest ride I ever have had having been now two nights with scarce any sleep. The interior of our coach presents a curious scene. Having exhausted all interesting topics of conversation, a dead silence ensues, each person fixes himself on his seat with a desperate resolution to wait the journeys end. But in a short time gets off his guard and commences a suspicious nodding and bowing to his neighbors, until some heavier plunge of the coach, completely upsets his gravity, and he is obliged to right himself and commence again.

MARCH 16:

However this morning about 3 o clock we arrived at Wheeling where we remained until about noon, and then took passage on the steamer Emily for Cincinnati.

MARCH 17:

On Ohio river.

MARCH 18:

Arrived at Cincinnati this morning, and stoped a few hours and then proceeded on the steamer Pike for Louisville—Passed Gen Harrison's *Log Cabin*—this afternoon.

MARCH 19, SUNDAY:

Landed at Louisville this morning, having arrived in the course of the night. Here we shall be compelled to wait several days, there being no boat ready to start.

MARCH 20:

Went out a short distance shooting. Louisville is situated on level ground—and extends along the river some distance, but is not very deep. There are no wharves but the banks are graded and paved—to low water mark.

Oweing to the use of bituminous coal—all the western cities are very dirty.

MARCH 21:

Went out hunting several miles from the city, but obtained nothing of value.

MARCH 22:

Took passage on the steamboat Gallant, and

MARCH 23:

This morning we started for St. Louis—down the Ohio river.

MARCH 24:

Still on our way.

MARCH 25:

Passed the famous city of Cairo, & entered the Mississippi—This place which is situated at the junction of the two rivers, was made the subject of a famous speculation several years since—It undoubtedly possesses great natural advantages—but is low and unhealthy—and contains at present but few houses.

MARCH 26:

Great quantities of ice floating down.

MARCH 27:

Our progress much impeded by ice.

MARCH 28:

Arrival at St Louis—This city is situated on the western bank of the Mississippi, about 20 miles below the mouth of the Missouri (1390 miles from Gulf of Mexico—). It is built on a limestone bluff, and extends along the river some 5 miles and is a mile or more in depth.

It was settled by the French, 1764, and contains now near 30 000 inhabitants. The limestone bluff rises to an elevation of about 80 feet above the usual height of the Mississippi, is covered by a deep deposit of alluvial soil of great fertility—

APRIL 4:

This morning we started for Edwardsville, Illinois, distance about 20 miles, and such was the state of the roads that we did not arrive there until near sunset. The roads were awfully rough and muddy and without exception it was the hardest ride I ever had.

APRIL 5:

In attempting to cross a creek on a log, while out hunting to day I lost my ballance and fell, in, and in trying to get out dropped my gun, which sank in the water about 10 feet deep.

APRIL 6:

Succeeded in fishing up the gun—to my great joy. Saw some deer running for the first time to day. They carry their tail erect—when running.

APRIL 8:

One of my days of pleasure, rambling in the still forests alone. Saw wild Turkeys for the first time to day.

APRIL 13:

Moved off to day some 18 or 20 miles to a small town called Bunker Hill. This is the pleasantest place I have seen. The village is situated on a swell in the centre of a large prairie, and commands a view of the country in every direction. While here we killed a number of Geese—Grous &c

APRIL 21:

Started this morning in a carriage for Alton—and from thence by steamboat down the Mississippi river for St Louis—where we arrived about sunset.

APRIL 25:

Left St Louis at 11½ o clock on Steamer Omega—We have on board several Indians, and about 100 men who are employed by the Fur Company. They are principally Canadians and Frenchmen, and the French language is spoken almost exclusively—The Indians are Ioways—and Sac's and Foxes—Arrived at St Charles in the night, distance 40 miles from St Louis.

APRIL 26:

Almost lost one day here.

APRIL 27:

Passed many high bluffs, some very fine scenery.

APRIL 28:

Passed Jefferson City the seat of Government. Pleasantly situated on a high bank and commands a fine view of the river and surrounding country. Being situated at a turn in the river, it appears directly in front as we come from below, and presents a noble appearance or at least will when the houses are built. Distance 155 miles from St Louis.

APRIL 29:

Passed Bonneville this morning—a flourishing village 204 miles from St Louis—

APRIL 30, SUNDAY:

A strong head wind, and the current, compelled us to stop this forenoon. In the afternoon passed a farm covered with water—

MAY 1:

The whole country appears to be overflowed except the bluffs—

MAY 2:

Passed Independence, the last town of any size—that we shall probably meet with for some time. The village is not in sight from the river—It is 377 miles from St Louis. From this place the Santa Fe traders take their departure. Their goods are conveyed in large wagons drawn by mules. Weather to day very fine.

MAY 3:

Passed Fort Leavenworth. A frontier post beautifully situated on a high bluff, 431 miles above St Louis. From this fort there is a fine view of the country for many miles around, consisting of immense forests and boundless prairies as far as the eye can reach.

This afternoon unfortunatly taking the wrong channel, we ran on a sandbar and stuck fast, where we remained through the night, in the course of which we had a shower—with thunder and lightning.

MAY 4:

This morning by carrying a cable to the shore and using great exertion, they at length succeeded in getting the boat afloat but owing to the violence of the wind were unable to proceed. We landed and shot a number of Parakeets &c—Mr Harris shot a finch which is probably new—Ran a few miles just at night passed a prairie—

MAY 5:

Passed Black Snake Hills, 115 feet above the river.

MAY 6:

Stopped by the wind, this forenoon. Bell shot a vireo that differs from any described. About sunset arrived at the Indian Landing. As soon as the boat touched the shore the old chief who had come with us from St Louis, started at once across the prairie for home, the village being about three miles distant. In a very short time the natives began to come in both on foot and on horseback and before we had been there half an hour there was upwards of a hundred. Some of them were fine looking fellows, others looked like the Devil, many of them having their faces painted with red, black or yellow, which did not add much to their beauty as I could see.

MAY 7, SUNDAY:

Beautifull day, had a fine run. The trees and prairies begin to look delightfull—

MAY 8:

Fine day, obtained a black Squirrel, probably the Scirus Audubonii— outlined it this evening.

MAY 9:

Made another outline this morning—Passed the mouth of the Platte river. This afternoon we stopped to put out freight at Belle vue, a small village on the western bank of the river.

Quite a number of Indians were assembled on the bank watching our movements with great interest. Many of them were ornamented and painted in the most fashionable style, I suppose. One of them had the figure of a hand painted with black on the lower part of his face. The women are rather short and robust. They are a part of the Ottoe tribe.

At this place is stationed an Indian agent whose duty it is to see that no ardent spirits are smuggled into the Indian Territory. The agent being absent we proceeded on our route, and stopped for the night several miles above.

MAY 10:

This morning we were boarded by a detachment from a party of U. S. Dragoons who are stationed in this vicinity. While they were inspecting the cargo our party went out shooting and obtained a number of saffron Headed Troupials, and several Finches—

Passed Council Bluffs this afternoon. This place is on the western side of the river and was formerly a military station, but is now abandoned. The river which used to flow above at the foot of the Bluffs is now more than a mile distant. These changes are very frequent on this river, and are occassioned by the drift collecting—and obstructing the channel, which works itself a new passage in another direction carrying away many acres of soil, and rendering the navigation extremely difficult.

Rain in the night with thunder—

MAY 11:

Rainy in the morning. Saw a wolf passing along the bank.

The character of the country, for several days past has been gradually changing—long ranges of Hills, and beautifull prairies are fast taking the place of these dark forests which border the banks of the river below.

MAY 12:

Dull in the morning—Splendid Sunset, fine evening.

MAY 13:

This morning we passed a celebrated mound known by the name of Blackbirds grave, so called from the fact of its having been the burial place of a noted Omaha chief by that name. The grave with a pole standing on it is still to be seen. He is said to have chosen this spot, so that he might see *the white man pass up and down the river.*

Passed a bluff covered with cedar, the first we have seen for some time. Mr. A. shot a turkey while we stopped to take in wood. This afternoon passed Floyd's bluff on the top of which is the grave of Sergeant Floyd, one of Lewis and Clarkes party, who died on the route Aug 20 1804, and was buried here. The post which was set to mark his grave still remains. Stopped for the night at the mouth of the big Sioux river.

MAY 14, SUNDAY:

Saw a black bear swimming the river, and passing quite close to the boat, several rifles were shot at him but no damage done. At noon the wind began to blow so hard that we were compelled to stop—and by night it blew a perfect hurricane—

MAY 15:

The wind continues to blow so hard as to prevent our moving, and there is no hunting, the bottom having been overflowed is now exceedingly muddy. Coloured a lithograph print for the Capt. In the afternoon the wind lulled, and we went on a few miles.

MAY 16:

Passed some fine bluffs this morning. These bluffs appear to be composed of hardened clay in regular strata, and of a variety of colours. Stop at a small Trading station near the mouth of the Vermillion river. This afternoon one of our boilers gave out and we were obliged to stop to repair it. Not much success hunting.

The wild plum is very abundant here and is now in full blossom. It is a shrubby bush or tree, not more than 6 or 8 feet high. I noticed some of those black bunches on the twigs, that are so troublesome to the plums at the East. There are also plenty of wild gooseberry bushes, and also a few of the buffaloe berry.

MAY 17:

Still engaged in repairing—Our party started out hunting this morning and someone killed a deer, the first we have had. While rambling about to day I came across the stump of a cotton wood tree that had been cut down, which measured five feet across, and by counting rings of growth I estimated its age to be upwards of 150 years.

A Number of dead buffaloes have passed us floating down stream.

MAY 18:

Not started yet. This morning 4 boats loaded with furs, from Fort Pierre passed us on their way down the river to St Louis. Shot one of the new vireos this afternoon—&c

MAY 19:

We are off again. Passed the mouth of the Vermillion river this morning. Saw a deer swimming the river to day. These animals swim quite well, notwithstanding the smallness of their feet. The river where this one crossed was more than half a mile in width.

MAY 20:

Passed the mouth of the Jaques river. Shot the *pipilo arcticus* [eastern towhee; Octavo plate 394] this afternoon.

MAY 21:

Sunday again, but here in the wilderness the days are all alike—Passed the *Running water* this morning. This is quite a large stream, and derives its name from its strong current. Near its junction with the Missouri are the ruins of an old fort or trading station—where we stopped to take wood. While on shore here I killed a Say's Fly-catcher. This afternoon the wind became so strong that we were compelled to stop.

Bell shot a deer—Saw an Elk—first I ever saw. We have also seen Buffaloes to day for the first time. These noble animals when they are quietly feeding, or reposing on the grass, so nearly resemble our domestic species that I could hardly realize their being wild.

MAY 22:

This morning a small party of Indians were seen crouching among the bushes on the bank. They made signs for the boat to stop, which being disregarded, they saluted us with a volley of musket balls—one of them passed through the pantaloons of one of the men, and several others struck the boat but fortunatly no one was injured—

Stopped for the night at an Island called Cedar Island—which is covered with an evergreen resembling our common red Cedar—Several of the men have gone ashore to hunt for buffaloe—

MAY 23:

The hunters came in this morning with some buffaloe meat having killed 3 or 4 while out—These animals are thus wantonly destroyed—but a very small portion of the flesh being used, often times none—and the skin being of no value at this season—they are left to be devoured by wolves and buzzards. This has been the hottest day we have yet had. Thermometer stood at 92° at 12 o clock. This afternoon they managed to run the boat on to a sandbar—and so admirably have they stuck her into the mud, that all efforts to start her have proved ineffectual—

MAY 24:

—and here in spite of all the tugging have we remained fast, until this afternoon, when by placing a spar, upright alongside the boat with the lower end resting on the bottom of the river—and applying a tackle from the deck to the upper end which rises several feet above the boat, passing the fall around the windlass—they have at length raised the bow of the [boat] sufficiently to let her swing clear, and we are once more afloat. Made a sketch of the bank of the river opposite where we lay—

MAY 25:

Stormy day—The hills appear more ragged and broken—the knobs not being so regularly rounded as below—although the general character of the country is still the same.

MAY 26:

Saw a buffaloe swimming across the river this forenoon—About 5 o clock this afternoon we arrived at the commencement of the Big bend as it is called, where 4 of our party myself included—accompanied by 3 hunters, left the boat to cross the bend.

Each man equiped with his gun, blanket &c—We first crossed a fine level prairie, on which we saw a prairie dog village, and afterwards a range of hills the tops of which appeared to have been exposed to the action of fire—arriving at the opposite bank just before sunset—the distance where we crossed not being more than two or three miles—while it is upwards of 25 by the river—

In a very short time we had arranged our camp, kindled a splendid fire—and despatched two hunters in quest of game. They returned in less than an hour bringing a fine black tailed deer—This was soon dressed and sundry portions thereof in the shape of joints & steaks after being roasted on sharpened sticks before the fire—were despatched with a keen relish and pronounced *delicious*.

Wrapped in his blanket each man laid himself on the ground—with his feet to the fire—and the whole party were soon sleeping as soundly—for aught I know, as if at home—To me the situation was one of novelty it being my first encampment in the open air. However, I slept quite as sound as usual—and experienced no inconvenience whatever.

Close to our encampment was a small creek, the water of which was clear as crystal, but so strongly impregnated with mineral salts as to be unfit to drink—Saw one of Townsends Hare—first that has been seen—

MAY 27:

After despatching a few more steaks, by way of breakfast—The party started in various directions in search of birds, quadrupeds &c—but unfortunatly were not successfull in procuring any of consequence. About 3 o clock the Steamer arrived, and we proceeded on our journey—

Just before we stopped for the night 4 Indians were seen [****]sting along the bank and as soon as the boat stopped they came on board. The Captain smoked with them, gave them something to eat, and some tobacco. Mr A gave them some powder when the chief cooly asked, "What is the use of powder without balls?" They appeared to divide the presents very equally—

MAY 28, SUNDAY:

Had a fine run this morning—Passed fort George, a trading station belonging to an opposition company—Soon after we passed this fort the boat was stopped—being unable to find a channel sufficiently deep to proceed.

MAY 29:

Still lying by the shore about 1½ miles above the fort—around which are a number of Indians encamped—parties of whom have visited the boat to day—Among others was a Chief named *Four Bears*. He was a fine looking man dressed in Indian style—having his dress ornamented with porcupine quills—grizzly bear claws &c—

Several specimens of the Black throated Grosbeak were shot this morning.

MAY 30:

Been engaged to day in makeing a drawing of the head of a buffaloe calf the size of life—The boat has made but little progress to day—having been obliged to sound most of the time—This afternoon several traders came down from Fort Pierre which is but few miles above us—Among others—Mr Picotte and Mr Chardon of whom Catlin gives some account.

MAY 31:

After having been grounded repeatedly they have at length forced the boat over the shoals, and are now lying at Fort Pierre—This is one of the principal forts or stations of the American Fur Company—1,274 miles above St Louis It is situated on the western bank of the Missouri just above the mouth of the Teton river—At this place will be left some 50 men, and several tons of freight—which will lighten the boat a great deal, and enable her to proceed much faster.

Around this fort are now about 25 lodges of *Sioux*—or as they call themselves *Dah cotah* Indians, a large number of whom were assembled on the bank to witness the arrival of the boat—Many of the men are really noble looking fellows—They are generally large, tall, and finely formed, and as they stood in groups, or pranced up and down the shore, on horseback—with their long hair and showy dresses streaming in the wind they presented a truly picturesque appearance—River mile and half wide—

They ride remarkably well—both sexes—dashing ahead at a furious rate—and clinging so closely to the horse as to appear a part of the same animal. Their huts or lodges are composed of skins of the buffaloe &c, dressed, and sewed together—They are of a conical form rounded at the base—and are supported by a number of poles which rise to one common centre.

JUNE 1:

This morning I visited the Fort for the purpose of making some sketches of buffalo calves, of which they have some 10 or 12. These animals were taken by pursuing them on horseback and throwing a rope with a noose over their heads—After a short time they become quite gentle and are allowed to run at large in the yard. Most of these have been captured three or four weeks—and are quite as tame as the common calves, of which they have several in the yard with them. They have a kind of grunt which resembles precisely that of the hog—They are supposed to be 6 or 8 weeks old—

Mr Picotte one of the agents here presented Mr A with a number of curiosities, Among which are a quantity of petrified shell—some Indian dresses, and a splendid pair of Elk horns—The horns measured as follows—

Length around the curve—4½ feet
Breadth between the tips—2¼
Circumference at base—10 inches
About noon we were again underweigh and had a fine run.

JUNE 2:

A beautifull day. The atmosphere in this country is very clear enabling one to discern objects at a great distance—I have frequently been deceived supposing things to be nearer than they really were. The immense extent of country visible at one time, also, aids the illusion—every thing being on so large a scale—that one need be accustomed—to judge correctly of size or distance.

JUNE 3:

Plenty of wolves have been seen to day—Bell shot one—who measured—

from nose to root of tail—41½ inches
Tail to end of hair—18½
Height at fore shoulder—27

Passed mouth of Grand river—

JUNE 4:

This morning we passed the ruins of an Indian village—It was formerly occupied by the Ricarees but they have now taken possession of the old Mandan village—and this place is wholly deserted—

JUNE 5:

Cool in the morning—Thermometer 47° to 60°—Passed Beaver river— and in the afternoon Cannonball river—Four buffaloe crossed the river to day—and passed quite near our boat—several rifles were shot at them but none were killed.

JUNE 6:

Cold and rainy. Thermometer 40° at sunrise—Some white frost seen *very* early this morning.

Met four boats loaded with furs—on their way from the Yellow Stone to St Louis—In this manner great part of the furs are carried down—Mr Kipp one of the Agents of the Co—was taken on board and goes back with us to the Yellow stone—he having started with the expectation of meeting the Steam boat—Mr K. is one of the oldest traders in the country—having been among the Indians more than 20 years—

JUNE 7:

About 8 o clock this morning we reached Fort Clarke—another of the A. F. Companies stations.

It is situated on the western bank of the river—and near the great Mandan village—This village is now occupied by the Ricarees—the Mandans having been almost totally destroyed by the small pox which raged among them a few years since—The small number that remain, supposed to be about 150, are now living at a small village a few miles above. The Mandans appear to have been a peculiar race—quite different from many of the tribes of Indians who reside around them—and before the introduction of that terrible disease were quite numerous. Their huts or lodges—as well as many of their other customs differ from those of the Sioux—and many other tribes—Instead of tents of skins, supported by poles, they build a permanent hut—commencing by digging a circular cellar about two feet deep, around the outside of this are set stakes or posts 5 or 6 inches in diameter quite close together, rising 4 or 5 feet above the ground, with a slight inclination inward—

On the top of these are placed other poles which rise to one common centre at an angle of some 30 or 40 degrees. The whole is then covered with a layer of willow twigs and small poles lashed across and over the whole a thick coat of mud or clay, which gives them the appearance of a hemispherical heap of dirt somewhat resembling in shape a bowl bottom up—the whole is supported by numerous posts in the inside around which are hung various domestic utensils clothing &c—The bedsteads which somewhat resemble those of the whites are rudely constructed of poles lashed together over which are spread one or two buffalo robes for a bed, the whole is surrounded by curtains of dressed skins. The fire is built on the ground in the centre, and an opening in the top serves to let out the smoke and to let in light—

Near the centre of the village is the great *Medicine Lodge*, where their ceremonies are performed. It is constructed in a similar manner but is much larger, being near 60 feet in diameter and 20 feet high. Around the walls inside of this lodge were hung a number of medicine rattles, skulls of various animals and other articles used in their ceremonies.

On the floor laid an old man quite blind, who, as our guide informed us by signs, had come there to die —he shook hands with all of us—Another Indian sat wrapped in his robe on one side of the lodge but he neither moved nor spoke.

A visit to their village such a day as this destroys all the romance of Indian life. It was stormy and cold, and they appeared anything but comfortable. The water dripped through the roof in many places so that the floor was quite wet.

This evening we had a kind of council on board the boat, consisting of 30 or 40 of the principal men of the tribe—and really it was a fine sight to see so many large noble looking fellows arranged around the cabin, all entirely naked with the exception of a bit of cloth around the waist and a buffaloe robe which they threw down to sit upon.

Coffee and biscuit were served to them and several of the traders and the Capt. of the boat addressed them through the aid of an interpreter, on the advantages of a friendly intercourse to both parties. To all of which they listened very attentively signifying their assent to each proposition by a kind of guttural "Agh," which amused me very much.

Some tobacco and other articles were divided amongst them. On leaving the boat they *all* shook hands with each of us in succession.

JUNE 8:

Cold thermometer at 37° this morning. After running a few miles we arrived at the Minnataree village, where we have had nearly the same ceremonies with the natives as yesterday—

JUNE 9:

Passed the mouth of the little Missouri—This afternoon a hare and a buffaloe cow were killed—

JUNE 10:

Outlined the head of the Cow this morning—size of life—Two Indian chiefs were taken on board yesterday, manifest a great deal of curiosity at seeing us drawing, and skinning the birds and quadrupeds—One of these Indians is a Ricaree named *Iron Bear*, the other a Minataree or Grosventres, and is called *Four Bears*. Squires wrote their names on a piece of paper which they examined very closely and like Leather Stocking in the Pioneers appeared to think it must all be right but they know nothing about the matter. This afternoon a number of Indians were seen on the shore—the boat was run in and one of them came on board—They were Assiniboins—

JUNE 11:

Passed some remarkable hills to day along the top of which I saw for the first time, several of the mountain sheep or bighorn—they bounded along in a surprising manner and appeared to be quite at home.

JUNE 12:

Passed the mouth of the Yellow stone this afternoon—About 7 o clock this evening we reached Fort Union—having been little more than 47 days working our way up from St Louis to this place—the quickest trip that has yet been made—

Fort Union is situated on the northern bank of the river three or four miles above the mouth of the Yellow Stone—

JUNE 13:

Engaged in unloading boat to day. We remained on board writing letters &c—

JUNE 14:

This morning we bade the Capt. good bye—and about 9 o clock the boat started on her reutrn to St Louis—To day has been a kind of holiday at the fort, the arrival of the steamboat being an event of much importance here. This afternoon a wolf was seen running on the prairie at a short distance. Mr Culbertson the principal Agent here immediatly mounted his horse and proceeded in pursuit of him—In a very short time he came up with, and shot him while running at full speed—and in less than 20 minutes the wolf was brought into the fort. Some of these men ride in a most surprising manner. Several of them rode out about ¾ of a mile from the fort starting from thence with unloaded guns, and while running that short distance at full speed they managed to load and fire from 9 to 11 times.

The horses are guided by inclining the body to either side—the reins being thrown loose upon the neck, leaving both hands free to use the gun.

JUNE 15:

Two Mackinaw boats loaded with furs from the Crow nation arrived here this morning on their way to St Louis—A squaw on board brought a fine young badger alive, which Mr [Audubon] purchased and intends carrying it home.

JUNE 16:

Engaged to day in making a drawing of a young wolf, which had been taken from a hole some time before—It is probably about two months old—The hunters to day brought in a young fawn—

JUNE 17:

At work on the drawing of the young deer. The hunters frequently bring the does within shot, by imitating the cry of the young—which is done by a kind of whistle made for the purpose.

JUNE 18, SUNDAY:

Two antilopes were brought in to day—these are the first I have seen. They are beautifull animals—but larger than I expected to see—[Measurements appear here.]

JUNE 19:

Working at the drawing of antilope head full size—Messrs Harris and Bell went out shooting and brought in a golden wing woodpecker with a red stripe on the cheek instead of black—and a small species of titlark which is probably new. Length 6 inches—alar extent 10¼.

JUNE 20:

Cold and Rainy—Finished drawing of antilopes head.

JUNE 21:

Cool—went out this afternoon and shot an Arctic blue bird—the first that has been killed—

JUNE 22:

About 12 o clock today Mr Chardon with his party of some 30 men started with their keel boat, for the Blackfoot country. The flag was hoisted on the fort and a salute fired from the six pounder, which was answered by a swivel on board the boat. They expect to be between 30 and 40 days working their way up to their place of destination. And a most tiresome life it must be pulling at the *Cordelle*—

Been engaged to day—Drawing a spike of beautifull flowers of a fine purple colour. This plant, as are most of the kinds about here, is quite new to me—

JUNE 23:

Went out shooting this morning but killed nothing—while out I watched one of the new titlarks for nearly an hour—as it sailed around over my head high in the air—singing its simple notes at intervals of about 10 seconds, the song itself occupying about 5 seconds—While singing they remain nearly still moving their wings in a rapid manner like a little hawk—and in the intervals between they sweep around in an undulating manner closing the wings to the body like the goldfinch—3 of these titlarks killed today.

JUNE 24:

Found the nest of the titlark—and shot the female as she rose from it. It was built on the ground, in a small cavity so that the top of the nest was even with the surface and slightly shaded by a small tuft of grass—The eggs five in number are pale brown thickly spotted with darker.

JUNE 25, SUNDAY:

Very warm—Ther 84°—

JUNE 26:

A kind of Sturgeon was caught to day—It had run into shoal water and could not get off and was taken by one of the men with his hands—It measured 57 inches from tip of nose to end of tail on the upper part, the lower fin being an inch shorter—weight 22 lbs—

JUNE 27:

Made a drawing of the plant called Pomme Blanch—the root of which is used in this country by the Indians and others, as an article of food. It is deep in the ground—and is near the size of a hens egg—The outside is covered with a thick woody skin of a dark brown colour. The inside is white and resembles a chestnut in appearance and taste except it is not so sweet—

JUNE 28:

Mr Harris shot a fine wolf last night, and I have been engaged to day in making a drawing—full length but smaller than life.

JUNE 29:

Mr Kipp started this afternoon on a Mackinaw boat for the Crow nation—with some 25 men and a freight of goods for trading.

JUNE 30:

Went out shooting this afternoon—killed a Red shafted woodpecker, a very pale specimen—It was a female—and was killed near the hole which contained young.

JULY 1:

At work on the drawing of a Doe's head—went out with Mr Harris this afternoon and obtained several more specimens of very pale Redshaft woodpecker—a golden wing with red stripe on the cheeks—and one with black cheeks precisely like our Eastern one—

These redshafts were marked with a black stripe on the cheeks whereas those of the true Redshaft are red, so these must either be several species very nearly allied—or an intermixture of the Red and Yellow—

JULY 2, SUNDAY:

Mr Culbertson presented me to day—a beautifull Indian dress consisting of a Shirt, leggins and Mantle all of which are made of skins of various animals and highly ornamented with porcupine quills, pieces of shells &c—

By cutting down the trees we obtained the young of the pale redshaft woodpecker which I shot a day or two ago. They resembled the old one in colour—and some of them had the rudiments of a *black stripe* on the cheeks.

JULY 3:

Very hot—thermometer at 90—sketched young buffaloes—

JULY 4:

Independence—not celebrated—The young buffaloe about the fort are now shedding their hair. It comes off in large flakes matted together so firmly as to be sometimes used by the Indians to make caps of—Spines of the vertebrae of the buffaloe measure 19 inches—

JULY 5:

Cold and stormy—A female Bighorn brought in—weight 140 lbs—

JULY 6:

Drawing flower—Crossed the river for the first time this afternoon—and made a sketch of the fort—Just at night a small war party of about 12 Indians came in singing, having the scalp of an enemy—They are and have been for a long series of years almost continualy at war with each other—

One great cause of this is want of other employment, almost all of the labour being imposed on the female—and it is impossible to keep the human mind in idleness—whether in a civilized or savage state they try to kill time by killing each other—

Another cause is the disgrace that is attached to those who have not been to war—It being a bitter taunt equivalent to calling a man a coward to say to him—You are of no account you have not been to war—consequently the young men are all anxious to signalize themselves as soon as possible.

They are accordingly forming war parties who proceed to the enemies village and either steal some horses, which is considered a great honour—or surprising some unfortunate persons at a distance from their tribe and kill and scalp them without mercy. (It does not matter much of what tribe.) They then return and celebrate their exploits by singing and yelling—of which they gave us a fair specimen this evening.

JULY 7:

The Indians performed a couple of hours this morning in fine style—The whole party in their war dresses—which by the way are but little more than nature gave them, with their faces painted with black and red looking like fiends—presented an appearance that would contrast strongly with a military company in Boston or N. York—but for aught I know there is as much reason in the one as the other.

The Indians place their dead on scaffolds or on trees of which there are a number about a mile from the fort, some of which I sketched this afternoon—

JULY 8:

Very hot—Ther 92°—Made a drawing of a Plant—the root of which is said to cure the bite of the Rattle Snake—[****] on hills.

JULY 9, SUNDAY:

Drawing wolfs head size of life—

JULY 10:

Drawing a Cactus—We find several species of these plants—and in many places so plenty as to be troublesome in travelling—Unless protected by stout soles—the spines pierce the feet and nothing will do but to set Down and pull off your Moccassins and extract them at once—There is also a kind of grass, the seeds of which being pointed and barbed proves exceedingly troublesome—

JULY 11:

Painting a view of the Fort—in oil—for Mr Culbertson—

JULY 12:

Work on Bucks Head.

JULY 13:

The wolves come around the fort and men are kept constantly on guard to protect the horses from them and the Indians—

JULY 14:

This afternoon Mr Culbertson, Owen—and Squires—arrayed in Indian Costume accompanied by two Blackfeet squaws in native dress made a grand display on horseback—They performed a number of evolutions on the prairie, and rode to the hills where they espied a wolf to which they gave chase and shot—and after returned to the fort at full speed—The women ride as fearlessly as the men—and astride their horses in the same manner—

JULY 15:

Mr A &c started on an expedition up the Yellowstone—

JULY 16, SUNDAY:

Drew a plant—Yucca filamentosa?—A Buffaloe bull killed to day measured [Measurements appear here.]

JULY 17:

View of the country from a hill back of the fort—Yellow stone in sight, *but I miss the Ocean—*

JULY 18:

Very hot. rain at night with thunder—

JULY 19:

Very hot again—view of the country near the fort consisting of hills & ravines.

JULY 20:

Started on grand Buffaloe hunt—I did not attempt to ride—They are hunted on horseback—the hunter selecting his animal—rides close up, and shoots him, loads again and proceeds—while, at full speed—In this way an expert hunter will kill 3 or 4 from one barrel—It sometimes happens that the shot is not mortal, and the wounded animal being closely pursued becomes furious and turns upon his pursuers causeing the horse to sheer aside to the no small dangers of the rider—Something of this sort happened to Squires—he having shot at, and wounded a bull—was pursuing at full speed—when the beast made a sudden turn upon the horse—who wheeling to avoid him—threw Squires to the ground, fortunately the buffaloe passed on—and Squires was but little injured—Bell, who is an expert rider, and a good shot, succeeded better.

JULY 21:

Gathering berries, from a species of {pyens} resembling our swamp one, and in absence of other fruit they are really good.

JULY 22:

At fort a party of Indians come in.

JULY 23, SUNDAY:

A party of Cree Indians of about 70 in number—stopped here a few hours on their way to smoke with the Grosventres—4 buffaloe killed—weight of one 1777 lbs. Went across the river for berries.

JULY 24:

Finish sunflower—severe squall with thunder.

JULY 25:

Antilopes brought in—one of the hunters killed two and wounded a third with a single bullet—as they said—

JULY 26:

Finish drawing of antilopes head size of life—Buffaloe hunters go out—

JULY 27:

A party of Indians come in singing and whooping.

JULY 28:

Buffaloe hunters come in having killed 7 since they left. The hunters always select the young cows—and when in good condition their flesh is excellent—The Indians eat the brain, the inner coat of the nostrils &c raw! and appeared to consider them as great delicacies. They also eat the liver—the coats of stomach, or tripe, in the same state—I tried a piece of the last, but did not relish it much. Though I could eat it about as well as any tripe—

JULY 29:

Sketched young buffaloes as they lay in a fine group.

JULY 30, SUNDAY:

Very fine flower—

JULY 31:

Gathered a few seeds.

AUGUST 1:

Went out with a party to gather a quantity of Pomme blanch (Psoralea esculenta)—The Indian women dig them with great facility by means of a pointed stick. In autumn the top of the plant dries—and separates from the roots near the ground and blows away. High prairies abound with prairie turnip.

AUGUST 2:

A small party of Chippeways come in to trade—and at night a party of Assiniboins shot [****].

AUGUST 3:

This morning the Crees come back that passed here July 23—it seems that instead of making peace, or smoking as they say—these Indians surprised a small [group] of Grosventres—and killed & wounded several. Went out to the hills with Mr Harris—

AUGUST 4:

Buffaloe hunters go out and

AUGUST 5:

bring in some fine cows but nothing else of importance.

AUGUST 6, SUNDAY:

In the ravines about Fort Union are found elm, maple and ash, dwarf chokecherry—a species of black currant the fruit of which is not very finely flavored and a wild gooseberry resembling the common one.

AUGUST 7:

Collected a few seeds—

AUGUST 8:

A short time ago—an Indian who had been long sick died—He had been hanging round the fort and received some medicine from the whites but it did no good—His brother who has now come, is very angry—because they did not cure him—women cry.

AUGUST 9:

A small party of Assiniboins come.

AUGUST 10:

Indian girls have fine voices—sweet and musical.

AUGUST 11:

Rather cool.

AUGUST 12:

Cold and stormy—wind blows a great deal here—

AUGUST 13:

Fine day—*Sunday.*

AUGUST 14:

Made a sketch of Indian woman in native costume—

AUGUST 15:

Preparing to start—all busy packing—Hammering and bustle heard throughout the Fort.

AUGUST 16:

Four or five Assiniboins come in on fine horses which they ride admirably—They are a part of a large band now in sight—

About 12 o clock we bade good bye to the inmates of Fort Union—and it is not probable we shall ever meet again—They have endeavoured as much as in their power to make us comfortable—and notwithstanding some disappointments—I have passed many pleasant hours here—It is with regret I take leave—but the hope of again seeing Home—Friends, and again enjoying scenes long familiar—render the parting less bitter—

Fort Union is situated on the northern bank of the Missouri—about 6 miles above its junction with the yellow stone, on a high plain close to the river—The plain extends along the river several miles and is about 1½ miles wide—The Fort was commenced in 1829—is 220 feet in front, and extends back 240 feet—surrounded with pickets 20 feet high made of hewn cotton wood 1 ft square—On the S.W. and N.E. are bastions built of stone 24 feet square and about 30 feet in height, projecting beyond the pickets in such manner as to rake them—The gate is 14 ft high by 12 wide. Flag staff in centre 60 feet—

The ground is very muddy when wet, but soon dries hard enough for bricks—The hills around the fort are composed of strata of sand—clay, coal, and contain remains of shells, petrified wood &c—The bottoms or points near contain cotton wood, with an undergrowth of buffaloe berries and wild rose bushes—Here far away from civilization, the traders pass the best of their days—some from a love of adventure, some for gain—and others for crime are driven from civilized society.

Soon after 12—our party consisting of 14 persons in all, embarked on board a mackinaw boat—a flat bottomed craft about 40 feet long and 8 wide—and started down the river—A few miles below found a fine pair of Elk horns, that had been pointed out to us. They have a remarkable prong projecting forward—Several young ducks that had been shot were cooked for supper—coal 3 feet thick bottom strato—close to the bank of the river—

AUGUST 17:

Fine day—very pleasant on calm still days floating along this wilderness—especially in those places where the banks of the river are overhung by the ancient forest. Occasionally the boatmen strike up one of their wild Canadian boat songs, keeping time as they row with their oars—These delightful days afford me an enjoyment much superior to anything I ever experienced in the bustle and turmoil of civilized life—

Passed the mouth of Muddy river about 11 o clock—Saw several swans, geese and ducks—also great number of deer and some Elk, one of which was shot at and wounded but made his escape—Two bulls killed at night—

AUGUST 18:

Started as usual at day break, weather very fine and warm—stopped to breakfast—made a sketch—

Saw a band of 15 or 20 Elk feeding and laying on the sand close to the waters edge but they bounded off before we came within shooting distance—Stopped a short time to hunt but obtained nothing—A part of our company landed and walked several miles just before we encamped for the night, but killed no game.

AUGUST 19:

Cool this morning—stopped to breakfast just above the mouth of Knife river—Messrs Audubon and Culbertson walked on in advance of the boat a few miles and shot a buffalo cow—Stopped about 4 o clock at an island to hunt and killed a deer, and a cow—

AUGUST 20, SUNDAY:

Ran a few miles, and stopped by the wind. Bell and Myself went ashore and endeavoured to get a shot at a large herd of buffaloes that were feeding on a level plain at a short distance. We walked and crept to within 100 yards when they began to take alarm—we fired but stopped none—

Seeing two buffaloe bulls leisurely approaching the river, Bell & myself concealed ourselves in some bushes that grew along the bank, and waited for them to come up—The first one—an animal of immense size and most formidable appearance—came directly toward me, and when within a dozen yards I leveled and blazed away. He sprange off a few yards and fell, but not dead, the shot taking effect rather too high on the shoulder to be immediately fatal. The other, fully equal in size and quite as ferocious in appearance, being 2 or 3 hundred yards behind was not alarmed at the report of the gun, but came directly on unconscious of danger in the same track until quite near, when I discharged the contents of my second barrel at him he bounded ahead a few paces and fell dead! The first one, startled by the second report and the sight of us as we came from our cover, started and ran a short distance when he staggered, fell and died!

This was good shooting, to kill two buffaloes weighing 2000 lbs each—with a single ball each—but the sight of them as they lay bleeding on the earth seemed to me too much like butchering—I immediately made a sketch of the nearest one as he lay, afterwards cut off the tails as trophys and left them for the wolves!

An old bull is one of the most frightful animals I ever saw and at times they are as savage as they look. Some of our men shot at and killed a cow, when an old bull stepped in front, and maintained his ground so well, that, although they shot at him several times they were obliged to leave without their meat.

AUGUST 21:

Ran a few miles this morning and again stopped by the wind which gives so much motion to the water that the steersman is unable to distinguish the sandbars &c.

AUGUST 22:

Still impeded by wind—reached the mouth of Little Missouri—While on shore this afternoon I espied a grizzly bear about a quarter of a mile from the boat and on informing Mr A. he in company with Bell &c crept along under cover of the bank until within gunshot—when the animal rose upon his haunches to reconnoitre they gave him a volley which laid him in the dust. He proved to be rather young and would probably weigh about 300 lbs.

AUGUST 23:

Windy this morning—close to our camp is a large prairie dog village.—These little animals which resemble our woodchucks but are smaller, are extremely difficult to obtain—They set at the mouth of their holes and bark like toy dogs and on the least appearance of danger—or when shot at and wounded, hide themselves at once. We procured but two, although we spent several hours watching for them. Ran a few miles at night—Saw a bear.

AUGUST 24:

Ran a good piece this morning when the wind arose—found petrified wood stones with leaves impressed—gathered seed of a plant with aniseate odour—wild cherries very good.

Our encampment presents a curious spectacle; composed as it is of wild looking figures in Indian-hunter's and civilized costume each engaged after his own fashion in cooking, eating, or preparing for the night—some talking English some French some Indian.

AUGUST 25:

Very fine day—went on without stopping until afternoon, when we arrived at Fort Clark where we stopped for the night—Plenty of Indians—one of the chiefs is put on board the boat to keep off thieves.

AUGUST 26:

Stopped some by wind, but made a good days work—made a sketch of the square hills—limestone begins—

AUGUST 27, SUNDAY:

Hardly remember it—Stopped about 12 o clock & killed a cow. In the afternoon the wind lulled and we went on—A buffalo bull crossed the river just as we passed—and came very near our boat—the whole party shot at him but of no use—after the boat passed he swam back again to the side he started from. Bear tracks plenty—Heart river on west side Apple river on East.

AUGUST 28:

Wind blew a gale from S.E. last night and continues so high that we have not started this morning—wild artichoke very abundant—

Our encampment is 1 mile above Apple? river.

AUGUST 29:

The wind still continuing to blow we have remained fast—

Time passes rather dull when tied up in such places—read some of Roger's poem Pleasures of Memory—reminded me of home—I think of home and am there although 2000 miles away.

Little striped snake 1 yellow line along the back, each side a red and black line—then a yellow and below that a black line on each side. moved a mile or two at sunset.

AUGUST 30:

Tied up all day—a cow and calf killed—The willows and rosebushes so thick along the bank that it is almost impossible to get through. Ran a few miles about sunset.

AUGUST 31:

Started early this morning with fine calm weather—about noon reached Cannon ball river, where we stopped to dine—This river receives its name— from the great number of rounded stones that abound on its banks—we searched a while but could find none small enough to take away.

Just before sunset we reached Beaver river where we encamped for the night.

SEPTEMBER 1:

Started rather late this morning in consequence of its being wet, and we have had slight showers through the day—notwithstanding which we have made good progress, stopping once to try to get a shot at some buffalo but without success—After we had stopped for the night a band of buffalo were seen—and several of the party started after, and shot one but could not get her—In the night we had a heavy shower.

SEPTEMBER 2:

Clear and cool—ran two or three hours when the wind became so high that we were obliged to stop.

The banks of the river where we are now encamped, are wholly different from those at Fort Union. The soil is composed almost wholly of clay generally of a dark colour, and containing many crystals of gypsum and great quantities of shells—most of the shells are broken and when exposed to the air crumble to dust—first noticed the change in the country yesterday—

Made a sketch of our encampment—we obtained a few pertifactions—first noticed a plant with variegated leaves—The high winds that prevail most of the time are exceedingly annoying not only impeding our progress, but filling our eyes and food with [word omitted] and dust—

SEPTEMBER 3, SUNDAY:

Remarkably fine calm day—passed old Ricaree village and the mouth of Grand river this morning—Saw some of the white flowers that abound at Fort Union—They appear to prefer a steep bank with a western exposure—In the afternoon a band of buffaloes were seen near the shore and several men went in pursuit of them but killed none—Camped at the mouth of Moreau river 18 miles below Grand river ? west side—

SEPTEMBER 4:

Went on until stopped by the wind some 8 or 10 miles—

The point where we are encamped is filled with buffaloe berry bushes—among them we find a few with berries of a yellow colour—Tracks of elk, and deer very plenty and also signs of beaver—At one place they had cut off trees 6 or 8 inches in diameter and afterwards cut them into pieces 3 or 4 feet long and carried them to the water—One fine deer killed.

SEPTEMBER 5:

Traps were set for beaver last night but none caught—This morning their lodge or house which we found along the bank of the river, was torn down and examined. It was made by digging an oven shaped hole into the bank—communicating with the river by a passage way, covered with willow brush.

After which we ran on a few miles until about 11 o clock when the wind compeld us to stop—Soon after we started this morning we saw a sparrow hawk in pursuit of a spotted sandpiper who escaped by diving repeatedly.

In the afternoon I went on to the hills back of the camp, and on attaining the summit level, had the finest view of the prairie country I have seen—on every side as far as the eye could reach, was spread out a boundless plain apparently perfectly level, but in reality consisting of an endless succession of wooded hills or swells like the billows of the ocean. Where I went the country was covered with bowlder drift. A young buck elk was killed.

SEPTEMBER 6:

Tied up all day—The wind blew hard all night and this forenoon we have had quite a gale—This is one of the days that pass heavily—The point being small all the game was started yesterday—and the surrounding country not being very interesting after seeing so much of the same character—we are compelled to pass the time as best we can—Found the nest of a dove on the ground one young and one egg—found another just Fledged young one dead, rather late for eggs & young—

SEPTEMBER 7:

Last night the wind which had been blowing from the south suddenly changed to Northwest and gave us a tremendous puff accompanied with thunder and lightning—Our boat sheltered before, was now exposed to its full fury, and for a few minutes there was quite a bustle and rush to get on shore and to secure the boat—The wind soon abated a little, and the rain began to fall so plentifully that all hands were glad to shelter themselves on board and let the storm have its own way. It blew through the night and has blown hard all day giving no chance to start until about 5 o clock this afternoon when the wind lulled a little and we moved on a mile or so to an Island where we stopped to hunt but killed nothing. Bell shot a whip-porwill that differs from the eastern one—found here a species of thorn with edible fruit.

SEPTEMBER 8:

Very cold and uncomfortable with a slight sprinkle of rain—notwithstanding which, we have made a good run and arrived at Fort Pierre about 5 o clock this afternoon—Messrs Laidlow, Bowis, Murray &c received us very kindly—

SEPTEMBER 9:

Cold and stormy—Mr A. has decided to take a larger boat which they have here—and which the workmen are engaged in fitting up. This will probably detain us here several days—

SEPTEMBER 10, SUNDAY:

uncomfortable, disagreable weather—1253 miles mouth Missouri—Made a sketch of the Fort.

SEPTEMBER 11:

The wind which continues to blow from the south, as it has for two days past, increased this forenoon to a perfect hurricane and raised such a commotion in the river, that we were obliged to unload our boat to prevent her being filled and sunk—

SEPTEMBER 12:

Fine this morning with a gentle west wind—but we soon had a breese which brought clouds and rain, and it has proved to be one of the most disagreable days we have had being cold, wet, and uncomfortable—

SEPTEMBER 13:

Weather same as yesterday however the boat is nearly ready.

SEPTEMBER 14:

Cloudy but no rain—Mr. Laidlow and party started for Fort Union this forenoon—Our boat being completed, loaded &c—at 3½ o clock this afternoon we bade adieu to Fort Pierre and its inmates—

Mr Culbertson who has accompanied us thus far on our way down—leaves us here and will start for the Platte in a few days—

At 5 o clock we arrived at a small Island where vegetables are grown for the Fort—Here we stopped to get some green corn—potatoes—&ᶜ—We afterward crossed the river and encamped for the night. We have now ten persons on board.

SEPTEMBER 15:

Did not start very early this morning being dark and foggy. Before 9 o clock we arrived at Fort George where we stopped a few minutes and again went on our way rejoiceing, but about 1 o clock the wind raised so much that we came to a stand still—and although the rowers did their best it was no go—here we were forced to pass the remainder of the day—

To day we have had fresh pork, potatoes, green corn, carrots and beets, and I have eaten heartier than I have for a long while before—At Fort Pierre the cooking was badly done—and the fare miserable—much poorer indeed than it need be. The men here, at Fort Union—and in fact everywhere—live almost exclusively on dried buffaloes flesh—

SEPTEMBER 16:

Lovely morning—perfectly calm and beautiful—and when we float along with the current—the men resting on their oars—not a sound is heard to disturb the profound repose that reigns over this vast wilderness.

About 11 o clock we arrived at the Great Bend, where we stopped to search for fossils, and hunt for black tailed deer—but without success as no deer were seen—and no fossils of any value found—so we lost a fine calm day—Bell killed six rattlesnakes—first we have seen—

SEPTEMBER 17, SUNDAY:

Last night we had a heavy squall from the N.W. It came on suddenly while the sky was quite clear but we afterward had a slight shower—Clear and cold this morning with a fresh breeze, that soon increased to a gale—and put a stop to our progress—having run 8 or 10 miles. In the afternoon the wind lulled—and we made quite a good run—nearly around the great Bend—Three sharp-tailed grouse killed—saw some oak trees—

SEPTEMBER 18:

Detained part of the day by wind. The banks of the river here are composed of apparently, hardened clay—some of it hard as stone—in strata of different thickness—the perpendicular face of which where it has been worn away by the river rises like a wall of mason work some 50 or 60 feet. Above this is a dark brown deposit of perhaps 40 or 50 feet thickness—

SEPTEMBER 19:

Wind has impeded our progress to day—It does not matter from which quarter it blows we are sure to have a fresh breeze the vast extent of level country on every side offering no resistance—and affording no shelter from its fury—

This morning we stopped a short time at an island which is said to have been once covered with cedar but is now nearly bare—Stopped at noon opposite white river—Yellow buffalo berries—

Wolves are very plenty—At night their long drawn dismal howling is anything but agreeable, and—associated as it is with the knowledge of our distance from civilized life, produces an indescribable feeling of loneliness— They seem to howl in concert, taking up their parts in succession—At times they appear to feed upon wild cherries and plums—not very digestible food for them.

SEPTEMBER 20:

Did not start until about 1 o clock—pale red shaft woodpecker—Saw an immense flock of ducks on a sandbar soon after we started—This afternoon we ran on a sandbar which is nothing remarkable—and all we have to do is to get into the water and push the boat off again—rather longer than usual doing it to day—Encamped nearly opposite Bijou Hills.

SEPTEMBER 21:

Dull and cloudy this morning—passed Burnt Hills—

About noon a band of buffaloes were seen lying on the prairies a short distance from the river—The boat was stopped and—Provost, Harris and

Bell went after, and shot one—They proved to be all bulls—In the afternoon we had showers, and 5 o clock when we reached Cedar Island it rained quite hard—Here we encamped—and cut several sticks for oars—The rain continued through the night.

SEPTEMBER 22:

and has still continued the whole of to day without intermission—We have only moved a mile or two to find better accommodation—The strata of clay not so high as above—

SEPTEMBER 23:

Passed the Tower as it is called—probably the Dome of Lewis and Clark—Encamped on an island—weather cloudy and cool—(Puncah Island). Met steamboat at night—on her way up to [blank space in MS].

SEPTEMBER 24, SUNDAY:

Cloudy—cold and uncomfortable—

Passed Puncah river, and soon after the Running river—Ruins of Fort Mitchel on the point. Stopped by wind a few miles below—In the hill back of our camp are a number of Indian graves. Large flocks of Pelicans, and one flock of ducks sittting on a sandbar. Making oars—Found some fine wild plums—5 wood ducks killed.

SEPTEMBER 25:

Rainy this morning—started about 10 o clock—Passed Bonhomme Island—ruins gone—probably washed away years ago—soon after passed a mound or more probably a natural hill.

SEPTEMBER 26:

Cool & cloudy—saw a large flock of geese near the mouth of the River Jacque—which we passed about 11 o clock—2 Ducks—1 goose—pelican—

SEPTEMBER 27:

Weather cloudy and cold—passed the mouth of the vermilion river this morning and about one o clock reached the vermillion Fort as it is called,

a small station in charge of Mr Pascal who received us very kindly—
Here we stop to finish our oars, and get provisions—potatoes, corn and
beef—Just above this fort we met a keel boat coming up belonging to the
A. F. Co—One of the men, a mulatto by name of Micheaux whom we
had known on the steam boat was desirous of going down [with] us and
Mr A. engaged him as a hunter. Before we arrived at the fort the weather
began to be stormy and this afternoon it has rained and continues to rain
till in the night—

SEPTEMBER 28:

A most superb morning—clear, calm, and delightful, the finest we have had
for a long while—but to me these fine autumnal days bring a recollection of
the pleasures I have enjoyed on similar days at home—and notwithstanding
every thing is more pleasant, I feel an approach to homesickness that is not
thought of when the weather is bad.

After breakfast we took leave of the vermillion Fort—and went on about
15 miles to a point opposite the Ioway river, where we stopped to hunt
for Elk—of which Mr A. is very anxious to obtain good specimens—but
unfortunatly without success. Towards night the clouds began to gather and—

SEPTEMBER 29:

This morning it is wet and rainy enough for any one—All we have done to
day is merely cross the river to find a better shelter and here we have passed
the day as best we could. We find in the bluffs minerals as alum—&c

SEPTEMBER 30:

About 8 o clock the sun made his appearance and we started and went on
untill two—when we had a sudden and severe squall, which obliged us
to put ashore, and there we remained the remainder of the day the wind
blowing a perfect hurricane from the Northwest. Our situation was rather
exposed and the wind and water made our frail bark tremble like an affrighted
deer—but her plank held together and we safely rode out the gale—which
continued to rage with unabated fury until near midnight.

OCTOBER 1, SUNDAY:

Showery this morning however we started and about 9 o clock passed the
Sioux river just below which we stopped and killed three turkeys—again
went on some 3 miles when the wind increasing we came to a stop nearly
opposite Floyds Grave—but in the afternoon it became calmer and we
again went on and stopped for the night at the mouth of Maha creek—

OCTOBER 2:

A clear cold northwester—blowing so hard as to detain us a part of the
day—We however made some 15 or 20 miles progress and camped for the
night a few miles below Blackbirds grave.

The country in many places where we have stopped is covered with wild
artichokes, upwards of ten feet in height—growing very thick together
and interwoven in every direction with a species of wild bean vine. Wild
turkeys along the shore.

OCTOBER 3:

To day the weather has been so fine and calm that we have run all day without
stopping, an event that has not occurred to us for a long time before—Saw
plenty of deer along the shore—and immense quantities of geese and
ducks on the sand bars—Encamped near the old Council Bluffs—Great
changes have taken place in the channel of the river here—since Lewis
and Clarke passed—

OCTOBER 4:

Underweigh this morning before sunrise—At 8½ passed Bowyers creek,
and about 10 we stopped near a small creek on the west side—the wind
having become so high as to prevent our travelling—Here we remained
the rest of the day—The water in river is rising—this helps us on our way.

OCTOBER 5:

Ran a few miles this morning when the wind, which had blown all night,
increased to such a degree as to stop our progress about 6 miles above
Fort Croghan—Messrs. Harris, Bell and Squires started on foot for the

garrison—where they arrived about 10 o clock—At 2 o clock the wind lulled somewhat and we went on with the boat and arrived at the fort in about 2 hours, and were received in the most hospitable manner by Capt. Burgwine—Lieuts—McCrate—Carleton—and Noble—The garrison consists of a company of dragoons—stationed here to protect the Potawattomies—and to prevent whiskey being smuggled into the Indian territory—

They have received orders to abandon the post—and will leave tomorrow, a part on a Mackinaw boat in company with us, the remainder on horses by land.

OCTOBER 6:

Between 8 and 9 o clock this morning we left the fort and dropped down to Bellevue—where we stopped a short time to purchase a few articles—Here they informed us that the Indians now around here are Pawnees—and that a large war party of 1200 Sioux had attacked this tribe this summer and killed 60 or 70 men—and taken a large number of horses. 4 to 11 hundred! We were shortly joined by the other boat and soon proceeded on our journey—9 miles below we passed the mouth of the Platte river which is now quite high. This will be of immense advantage to us as it covers the snags, and increases the current—About 5 encamped for the night some 20 miles below the Platte—our men bivouac—the soldiers camp—Limestone in place to day—

OCTOBER 7:

Cloudy and cool—But having a fair wind we have kept on all day—camped for the night at the mouth of the Nishnabotona river (128 miles above Ft. Leavenworth). Here we found a squatter and his wife—the first white woman I have seen for five months—They made us a corn cake—grinding their meal by rubbing the unshelled ears on a kind of tin grater—

OCTOBER 8, SUNDAY:

Started as usual this morning and went on a short time when it began to rain and blow—consequently we landed and remained a while but the weather again brightened and we pursued our course—passing the mouth of the Tarkeo and great Nemaha rivers.

OCTOBER 9:

Beautiful clear calm morning. About 12 o clock we arrived at the commencement of a bend in the river about 4 miles across and some 8 or 10 around. Here Messrs Harris, Bell—and myself landed to walk across. The point was covered with heavy growth of black walnut, oak &c—noble trees that have stood here for ages. I delight to roam through these old forests alone—on such fine, calm autumnal days as this—every thing is so quiet—and there is nothing to disturb the profound silence that reigns around save the occasional chatter of a squirrel—the tap of the woodpecker—or the scream of the parrots as they sport among the branches of these ancient trees—

But in the midst of this, and quite out of place, I thought, we found that curse of the red and white man—a distillery!

Killed 6 squirrels—several paraqueets, quails &c—found long acorns— also fruit of the Papaw—of which I ate one and was made quite sick—They were not perfectly ripe however.

OCTOBER 10:

Another fine day which has been so well improved that at 12 o clock we passed Cow Island 9 miles above Ft. L—and about 4 we arrived at Fort Leavenworth—Here are stationed three companies of infantry—to which will now be added the company of dragoons from above—Major Wharton is now in command—A Remarkably fine clear moonlight evening gave to the scenery around the fort a most beautiful apperance—music by a Military Band—

OCTOBER 11:

Cloudy and somewhat windy but not enough to stop us—at 12½ we passed the mouth of the Kansas river and are now in the state of Missouri—A short distance below we passed Madame Chouteau's place 40 miles from fort Leavenworth. About 4 o clock we reached Liberty landing—and just before sunset encamped a few rods above Independence landing—This morning as we floated along our voyagers sang some of their boat songs—The scenery along the banks of the river is beautiful—the golden yellow of the tall black walnut intermixed with the deep green oak, and majestic cotton wood trees that have stood here more than a century—their lofty stems covered with a profusion of creepers now of a brilliant scarlet colour—with wreaths and festoons of the same swing from tree to tree.

This evening the steamer Lebanon stopped a short time at the Independence landing—Mr Harris and myself went on board a few minutes—she is going as far as Leavenworth.

OCTOBER 12:

A very fine but rather cool day. The river is now sheltered somewhat by the timber that grows along the banks—so that we are not obliged to stop so much for wind—To day we have made a fair run—at 4 o clock we reached Lexington where we stopped a few minutes—Just below we met the steamer Tobacco Plant—and about 5 miles below camped for the night—

OCTOBER 13:

Cold but pleasant—kept on our journey all day—Just before sunset we passed grand river—near the mouth of which is the town of Brunswick—encamped half mile below—

OCTOBER 14:

We had this morning a fine breeze which by eleven o clock had increased so much as to stop our progress about 6 miles above Glasgow—near where we stop is a farm where they grow Hemp—Tobacco—and Corn—Tobacco when the plant is fully grown is cut up at the root and supended by the base on poles in a large open building and slow fire is kindled under it until dry—when the leaves are striped from the stalk and packed in hogshead for market—

About a mile distant is another farm and a fine young orchard—on which the owner reckoned there was a mighty smart chance of apples

In the woods saw some very large mushrooms 12–15 inches diameter—

OCTOBER 15, SUNDAY:

Hazy and cold in the morning—afterward clear—having fair wind we go on well—at 7½ we passed Glasgow—about 11 we passed arrow rock—at 2 Boonsville, at 3½ Rockport a few miles below which we encamped—

OCTOBER 16:

Cold—Passed Nashville—Marion—and about 12 arrived at Jefferson City where we stopped a short time—The ground about town is very uneven—wild geese abundant—

OCTOBER 17:

Fine October weather.

OCTOBER 18:

Passed this morning a remarkable cave called the Tavern—Arrived at St Charles—

OCTOBER 19:

Entered the Mississippi at 11 o clock—and at 3 arrived at St. Louis—water much Darker. Missouri gives character to Mississippi. The average rapidity of the Missouri is nearly twice that of the Mississippi.

OCTOBER 22:

Left St. Louis about one o clock on steamer Nautilus—St. Louis appears flat from want of spires—

OCTOBER 23:

Arrived at Cairo about one o clock—P.M.

OCTOBER 25:

Reached Louisville about 9 o clock this evening—and this morning

OCTOBER 26:

passed through a canal cut to avoid the rapids—when the river is low—

OCTOBER 27:

Arrived at Cincinnati at 8 o clock this morning and left at 12—snow on the ground.

OCTOBER 29:

Sunday—

OCTOBER 30:

At Wheeling at 6 this evening.

OCTOBER 31:

Pittsburgh this afternoon—went on board a canal boat.

NOVEMBER 4:

Arrived at Philadelphia this evening—

NOVEMBER 5, *SUNDAY*:

Left for New York and reached there about 12—

NOVEMBER 6:

Left New York at 4 o clock P.M.

NOVEMBER 7:

At Boston this morning from thence to Hingham this afternoon where I arrived—safely—Home—

[In the final twelve pages of the diary, the following undated notes are written in Sprague's hand.]

Burnt hills—This phenomona manifests itself by the occasional appearance of a dense smoke at the top of some conical hill—or along a line of country bounded by the horizon. The smoke from these hills and the crevices in the plastice clay is said to last at the same spot for a long time—say two or three years; indicating at them a large accumulation of combustible materials, but is not accompanied by the emmission of flames. It is evidently due to the decomposition, by the percolation of atmospheric waters to them of beds of pyrites, which reacting on the combustible materials such as lignites and other substances of vegetable nature in their vicinity, give rise to a spontaneous combustion.

All the high prairies abound with the Psoralea esculenta—which is the prairie turnip of the Americans, Pomme de prairie and Pomme Blanche of the French—

Buffalo were seen feeding in large detached herds scattered over the prairies like huge droves of cattle.

The flesh of a fat buffalo-cow is perhaps the best beef that can be eaten, it is at once juicy, tender, nutricious, and very digestible, added to which it has a game flavour which renders it far superior to the very best beef of the States. It may, in fact, be not improperly denominated "*game beef.*"

The hump, the toungue—tender loin, [****], and marrow bones are considered the choice parts of the carcass, the residue being left on the ground for the wolves—Stan.

The appearance of the water in the Platte is precisely that of Mississippi and Missouri of a muddy white, and its current is, like theirs, constantly boiling and eddying in restless turbulence—

Buffalo berries grow upon a shrub fifteen feet high.

From the mouth of the Platte river the forests are narrower. The principal trees are the American and red elm, the soft maple, Canadian poplar, white and red ash; the most common under growth, horse briar, fox and false grape, red root, gray dogwood, currant, and gooseberry, with shrubs and dense rushes along the banks of the river. The same trees and shrubs grow on the numerous islands that are generally bordered with black and long-leaved willows. In the higher situations, and at the heads of creeks are sweet with the black walnut and mulberry, basswood, nettle wood, intermingled with the common hawthorn, prickly ash &c. On the high grassy or rocky banks the black and bar oaks constitute the principal growth but occasionally intermixed with wild cherry, red cedar, hornbeam, wild roses, and sumach. The low prairies bordering the rivers have a deep fertile soil—and abound with sedge-grasses and leguminous plants. Nicollet.

Leaving Council Bluffs the hills on either side are observed to be at greater distance from the river, which is itself twice its preceding width—The valley is fully fifteen miles wide;—and the broad prairies that carpet it exhibit the same richness of soil and luxuriance of vegetation as below—The width of the river varies from one fifth of a mile to two miles. In its widest parts, the navigation is frequently impeded by sandbars and driftwood; but where it is narrow, the current flows in a straight onward direction.—N.

In many places on the upper Missouri the soil is composed of clay, in strata of different thickness—of different degrees of hardness—and varying in colour—It is of a character to be so acted by the weather as to exhibit a great variety of fanciful form summits—as domes, cupolas, towers, colonades imparting a singular and picturesque appearance.

Audubon's "George Catlin" Powder Horn from the Missouri River Expedition

A delightful footnote to the Missouri River expedition—and one offering a challenging mystery with an ironic twist as well—is the engraved powder horn made from a bison horn that Audubon brought home with him from his western travels in November 1843.

In addition to their journals and natural history collections, Audubon and his companions returned with large collections of Native American and other artifacts and souvenirs. Among Audubon's artifacts was an intricately and precisely constructed powder horn made from the horn of a bison bull.[1] (See figs. 34 38.) Several similarly constructed "St. Louis–style" horns are known to exist and are held in private and public collections.[2] What distinguishes the Audubon powder horn from all others except the fourth horn discussed here are the scenes of northern plains Indian life carefully engraved into the outer shell of the horn. In a personal communication, journeyman horner Rick Sheets explained that the "alternating light and

1 This horn has made only two public appearances. At the Charleston Museum, it was included in the exhibition *Audubon: The Charleston Connection*, September 8–November 17, 1985. Photographs of it appear in the published catalog, as does mention of the Audubon family's belief that Audubon engraved the horn himself (Sanders and Ripley 114–15). It was also included in the traveling exhibition *John James Audubon in the West: The Final Expedition*, June 23, 2000–September 30, 2001. To my knowledge, no other reference to this powder horn has appeared in print.

2 The Missouri History Museum holds two powder horns of the same style of construction as the Audubon horn, one of which is said to have belonged to William Clark, who began his appointment as superintendent of Indian affairs at St. Louis in 1822. Neither of these horns bears any engravings.

dark pie shapes in the butt are made up of white bison bone and flattened dark bison horn." While we do not know who made these powder horns, "the complex construction suggests they came from a commercial shop."

Audubon's powder horn has remained in the hands of family descendants ever since the patriarch's death in 1851, as has the tradition that Audubon did the engravings himself. Various circumstances, however, argue against this attribution, and the available evidence suggests that the most likely artist of the engravings is George Catlin, the painter of plains Indian scenes whose representations Audubon regularly disparaged as highly romanticized and unrealistic.

The first reason to doubt that Audubon did the engraving is that all the scenes depict Native American cultural ways: skinning a bison, dressing a bison skin, scalping an enemy, hunting bison and pronghorn, performing a buffalo dance, smoking a peace pipe. On this trip, Audubon's interests in Natives were opportunistic, not those of someone genuinely interested in Native culture. Yet the extreme degree of detail in the scenes suggests an intimate knowledge acquired over a long period of time as well as a desire to record and report the detail.

Another consideration is that at least three other bison horn powder horns exist that bear practically identical engravings. The first of these I found among Edward Harris's souvenirs from the 1843 expedition housed at the Alabama Department of Archives and History in Montgomery. The Harris powder horn is of much simpler and plainer construction, but the images are just as elaborately engraved as are those on the Audubon horn. (See figs. 39–42.) To believe that Audubon engraved all of these horns requires believing that he first acquired his intricately constructed powder horn without engravings, cut in the scenes, and then engraved identical scenes onto a plainer powder horn for his friend Harris and onto another, and much finer, horn for someone else.

However, the existence of another powder horn with nearly identical engravings on it is probably the most persuasive reason to believe that some-one other than Audubon engraved these horns. The third engraved horn was reported in a publication of the Wyoming Historical and Geological Society in 1930 by A. C. Parker and is referred to as the "Catlin powder horn." (See figs. 43–44.) Parker draws his evidence that Catlin engraved the horn from two sources: first, an oral tradition that Catlin presented the powder horn to "John Blacksmith (or Red Jacket) about 1825" when Catlin painted Red Jacket's portrait near Niagara Falls and that the horn then passed through several hands before being donated to the society (6); and, second, the clear similarities of style and content between the scenes depicted on the horn and many of Catlin's drawings.[3] (See fig. 45.)

The fourth St. Louis–style powder horn bearing practically identical "Catlin" engravings was acquired in St. Louis in the 1960s and is in a private collection. (See fig. 46.) As Rick Sheets wrote to me, "The powder horn is so similar to the one in the Audubon Collection in both form and embellishment that it has to come from the same engraver's hand and the same shop." Rick further explained, "The engraving cuts making up these designs are very simple—a small blade or possibly an awl or other pointed tool was used to lightly scratch the horn. Yet the art itself is very well executed without hesitation or wasted lines. Since professional engraving tools such as a burin were not used and the art is so consistent, these horns could have been engraved by a trained artist without knowledge of engraving techniques (or access to engraving tools)."

More evidence is needed before we can claim with certainty that Catlin devoted a good bit of time to engrave several horns on the prairies. Still, the Audubon powder horn and the others are rich emblems of the exchange of gifts, the collecting of artifacts, the intersecting of many lives that occurred at St. Louis and of the desire to bring evidence of western experiences "back East" in the first half of the nineteenth century.

3   A convenient edition of Catlin's work to consult with regard to similarities to the Catlin horn is that prepared by Michael Mooney; see especially pages 112, 122, 284, and 299.

APPENDIX

"THE PET BEAR," AN UNPUBLISHED EPISODE

Among the Audubon Family Papers, 1805–1938, held in Special Collections at the University of Kentucky Library is a five-page manuscript entitled "The Pet Bear." It appears to be a story that Audubon was told on one of his trips through Pennsylvania and that he subsequently wrote up as an "episode," probably for publication in a newspaper or magazine. Although Audubon's name does not appear on the manuscript and the hand seems to be that of someone other than Audubon, its inclusion in this collection of manuscripts and internal evidence suggest that "The Pet Bear" is a previously unpublished tale by Audubon. The narrator's tone is marked by Audubon's characteristic playful, wry humor, and the narrator's remark that the state library of Pennsylvania has not yet purchased a copy of Audubon and Bachman's *Quadrupeds of North America* points to Audubon as the likely author.

The framing narrative is set in Harrisburg, through which Audubon passed several times in his decades of travel.[1] He may have gathered the basic tale about the pet bear any time between his first trip from Louisville, Kentucky, to Philadelphia, around 1810, and his later interior travels in the mid-1820s. He regularly gathered and recorded anecdotes of birds and mammals as he traveled for possible use in various publications. The episode "Scipio and the Bear," for example,

---

1   His route from Henderson, Kentucky, to Philadelphia took him through Harrisburg, as he explains in the episode "A Wild Horse" (S 3:272).

first published in volume 1 of *Ornithological Biography* in 1831, became a large portion of the account of the habits of the black bear (*Ursus americanus*) in the octavo version of *The Quadrupeds of North America* in 1854 (3:187–97). He may have intended a similar use of this bear story, or he may have meant to publish "The Pet Bear," with its special recommendation of the *Quadrupeds,* in a newspaper or magazine to publicize the new work. Over the years, he regularly hired scribes to make copies of his manuscripts to use in publications.

The Pet Bear

It might have been about half Past 10. PM, on a cold snowy night in the middle of January, that I was comfortably seated at one Corner of the Capacious chimney of the bar-room at "Mine Inn," in the Capital of the Glorious and patriotic Key-Stone State—a most excellent fire of her own anthracite, sent a genial warmth to my heart as well as throughout the room, and thought of this blessing of nature to mankind, on the Comfort it carried to the firesides of so many.—but I am proving either pathetic or wiley unless—I was, I said, Comfortably seated; a pretty good Noriega between my lips, which last, I occasionally damped with a mixture of old Rye and water, that the bar keeper had placed in a glass on the mantel piece within my reach.

During the day I had attended the session of the legislature and made the acquaintance of some members thereof of high standing in the State. Judge Johnston (I will call him by that name) was one of these gentlemen—and a fine specimen he was of that character which formed the bodies of the *men* of some "60 years since"[2]—our early explorers, settlers, & pioneers. He had like most of them a "*physique*"—in plain terms, he had been, and still Continued to be, a man of great strength, and shewed by his deep-set eye and thoughtful brow, that the intellect was there to match the tall and well proportioned frame.

We had been talking for some time (and the Judge supplied with the same "Creature Comforts" as myself). There was no one but ourselves and the sleeping Bar Keeper in the room, and it was a pleasure to see the Kindling

2 This refers to the subtitle of Walter Scott's *Waverley; or, 'Tis Sixty Years Since* (1814).

eye & observe the half smile—as he detailed some of his "life scenes," in our friendly talk: One of his anecdotes you shall have.

You are acquainted, said the Judge, with the habits of the Common American bear, or if not, you can learn all about his habits & habitat, and see his portrait in Audubon and Bachman's fine work, the "Quadrupeds of North America"—Yes, said I, interrupting, they want a copy of that work in your state library yet, I believe; but please go on—

Well then, Continued the Judge, here goes for my Bear experience in the northwestern part of the state of New York many years ago—Yes, a devilish fine fun I had with one too, a pet bear, he was.

One fine morning in September, I had made an early start and had been riding for some hours through the dense woods, in a district almost beyond the settlements; only at intervals was the shade of the forest broken by {several} "Clearings," in which the "girdled" trees looked ghastly and black, having been schorched when the brush wood was burned off the ground. Some of the scattered log Cabins were abandoned, and the whole scene bordered on the desolate. The road was merely a trail with a few waggon tracks marking its course, and was well adapted for upsets & slow travelling generally—being stumpy, lumpy, and supplied with an assortment of mud holes of various sizes and dubious depths.

My horse being tired, and myself hungry, I determined to stop and take a breakfast at the first place where I could obtain one, and on getting to the top of a long red-clay hill, I found a log house that seemed to offer what I wanted. It was a one story building, clap-board roof, a mud daubed chimney adorned each gables end, [****atel****] the walls of the mansion which were "chinked up" with the same material to "keep the wind away."[3] There was a room on either side of the "Hall," or passage that led back to the yard, as is usually the case in the frontier houses.

Jumping off my horse, I looked into the passage, & saw the occupant just Comming out of a [****ie] with a large tin, or pewter platter full of broken

3 *Hamlet* V,i,214: "Imperious Caesar, dead and turn'd to clay, / Might stop a hole to keep the wind away."

victuals in his hand. I hailed him, and was informed I could feed my horse, and that he would set some breakfast presently for me. I led my horse round the house to a shed, tied him up to the rack, and gave him some fodder, and then going back to the house through the yard, found the man had a pet Bear, which was chained to a big hollow log that lay on the ground, and into the cavity of which Bruin could retreat as it suited him—for shelter or repose.

This Bear was finishing the Victuals his master had given him, and I stopped for a moment to look at him. He was of large size and quite playful. As soon as his repast was ended, the man held out his hands to him, and the animal raised himself on his hind legs, & put his fore-paws on his master's shoulders, he then licked his enormous lips, and carelessly shewed his teeth in indolent good humour.

The owner of this pretty pet now patted him on the cheeks and then pushed him backwards, on which the Bear fell upon his back with all the grace common to his race. His Keeper put his foot on the animal's belly, and rolled him about a little, while he thus lay with his legs uppermost. I went into the passage, and taking off my coat & cravat, began to wash my face and neck with the aid of the basin, brine-stone, water and hard soap.

While I was performing this refreshing duty, my host came in, and told me that as his wife had the "shakes" (or as some wag calls it, "ague in the wilderness") he would cook me some breakfast himself; and accordingly he entered the Kitchen.

I now put on my cravat quicker than ever did Beau Brummel[4]—jerked on my coat and went out in the yard to play with the pet bear, as I had seen the owner do, and going up to his Bearship, held out my hands in the most friendly manner, but to my horror the beast immediately seized my right leg, and with a sudden jerk I found myself on my back, and the Bear dragging me by the boot, so that each of us were going *stern foremost* rather fast towards the hollow log.

4   George Bryan Brummel (1778–1840), called "Beau," was the infamous friend of George, Prince of Wales, the future George IV, and a leader of fashion in London. As a prominent dandy, he was the target of much satire and ridicule, especially noted for his intricately folded cravat.

I Kicked with all my might, and would have halloed for help had I not been too much ashamed of my ridiculous position.

It was however no time for "Swapping Knives" so I struggled desperately, and by good luck my boot together with some part of the leg of my breeches came off, and finding myself loose, I rolled, shuffled, and scuffled backwards as fast as I could—the Vile Bear giving chase in the rear—but fortunately he was unable to catch me again, and was "brought up" as the sailors say by his chain before he could do me any further mischief. I now arose, examined into my Condition, and looked all around to see if my disgrace had been witnessed—I was comforted by thinking it had not, but my boot was still in possession of the enemy. Presently I saw a good stick of wood about 8 feet long in a corner of the yard and possessing myself of it, advanced on "Mon Cher Bear"—He made fight, Came fiercely to the limit his chain admitted, and exhibited considerable ability as a "boxer," but I soon made his head feel rather sick—he retreated into his hole, and I recovered my property—so that in our trade he got *no boot*.

Perceiving that all this time I had been unobserved I "smoothed down the wrinkled front of war,"[5] pulled on my boot—stuffed the remainder of the torn leg of my breeches into it to hide any traces of the "scrimmage," & went into the house determined to say "nothing to nobody" about the Bears in that region.

My host met me at the door, the bacon and corn bread were hot, and I enjoyed a hearty meal, paid my entertainer, got on my horse and rode off—with a fierce determination for the future to remember the Baron of Bradwardine's motto "Bewar the Bar."[6]

The Judge now gave me a wink—we tossed off the rest of our drink and retired to roost.

[*S*dle.*]

---

5   Compare *Richard III* I,i,9: "Grim-visaged war hath smooth'd his wrinkled front."

6   In Scott's *Waverley*, "Bewar the Bar" is the family motto of the Baron of Bradwardine, whose mansion and grounds are ornamented with carvings and sculptures of bears, several rampant like the bear in Audubon's tale.

WORKS CITED

Adams, Alexander B. *John James Audubon: A Biography*. New York: G. P. Putnam's Sons, 1966.

Andress, David. *The Terror: The Merciless War for Freedom in Revolutionary France*. New York: Farrar, Straus and Giroux, 2005.

Arthur, Stanley Clisby. *Audubon: An Intimate Life of the American Woodsman*. New Orleans: Harmanson, 1937.

Ashton, Leonora Sill. "Some Recollections about the Granddaughters of John James Audubon." *Audubon Magazine* 53.4 (1951): 244–47.

Audubon, John James. "The American Woodcock." MS. American Museum of Natural History. New York.

———. Archives. Stark Museum of Art, Orange TX.

———. *Audubon in the West*. Ed. John Francis McDermott. Norman: University of Oklahoma Press, 1965.

———. *The Audubon Reader*. Ed. Richard Rhodes. Everyman's Library 284. New York: Alfred A. Knopf, 2006.

———. *The Birds of America from Drawings Made in the United States and Its Territories*. 7 vols. New York: J. J. Audubon; Philadelphia: J. B. Chevalier, 1840–44. [Octavo edition.]

———. *The Birds of America, from Original Drawings by John James Audubon*. 4 vols. Colored plates, double elephant folio. London: Author, 1827–38.

———. Collection. Missouri History Museum Archives, St. Louis.

———. Collection. Princeton University Library, Department of Rare Books and Special Collections.

———. *John James Audubon's Journal of 1826: The Voyage to "The Birds of America."* Ed. Daniel Patterson. Lincoln: University of Nebraska Press, 2011.

———. *Journal of John James Audubon Made during His Trip to New Orleans in 1820–1821*. Ed. Howard Corning. Boston: The Club of Odd Volumes, 1929.

———. *Journal of John James Audubon Made While Obtaining Subscriptions to His "Birds of America," 1840–1843*. Ed. Howard Corning. Boston: The Club of Odd Volumes, 1929.

———. "Letter from Audubon to the Editor, St. Augustine, East Florida, Dec. 7, 1831." *Monthly American Journal of Geology and Natural Science* 1.8 (1832): 358–63.

———. "Letter from J. J. Audubon, to the Editor." [No. 2.] *Monthly American Journal of Geology and Natural Science* 1 (1832): 407–14.

———. *Letters of John James Audubon.* Ed. Howard Corning. 2 vols. Boston: The Club of Odd Volumes, 1930.

———. Manuscript journal. 1820–21. Ernst Mayr Library, Museum of Comparative Zoology Archives, Harvard University.

———. Manuscript journal: The original field notebook. July 26–28 and August 16–November 6, 1843. Private collection. [Reprinted here in part 3.]

———. Manuscript journal. "Journal No 3." August 5–13, 1843. Box 8, folder 446. Morris Tyler Family Collection of John James Audubon. Gen MS 85. Beinecke Rare Book and Manuscript Library, Yale University. [Reprinted here in part 3.]

———. Manuscript journal. August 16–October 19, 1843. "Copy of my Journal from Fort Union homeward." Everett D. Graff Collection of Western Americana, Vault Graff 109, Newberry Library, Chicago. [Reprinted here in part 3.]

———. *My Style of Drawing Birds.* Introduction and transcription by Michael Zinman. Ardsley NY: Overland Press for the Haydn Foundation, 1979. 15–19.

———. *Ornithological Biography.* 5 vols. Edinburgh: Adam and Charles Black, 1831–39.

———. Papers, 1821–45. American Philosophical Society, Philadelphia.

Audubon, John James, and John Bachman. "Descriptions of New Species of Quadrupeds Inhabiting North America." *Proceedings of the Academy of Natural Sciences* 1.7 (October 1841): 91–104.

———. *The Quadrupeds of North America.* 3 vols. New York: Victor Gifford Audubon, 1854. [Royal Octavo edition.]

———. *The Viviparous Quadrupeds of North America.* 3 vols. New York: John James Audubon and Victor Gifford Audubon, 1846–1854.

Audubon, Lucy, ed. *The Life of John James Audubon, the Naturalist.* New York: G. P. Putnam and Son, 1869.

———. Manuscript letters. Box 1, folder 13-35. Morris Tyler Family Collection of John James Audubon. Gen MS 85. Beinecke Rare Book and Manuscript Library, Yale University.

Audubon, Maria Rebecca, ed. *Audubon and His Journals.* 2 vols. New York: Charles Scribner's Sons, 1897.

———. Letter to Frank Chapman. March 27, 1901. Archives, Library, American Museum of Natural History, New York.

———. Letter to Ruthven Deane. July 29, 1876. Archives, Missouri History Museum, St. Louis.

"Audubon's Expedition to California and the Rocky Mountains, &c." *Monthly American Journal of Geology and Natural Science* 1 (November 1831): 229.

Audubon, Victor. Manuscript letters. Box 2, folder 50. Morris Tyler Family Collection of John James Audubon. Gen MS 85. Beinecke Rare Book and Manuscript Library, Yale University.

Bachman, John. *John Bachman: Selected Writings on Science, Race, and Religion.* Ed. Gene Waddell. Athens: University of Georgia Press, 2011.

Baird, Spencer Fullerton. Manuscript letters. Box 3, folder 95. Morris Tyler Family Collection of John James Audubon. Gen MS 85. Beinecke Rare Book and Manuscript Library, Yale University.

Bakewell, William. Manuscript letters. Box 2, folder 64-66. Morris Tyler Family Collection of John James Audubon. Gen MS 85. Beinecke Rare Book and Manuscript Library, Yale University.

Barrow, Mark, Jr. *A Passion for Birds: American Ornithology after Audubon*. Princeton NJ: Princeton University Press, 1998.

Bell, John G. Manuscript journal. "Diaries of an Expedition with John James Audubon, 1843 Mar 11–Dec 31." 3 vols. Beinecke Rare Book and Manuscript Library, Yale University. [Reprinted here in part 5.]

Berkeley, Edmund, and Dorothy Smith Berkeley. *George William Featherstonhaugh: The First U.S. Government Geologist*. Tuscaloosa: University of Alabama Press, 1988.

"Biographical Sketch of John James Audubon." *The Animal Kingdom, Arranged According to Its Organization, Serving as a Foundation for the Natural History of Animals, and an Introduction to Comparative Anatomy*. By Georges Cuvier, baron. 4 vols. London: G. Henderson, 1837. [Second title page shows 1834.] 1:197–204.

Brewer, Thomas M. "Reminiscences of John James Audubon." *Harper's New Monthly Magazine* 61. (October 1880): 665–75.

Buchanan, Robert, ed. *The Life and Adventures of John James Audubon, the Naturalist*. London: Sampson Low, Son, and Marston, 1868.

Burroughs, John. *John James Audubon*. Woodstock NY: Overlook Press, 1987; Boston: Small, Maynard, 1902.

Butterworth, Hezekiah. *In the Days of Audubon: A Tale of the "Protector of Birds."* New York: D. Appleton, 1901.

Catlin, George. *Letters and Notes on the Manners, Customs, and Conditions of the North American Indians*. 2 vols. New York: Wiley and Putnam, 1841.

———. *Letters and Notes on the North American Indians*. Ed. Michael MacDonald Mooney. New York: Gramercy Books, 1975.

Chalmers, John. *Audubon in Edinburgh*. Edinburgh: NMS, 2003.

Chancellor, John. *Audubon: A Biography*. New York: Viking, 1978.

Chapman, Frank M. "Notes and News." *Auk* 7.1 (1890): 98–99.

Chardon, Francis A. *Chardon's Journal at Fort Clark 1834–1839*. Ed. Annie Heloise Abel. 1932. Lincoln: University of Nebraska Press, 1997.

"Charles Wilkins Webber." *Universal Cyclopædia & Atlas*. Ed. Rossiter Johnson. New York: D. Appleton, 1902.

Chittenden, Hiram Martin. *The American Fur Trade of the Far West*. 3 vols. New York: Francis P. Harper, 1902. 2 vols. Stanford CA: Academic Reprints, 1954.

———. *History of Early Steamboat Navigation on the Missouri River: Life and Adventures of Joseph LaBarge*. 2 vols. New York: Francis P. Harper, 1903.

Coues, Elliott. "Dr. Coues' Column." *Osprey* 1.10 (June 1897): 150.

———. "Dr. Coues' Column." *Osprey* 1.11–12 (July–August 1897): 135.

Cutright, Paul Russell, and Michael J. Broadhead. *Elliott Coues: Naturalist and Frontier Historian*. Urbana: University of Illinois Press, 1981.

Dall, William Healey. *Spencer Fullerton Baird: A Biography*. Philadelphia: J. B. Lippincott, 1915.

Deane, Ruthven, ed. "Unpublished Letters of Introduction Carried by John James Audubon on His Missouri River Expedition." *Auk* 25 (1908): 170–73.

———. "Unpublished Letters of John James Audubon and Spencer F. Baird." *Auk* 21 (1904): 255–59.

Drake, Daniel. *Discourse on the History, Character, and Prospects of the West.* Ed. Perry Miller. 1834. Gainesville FL: Scholars' Facsimiles and Reprints, 1955.

Durant, Mary, and Michael Harwood. *On the Road with John James Audubon.* New York: Dodd, Mead, 1980.

Ford, Alice, ed. *Audubon, by Himself: A Profile of John James Audubon, from Writings Selected, Arranged, and Edited.* New York: Natural History Press, 1969.

———, ed. *Audubon's Animals: The Quadrupeds of North America.* New York: Studio Publications, in Association with Thomas Y. Crowell, 1951.

———, ed. *The Bird Biographies of John James Audubon.* New York: Macmillan, 1957.

———. *John James Audubon: A Biography.* Norman: University of Oklahoma Press, 1964; New York: Abbeville Press, 1988.

Frémont, John Charles. *Report of the Exploring Expedition to the Rocky Mountains in the Year 1842, and to Oregon and North California in the Years 1843–'44.* Washington DC: Gales and Seaton, Printers, 1845.

Fuller, Errol. *The Great Auk.* New York: Harry N. Abrams, 1999.

Goetzmann, William H. *Exploration and Empire: The Explorer and the Scientist in the Winning of the American West.* 1966. Francis Parkman Prize Edition. New York: History Book Club, 2006.

Grieve, Symington. *The Great Auk, or Garefowl (Alca impennis, Linn.): Its History, Archaeology, and Remains.* London, 1885. Landisville PA: Coachwhip, 2007.

Grinnell, George Bird. "Memoirs" [1915]. George Bird Grinnell Papers. Manuscripts and Archives, Yale University Library.

———. "Recollections of Audubon Park." *Auk* 37.3 (July 1920): 372–80.

Hafen, LeRoy R. "Etienne Provost." *Fur Trappers and Traders of the Far Southwest.* 1968. Ed. LeRoy R. Hafen. Logan: Utah State University Press, 1997. 79–94.

Harris, Edward. *Up the Missouri with Audubon: The Journal of Edward Harris.* Ed. John Francis McDermott. Norman: University of Oklahoma Press, 1951.

Hart-Davis, Duff. *Audubon's Elephant: America's Greatest Naturalist and the Making of "The Birds of America."* New York: Henry Holt, 2004.

Harwood, Michael. "Mr. Audubon's Last Hurrah." *Audubon: The Magazine of the National Audubon Society* 87.6 (November 1985): 80–117.

Herrick, Francis Hobart. *Audubon the Naturalist: A History of His Life and Time.* 2 vols. New York: D. Appleton, 1917.

"Interesting Letter from Audubon." *New York Herald* August 3, 1843, col. D.

Irving, Washington. *A Tour on the Prairies.* Ed. John Francis McDermott. Norman: University of Oklahoma Press, 1956.

Isenberg, Andrew C. *The Destruction of the Bison: An Environmental History, 1750–1920.* Cambridge: Cambridge University Press, 2000.

Judd, Richard W. *The Untilled Garden: Natural History and the Spirit of Conservation in America, 1740–1840*. New York: Cambridge University Press, 2009.

Knott, John R. *Imagining Wild America*. Ann Arbor: University of Michigan Press, 2002.

Leopold, Aldo. *Game Management*. New York: Charles Scribner's Sons, 1933.

Lott, Dale F. *American Bison: A Natural History*. Berkeley: University of California Press, 2002.

Lownes, Albert E. "Notes on Audubon's Ornithological Biography." *Auk* 52.1 (January 1935): 103–4.

Marsh, George Perkins. *Man and Nature; or, Physical Geography as Modified by Human Action*. New York: Charles Scribner, 1864.

McCullough, David. *Mornings on Horseback*. New York: Simon and Schuster, 1981.

McDermott, John Francis, ed. "Editor's Introduction." *Up the Missouri with Audubon: The Journal of Edward Harris*. Norman: University of Oklahoma Press, 1951. 1–41.

Mitchell, J. K. *Indecision, a Tale of the Far West; and Other Poems*. Philadelphia: E. L. Carey & A. Hart, 1839.

Morton, Samuel George. *Crania Americana; or, A Comparative View of the Skulls of Various Aboriginal Nations of North and South America*. Philadelphia: J. Dobson, 1839.

Mossner, Ernest Campbell. "Deism." *The Encyclopedia of Philosophy*. Ed. Paul Edwards. 8 vols. New York: Macmillan, 1967. 2:326–36.

Moulton, Gary E., ed. *The Definitive Journals of Lewis & Clark*. 8 vols. Lincoln: University of Nebraska Press, 1983–2001.

"Mr. Audubon." *North American and Daily Advertiser* (Philadelphia PA) August 31, 1843; col. D.

Nash, Roderick Frazier. *The Rights of Nature: A History of Environmental Ethics*. Madison: University of Wisconsin Press, 1989.

Nuttall, Thomas. *A Manual of the Ornithology of the United States and Canada: Water Birds*. Boston: Hillard, Gray, 1834.

Palmer, T. S. *Legislation for the Protection of Birds Other Than Game Birds*. Washington DC: Government Printing Office, 1902.

Parker, A. C. "The Catlin Powder Horn." *Indian-Loving Catlin and His Buffalo Powder Horn*. Wilkes-Barre PA: Wyoming Historical and Geological Society, 1930. 5–9.

Peattie, Donald Culross, ed. *Audubon's America: The Narratives and Experiences of John James Audubon*. Boston: Houghton Mifflin, 1940.

———. *Singing in the Wilderness: A Salute to John James Audubon*. New York: G. P. Putnam's Sons, 1935.

Peck, Robert McCracken. "Audubon and Bachman: A Collaboration in Science." *John James Audubon in the West: The Last Expedition*. Ed. Sarah Boehme. New York: Harry N. Abrams, in association with the Buffalo Bill Historical Center, 2000. 71–115.

Randolph, Emanuel D. "Isaac Sprague, 'Delineator and Naturalist.'" *Journal of the History of Biology* 32.1 (Spring 1990): 91–126.

Rathbone, Hannah Mary. Letter to Theodore Woolman Rathbone. August 20, 1826. Collection of Hannah Royle, Edinburgh.

Reiger, John. *Escaping into Nature: The Making of a Sportsman Conservationist and Environmental Historian*. Corvallis: Oregon State University Press, 2013.

————. *The Passing of the Great West: Selected Papers of George Bird Grinnell*. New York: Charles Scribner's Sons, 1972.

Rhodes, Richard. *John James Audubon: The Making of an American*. New York: Knopf, 2004.

Sage, John H., ed. "Description of Audubon." *Auk* 34 (1917): 239–40.

Sanders, Albert E., and Warren Ripley, eds. *Audubon: The Charleston Connection*. Charleston: Charleston Museum, 1986. Exhibition catalog.

Shufeldt, Robert W. *On a Case of Female Impotency*. Washington DC, 1896.

Sire, Joseph A. "Journal of a Steamboat Voyage from St. Louis to Fort Union." Appendix H. Ed. and trans. Hiram Martin Chittenden. *The American Fur Trade of the Far West*. By Hiram Martin Chittenden. New York: Francis P. Harper, 1902. 3:984–1003.

Souder, William. *Under a Wild Sky: John James Audubon and the Making of "The Birds of America."* New York: North Point Press, 2004.

Sprague, Isaac. Manuscript diary. "Diary, 1843." Boston Athenaeum. [Reprinted here in part 5.]

Street, Phillips B. "The Edward Harris Collection of Birds." *Wilson Bulletin* 60.3 (September 1948): 167–84.

Streshinsky, Shirley. *Audubon: Life and Art in the American Wilderness*. New York: Villard, 1993.

Taylor, Alan. *The Civil War of 1812: American Citizens, British Subjects, Irish Rebels, & Indian Allies*. New York: Vintage Books, 2010.

Tyler, Alice Jaynes. *I Who Should Command All*. New Haven CT: Framamat, 1937.

Tyler, Ron. "The Publication of *The Viviparous Quadrupeds of North America*." *John James Audubon in the West: The Last Expedition*. Ed. Sarah Boehme. New York: Henry N. Abrams, 2000. 119–82.

Viola, Herman J., and Carolyn Margolis, eds. *Magnificent Voyagers: The U.S. Exploring Expedition, 1838–1842*. Washington DC: Smithsonian Institution Press, 1985.

Webber, Charles Wilkins. *The Hunter-Naturalist*. Philadelphia: J. W. Bradley, 1851.

Weidensaul, Scott. *Of a Feather: A Brief History of American Birding*. New York: Harcourt, 2008.

Winterfield, Charles. "About Birds and Audubon." *American Whig Review* 1.6 (April 1845): 371–83.

————. "American Ornithology." *American Whig Review* 1.3 (March 1845): 262–74.

————. "A Talk about Birds and Audubon." *American Whig Review* 2.1 (September 1845): 279–87.

INDEX

Italicized references to Fig. 1, Fig. 2, and so forth refer to illustrations in the gallery.
Page numbers in bold refer to journal or diary entries.

Academy of Natural Sciences (Philadelphia), 32

Academy of Natural Sciences (St. Louis), 59

Adams, Alexander B., 222–24

Adams, Charles Coffin, 5

Alabama Department of Archives and History, 59, 424

Alice Ford Papers (Audubon Center at Mill Grove), xv

alligator, 213, 274

American Fur Company, 36, 55, 60, 61, 388

American Ornithologists' Union, 7, 10

American Whig Review, 129

Arthur, Stanley Clisby, 219–20

Ashton, Leonora Sill, 8

Audubon, Caroline Hall (daughter-in-law), 6, 36

Audubon, Delia, 6

Audubon, Florence (granddaughter), 6, 9–10, *fig. 5*

Audubon, Georgiana Mallory (daughter-in-law), 45, 130

Audubon, Harriette Bachman (granddaughter), 6, 8

Audubon, Jean (father), 223, 235, 270

Audubon, John James: and affinity for wilderness, 66; argues for the ethical considerability of birds (see *Ornithological Biography*); aware of shaping his public image, 72, 253, 256; belief in great auk's continuing existence, 15; business failure of (1819), 51, 236; "Catlin" powder horn of, 423–25; and conservation, 238–46, 248–52, 257, 259–62, 264–65; desire for a western expedition, 31–38; desiring to write autobiography, 262–63; and difficulty with Sprague, 118–20; fabricates a hunting story, 92–96; hoping to encourage readers to nature study, 262–63; hunting ethic of, as represented in biographies, 211–33; as ideal of hunter-naturalist, 127–29; inability to foresee bison extinction, 14, 19–22; interview with, in St. Louis, 60–61; Labrador expedition of (1833), 215, 257–62; Mississippi River journey of (1820–21), 236–40; Missouri River expedition of (1843), 25–130, 264, 301, 302–3; multiple proposed projects of, 262–63; occasionally restraining his gunning, 238–40; and practical

Audubon, John James (*continued*)
necessity to kill birds, 240, 246–48,
252–56, 257–58, 264–65; *Quadruped*
plates by, *figs. 10, 13, 14, 17, 23*; said to
have warned against overhunting, 223,
229–30, 233; sketches by, *figs. 19, 21, 31–*
33; as sportsman, xvii, 29, 105, 213, 218,
219, 222, 231, 231–32, 242–43, 255, 256–
57, 283–84, 303–4; Western Museum
employment of, 236; as witness to
avian abundance, 237–38, 257–59
—journals, 3, 4–11, 80–81, 81n10, 90, 95;
journal of 1820–21, 237–40; journal
of 1826, 240–46; Labrador journal,
257–61
—manuscripts: "American Woodcock,"
275–77; Beinecke partial copy, xvi,
12–17, *figs. 26, 27*; Beinecke partial
copy, edited text of, 133–58; "Buffalo,"
21–22; Newberry partial copy, xv, *fig.*
30; Newberry partial copy, edited text
of, 159–208; original field notebook,
xvi, 5, 103, *figs. 19–20, 28–29, 31–33*;
original field notebook, edited text of,
159–208; "The Pet Bear," 427–31
—works: *The Birds of America*, 23, 33,
214, 225, 231, 243, 246, 248, 263, 265,
268, 282; *The Birds of America* (Octavo
edition), 34, 36, 89, 212, 267, 271, 274;
The Viviparous Quadrupeds of North
America, xvi, 29, 34, 34n1, 60, 81n10,
90, 94–95, 230n2, 264, 427, 429, *figs.*
10, 12–14, 17, 23. See also *Ornithological*
Biography (Audubon)
Audubon, John Woodhouse (son), 5,
34n1, 37, 38, 110, 130, 135, 258; paintings
by, *figs. 2, 25*; *Quadruped* plate by, *fig. 12*
Audubon, Lucy Bakewell (wife): allow-
ing George Grinnell to read Missouri
River journals, 4; assisting and
supporting Audubon, 33, 252–53, 256,
268, 282; and Audubon's journals, 4–5,
81n10; Audubon Society named for,
302; caring for family in Audubon's
absence, 50, 110, 236, 244, 253, 256;
concerns for Audubon's safety during
expedition, 50–51; and deaths of two
daughters, 49, 235–36, 237; diaries
of, 6n3, 10; early concerns about
Audubon's proposed expedition,
33–34; early difficulties endured with
Audubon, 235–37, 246–47; *The Life of*
John James Audubon, the Naturalist, 5;
quoted, 257–61; regard of, for Edward
Harris, 40, 50; as teacher of Grinnell, 4n2; uncomfortable conditions
endured in brother's home, 255–56;
vowing never to discuss politics, 242;
wedding of (1808), 31
Audubon, Maria Rebecca (granddaughter), xv, xvi, 5–11, 303–4, *fig. 5*;
Audubon and His Journals, xv, xvi, 6–11,
12–20, 303; destruction of Audubon's
journals by, 4, 6–8, 10–11; editing of
Audubon's journals by, xv, 4, 6–11,
22–23; and the "Great Auk Speech,"
12, 15–17, 19–23 (*see also* "Great Auk
Speech"). *See also* Audubon, John
James: journals
Audubon, Maria Rebecca Bachman
(daughter-in-law): death of, 35
Audubon, Mary Eliza (granddaughter), 6
Audubon, Mary Eliza Bachman
(daughter-in-law): death of, 35
Audubon, Rosa (daughter), 236
Audubon, Victor Gifford (son), 5, 45,
50, 81, 241, 247, 252; painting by, *fig. 2*,
publicizing Audubon's travels, 122
Audubon Society, 213, 302
Aumack, Jacob (captain), 237

Bachman, John: as coauthor of *Quad-*
rupeds, 21, 29, 35, 38, 114, 230n2, 427,
429; criticizing the Missouri River
expedition, 81n10, 90, 302; as father
of Audubon's daughters-in-law, 35; as

friend of Audubon, 253, 259, 272, 302; as naturalist advising Audubon, 43–44

badger, 115, 129, 156–57, 194, 320

Baird, Spencer Fullerton, 20, 37n4, 43, 85, 86n11; recruited for Missouri River expedition, 38–40

Bakewell, Lucy (daughter), 235

Bakewell, Lucy Green (mother-in-law), 235

Bakewell, Thomas Woodhouse (brother-in-law), 235

Bakewell, William Gifford (brother-in-law), 38, 49, 50, 252

Bayou Sarah, Louisiana, 247

bear, grizzly: abundance of, 81, 164–65, 172, 355, 356, 358; killed, 118–19, 166–67, 357, 405, 406; observed, 344–45, 355; stories of, 111, 138–39, 140, 168

beaver, 55, 161, 173, 177–81, 303, 359, 408, fig. 17

Beinecke Library (Yale University), xvi

Bell, John G., 40; and Bell's vireo, 66–67, 96, 313, 381; bison-killing reports of, 103, 105–7; bison-population reports of, 75, 107–9, 110; competitive spirit of, 116, 169; diary of (copied into Audubon's journal), 145–47; diary of (quoted), 19–20, 61–62, 105–9; 1843 diary of, 307–74; as naturalist, 71; as prairie hunter, 18, 27, 79, 84, 88, 97–100, 101, 111, 114; saving life of Edward Harris, 93–96; as skinner, 40–41, 59, 60, 66, 85, 167; as taxidermist, 41–42, 105

Bell's vireo. See Bell, John G.

Berthoud, Eliza Bakewell (sister-in-law), 51, 56, 81n10, 252

Berthoud, Nicholas (brother-in-law), 51, 55, 59, 126, 252

bighorn sheep, 12, 79, 81, 84, 111, 112, 144, 155–56, 303, 392. See also Mauvaise Terre (badland)

"Bird of Washington," 102, 102n16

bison, American, fig. 14; accounts of killing of, 22, 70, 100, 103, 104–9, 112–13, 117–18, 119; Audubon's field sketches of, figs. 32–33; horseback hunting of, 27–28, 88, 92–95, 96–99, 112–13; large numbers of, 18, 19, 20, 21–22, 73, 75, 79, 103, 107–8, 109–10, 112, 113, 233; natural causes of death of, 13, 69; near extinction of, 16, 23, 213, 218, 219; pastoral views of, 69, 104–5, 107–8; recovery of populations of, 219

Bodmer, Karl, 45

Bonaparte, Napoleon. See Napoleon Bonaparte

booby gannet, 293

Bowen, Samuel, 235–36

Brewer, Thomas Mayo, 37

Buchanan, Robert, 5

buffalo. See bison, American

Buffon, Georges-Louis Leclerc, Comte de, 61

Bunker Hill, Illinois, 58, 379

Burgwin, John, 123, 416

Burroughs, John, 214–15

Butterworth, Hezekiah, 214

caracara eagle, 214

Carleton, Lieutenant James Henry, 123–24, 200, 201, 202, 369

Carolina parakeets, 121, 213, 215, 238, 268, 381

Carson, Rachel, 220

Catlin, George, 19, 45, 78, 111, 111n19, 143, 361, 424–25

Chancellor, John, 224–25

Chapman, Frank, 8

Chardon, Francis A., 76–77, 78n9, 86, 122, 170, 317, 322, 388, 394

Charleston, South Carolina, 35, 36, 43, 252, 253, 257, 272

Charleston Museum, 423n1

Chittenden, Hiram, 76

Chouteau, Pierre, Jr., 36–37, 55

Cincinnati, Ohio, 31, 49, 236, 244n5
Civilian Conservation Corps, 217
clapper rail, 261, 285–86
Clark, William, 423n2; original journal of, given to Audubon, 59. *See also* Corps of Discovery (Lewis and Clark expedition); Mitchell, D. D.
Clean Air Act, 222
Comstock, Mary Louise, 6
Cooper, James Fenimore, 249, 251, 300
Cooper, Susan Fenimore, 117, 246, 300
Corning, Howard, 216–17
Corps of Discovery (Lewis and Clark expedition), 31, 44, 263, 299, 300
cottontail rabbit, 35
Coues, Elliott, 7, 9, 10, 10n7, 16, 134n1
coyote (prairie wolf), 145, 168, 198
crow, 292
Culbertson, Alexander, 18, 19, 27, 80, 83, 84, 88, 90–91, 92, 97–99, 101, 112, 115, 115n20, 320, 325, 331, 334–36, 337–38, 396, 398
Culbertson, Natawista (Mrs. Culbertson), 90–91, 115, 121
Cumberland, Maryland, 47
Cummings, Samuel (captain), 130

Deane, Ruthven, 8, 9
deer: black-tailed (mule deer), 66, 69, 73, 74, 115, 129, 173, 187, 190, 203, *fig. 12*; white-tailed, 146, 352
Drake, Daniel, 31, 236, 244n5
Durant, Mary, 23n11, 225–27

eagles, bald, 102n16, 109, 134, 145, 188, 349; Audubon's representation of, 291; Audubon's shooting of, 256
eagles, golden: Audubon's account of killing of, 212, 214, 230, 232, 273
Edinburgh, 216, 227, 246, 252, 268, 274, 282, 301
elk: hunting of, 123, 180–81, 318, 345, 347–50; observations of, 79, 144–45, 163, 178, 250, 353

Émile; or, Treatise on Education (Rousseau), 290

Featherstonhaugh, George, 253, 254, 255
flicker, northern, 89, 89n12
Florida Keys, 222, 226, 269, 283
Floyd's Bluff, 383
Ford, Alice, 103n17, 221–22
Fort Chardon, 77
Fort Clark, 75, 77–78, 169, 318, 357–58, 390–92
Fort Colville, 36, 37
Fort Croghan, 123, 199, 200, 415–16
Fort George, 75, 187, 316–17
Fort John, 115n20, 122, 361
Fort Leavenworth, 68, 124, 202, 369, 381, 417
Fort McKenzie, 111
Fort Pierre, 69, 75, 121–22, 317, 360–61, 388, 410, 411
Fort Union, 77, 83–114, 232, 319, 393, 402–3, *fig. 16*
Fort Vermillion, 123, 194
fox, swift, 122, 129, 133, 170, 339, *fig. 23*
Frémont, John C., 57, 64

Gallant, 49, 51, 52
Gannet Rocks (Labrador), 258
Glasgow House, 55, 126
golden plovers, 121, 188–89, 217, 238, 310, 361
gopher ("pouched rat"), 57, 59, *fig. 10*
Gray, Asa, 43
great auk, 12, 14, 15–16, 213
"Great Auk Speech," 13, 15, 16–17, 21, 23, 216, 219, 222, 223, 228, 230, 301
Great Bend. *See* Missouri River
Great Fire of 1835 (New York), 3–4
Great Depression, 217–18
Grieve, Symington, 15–16, 213
Grinnell, George Bird, 4, 4n2, 8, 213, 302
Griswold, Rufus, 56n6
guillemots, 259, 273–74

gulls: great black-backed, 225; laughing, 279–80, 295

Hall, Basil, 292, 294
hare, Townsend's, 81, 387
hare, wormwood, 89–90
Harlan, Richard, 3, 35, 262, 263
Harris, Edward: as Audubon's friend, 12, 37, 40–41, *fig. 6*; and Harris's finch, 66, 201, 313, 381; journals of (quoted), 17–19, 47, 61, 67, 70, 73–74, 76–77, 83–84, 88, 89–90, 93–94, 97, 100, 101; and lament about hunting, 18, 100, 227, 264; life saved by John Bell, 93–94, 332; as naturalist, 67, 70–71, 89–90; as prairie hunter, 18, 27, 78–79, 84–85, 88, 91–92, 97–99, 101, 111, 112–13, 114, 334–36; and souvenirs from 1843 expedition, 424, *figs. 18, 39–42*
Harris's finch. *See* Harris, Edward
Hart-Davis, Duff, 230
Harwood, Michael, 23n11, 225–28, 229, 303
Havell, Robert, Jr., 246, 252, 257
Henderson, Kentucky, 31, 51, 235, 236, 243, 249, 284
Herrick, Francis Hobart, 8, 215–16
Hingham, Massachusetts, 42, 76, 118
hoax letter ("Ke-ko-ka-ki, or jumper"), 101–3, 103n17
Holman, John P., 4, 4n2
Humboldt, Alexander von, 39
hunting, American attitudes toward, 12

ibis: white, 239–40, 274; wood, 293
Independence, Missouri, 64
Irving, Washington, 45, 100n15, 117, 118, 251

James, Edwin, 44
Jessup, Augustus, 32

Kirkland, Caroline, 300

Labarge, Joseph, 63, 79–80
Labrador. *See* Audubon, John James: Labrador expedition of (1833)
Laidlaw, William, 69–70
Laurel Hill, 48, 377
Lehman, George, 252, 253–54
Leopold, Aldo, 217
Liverpool, 230, 240, 241n4, 242, 268, 300
London Fur Company, 37
Long, Stephen H. (major), 31
Louisville, Kentucky, 31, 378; Audubon's 1843 visits to, 49–50, 51, 56, 127; Audubon's first residence in, 31, 235, 236; Lucy's 1831 residence with brother, 33, 252, 255–56; Shippingport canal lock near, 51, 249

MacGillivray, William, 268, 282, 282n8, 301
Madigan, Patrick Francis, 11, 11n8
Mandan village, 77–78, 119, 169, 318, 357–58, 390–92
Marsh, George Perkins, 246, 300
Mason, Joseph, 229, 236
Mauvaise Terre (badland), 112, 151–55
McCullough, David, 96
McCullough, Thomas, 34
McKenzie, Owen, 17, 19, 87, 90–91, 92, 104, 105–9, 112, 134, 141, 145, 323–24, 330–32, 343–51
meadowlark, 71
Michaux, Jean Baptiste, 73, 140, 194, 196, 200
Miller, Alfred Jacob, 45, 57
Minnie's Land, 5, 34, 35, 38, 72, 76, 110, 124, 130, 223
Mississippi kite, 239–40, 283, 290
Missouri River, *fig. 11*; difficulties of navigation, 63, 65, 71–72, 77, 120–21; Great Bend of, 66, 69, 73–75, 122, 173, 188, 316
Mitchell, D. D., 59. *See also* Clark, William

Mitchell, J. K.: poem about Audubon, 56, 56n6, 112, 147–48

Monthly American Journal of Geology and Natural Science (Philadelphia). *See* Featherstonhaugh, George

Napoleon Bonaparte, 243

Nash, Roderick, 245

National Audubon Society, 224

Native Americans: accounts of conflict with, 71–72, 76–77, 111; Audubon's defense of, 259–60, 290; derogatory views of, 86–88; Four Bears, 169, 318, 357, 387–88; Iron Bear, 169, 318, 392; list of "Indian Nations" traveled through, 208; romanticized views of, 76, 111n19; skulls of, collected by Audubon, 86–89; White Cow, 87–88. *See also* Mandan village

Nautilus, 127

Newberry Library (Chicago), xv

New Haven, 122, 192

New Orleans, 32

Nicollet, Joseph Nicolas, 38, 57

Nuttall, Thomas, 45, 183, 184

Omega, 55, 57, 61–64, 66–68, 69, 75, 78–79

opossum, 220

Ord, George, 103, 103n17

Ornithological Biography (Audubon): xvi, xvii, 211, 252, 263, 267–68; as argument for the ethical considerability of birds, 290–97; Audubon's persona in, 268–90

Ornithological Biography, essays cited: "American Avoset," 282; "The American Crow," 271–72, 292; "American White Pelican," 287–89; "American Woodcock," 275–77; "Anhinga or Snake-Bird," 281; "The Arctic Tern," 286–87; "The Black Guillemot," 273–74; "Black-Headed, or Laughing Gull," 278–80; "The Canada Goose," 221,

283–84; "The Clapper Rail," 285–86; "The Eggers of Labrador," 261–62; "The Florida Cormorant," 287; "The Golden Eagle," 273; "Knot or Ash-Coloured Sandpiper," 280–81; "The Mallard," 293–94; "The Mississippi Kite," 283; "The Ohio," 248–52, 261, 268; "The Opossum," 220; "The Pied-billed Dobchick," 277–78; "The Raven," 270; "The Roseate Tern," 277; "Scipio and the Bear," 427–28; "Surf Duck," 281–82; "The White-Crowned Sparrow," 271; "The White Ibis," 274; "The Whooping Crane," 274–75; "Winter Wren," 289–90; "The Zenaida Dove," 272

Parker, A. C., 425

passenger pigeon, 121, 167, 178, 213, 215, 324, 359

Peale, Charles Willson, 32

Peale, Titian, 32

Peattie, Donald Culross, 217–19

pelicans: brown, 224, 255, 295; white, 193, 198, 203, 287–88, 363, 365, 370, 413

Philadelphia, 40, 47, 240, 243, 252, 278, 376, 427

Philadelphia Saturday Courier, 56, 112

Pickering, Charles, 36–37

Picotte, Honoré, 75, 317, 388, 389

Pike, 49

pintail duck, 121, 175, 190

Poinsett, Joel R., 263

porcupine, 109, 134, 145, 350

prairie dog, 73, 167, 316, 357, 360, 405–6, *fig. 13*

pronghorn (antelope), 84, 135, 144, 160, 303, 320, 334; Sprague drawing head of, 321

Provost, Etienne, 57, 83, 109, 115, 125–26

puffin, 221, 257, 258, 284n9

purple grackle, 291

Rathbone family (Liverpool), 230

rattlesnake, 411

raven, 67, 296, 363

razor-billed auk, 296

red-headed woodpecker, 291

Reiger, John, 4n2

Reign of Terror (French Revolution),
    243, 286, 287n10

Rhodes, Richard, 231–33

Ripley, 257

Rogers, Samuel: "The Pleasures of
    Memory," 120

Roosevelt, Theodore, 42, 96

Rozier, Ferdinand, 235

San Domingo, 223

Say, Thomas, 31

Scott, Walter, 227, 244, 248

sentimental deism, 290

Seymour, Samuel, 32

Sheets, Rick, 423–24, 425

Shippingport. *See* Louisville, Kentucky

Shufeldt, Robert W., 9–10

Sire, Joseph A. (captain), 55, 57, 61, 63–64,
    68, 71–72, 75, 77, 79, 80

Smithsonian Institution (Washington
    DC), 39–40

Souder, David, 231

Spady, Matthew, 7, 9n6

Spark, 256

Sprague, Isaac, 12, 13, 40, 42   43, 58, 68, 75,
    84–85, 95, 103–4, 113–14, 42n5; diary of
    (quoted), 48, 58, 64, 69, 76–78, 117–18,
    122–23, 124–25; and difficulty with
    Audubon, 110, 117–20; 1843 diary of,
    375–422; and his portrait of Audubon,
    47, *fig. 8*; and his sketch of a camp, 122,
    fig. 24; as a writer, 77–78, 117

Squires, Lewis M., 40, 43, 47; as Audu-
    bon's scribe, 15, 16, 56n6, 60, 76, 80, 111,
    112, 133, 157, 159; as prairie hunter, 27,
    84, 88, 91, 97–98, 99, 335–36, 338; wish-
    ing to remain at Fort Union, 103–4

St. Augustine, 33, 214, 252, 253

St. Charles, Missouri, 62, 125–26

Ste. Genevieve, Missouri, 31, 64

Stewart, William Drummond, 37, 45, 57,
    58, 125

St. Louis, 51, 55–57, 59–62, 299, 419, 425;
    1840 view of, *fig. 9*

Street, Phillips B., 96n13

Streshinsky, Shirley, 23n11, 228–30, 230n2

Thoreau, Henry David, 117, 245, 245n6,
    246, 300

Three Mammelles (campsite), 103, 160,
    figs. 19–21

Townsend, John Kirk, 38, 45, 278, 282n8

tree sparrow, 295

turkey, 58, 197, 242

Tyler, Alice Jaynes, 6

War of 1812, 235, 243

Ward, Henry, 252, 253, 254

Webber, Charles Wilkins, 127–29, 207n11

Weidensaul, Scott, 96

Western Museum (Cincinnati), 31, 49,
    236

Wheeling, West Virginia, 48, 49, 253

white-crowned sparrow, 272

whooping crane, 221, 274–75

Wilkes expedition, 36

Wilson, Alexander, 42, 102,

wolf: as creature to be killed, 85, 145, 317,
    319, 323–24, 325, 327, 328, 329; devour-
    ing dead game, 145, 322, 338, 348, 350;
    drawn by Sprague, 394, 396, 398; head
    of, drawn by Audubon, 328; as pet,
    187, 319; pursued on horseback, 83, 91,
    320, 323–24, 330, 393; sport shooting
    of, 85, 320–21, 329

woodcock, 22, 275–77, 293

Wyoming Historical and Geological
    Society (Wilkes-Barre), 425

Zenaida dove, 294